The Low-Code AI Maturity Model

Leading Responsible AI Transformation with Microsoft Power Platform

Steve Jeffery

Apress®

The Low-Code AI Maturity Model: Leading Responsible AI Transformation with Microsoft Power Platform

Steve Jeffery
Cambridge, UK

ISBN-13 (pbk): 979-8-8688-1729-8 ISBN-13 (electronic): 979-8-8688-1730-4
https://doi.org/10.1007/979-8-8688-1730-4

Managing Director, Apress Media LLC: Welmoed Spahr
Acquisitions Editor: Shivangi Ramachandran
Development Editor: James Markham
Editorial Assistant: Jessica Vakili

Cover designed by eStudioCalamar

Cover image designed by Freepik (www.freepik.com)

Distributed to the book trade worldwide by Springer Science+Business Media New York, 1 New York Plaza, New York, NY 10004. Phone 1-800-SPRINGER, fax (201) 348-4505, e-mail orders-ny@springer-sbm.com, or visit www.springeronline.com. Apress Media, LLC is a Delaware LLC and the sole member (owner) is Springer Science + Business Media Finance Inc (SSBM Finance Inc). SSBM Finance Inc is a **Delaware** corporation.

For information on translations, please e-mail booktranslations@springernature.com; for reprint, paperback, or audio rights, please e-mail bookpermissions@springernature.com.

Apress titles may be purchased in bulk for academic, corporate, or promotional use. eBook versions and licenses are also available for most titles. For more information, reference our Print and eBook Bulk Sales web page at http://www.apress.com/bulk-sales.

Any source code or other supplementary material referenced by the author in this book is available to readers on GitHub. For more detailed information, please visit https://www.apress.com/gp/services/source-code.

If disposing of this product, please recycle the paper

To my wife and children: thank you for being the steady voices that keep me grounded and focused and remind me every day what the future is for.

To my parents - thank you for giving me curiosity, stubbornness, and a sense of humor. I've used all three.

To Scott - whose absence is felt by many and whose memory quietly remains present in all who knew you.

To Phil - thank you for your unwavering encouragement and for making space for work that matters. Thank you for being a true reviewer and a trusted leader. There is no way I could have written this book without your help and mentorship.

To Sarah, Mark, and Mark - past managers who helped shape my thinking and trusted me with big challenges and for helping me grow into someone who could write this book.

To Microsoft - where I've spent over 14 years working alongside incredible colleagues and customers. The people, the challenges, and the platform itself continue to inspire and push me forward.

And finally, to customers - the people who ask hard questions, build amazing things, and keep pushing this space forward with curiosity and care.

A note to the reader

This book is the product of many conversations, mistakes, discoveries, and years spent helping organizations navigate change.

You won't find sweeping generalizations or one-size-fits-all answers here. What I've tried to share is something practical – a path you can shape to fit your organization, your data, and your values.

If at any point you felt like someone understood what you were wrestling with – and helped you move forward – then this book did what I hoped it would.

Table of Contents

About the Author

Steve Jeffery has been an IT professional since 1996, with a career that spans hardware, software, consulting, and architecture. His journey started in hardware, evolving to software roles in the early 2000s and ultimately leading to his current role as Principal Program Manager in Microsoft's Power CAT team. During the past 13 years at Microsoft, he has gained extensive experience working directly with large enterprise customers, guiding them in adopting Microsoft's Power Platform and helping them overcome the complex challenges of integrating new technologies at scale.

One of his most significant contributions has been to Microsoft's Center of Excellence (CoE) Starter Kit and the Business Value Toolkit – resources that have empowered organizations globally to deploy, manage, and drive value from low-code and AI solutions. As a recognized expert in low-code AI, he has a deep understanding of the practical and strategic elements required for successful adoption, from technical implementation to cultural alignment.

About the Technical Reviewer

Phil Topness is part of Microsoft's Power CAT team, helping the largest organizations around the globe transform with AI and low-code solutions.

During the course of his 30-year career, he has worked across industries including energy, manufacturing, technology, marketing, government, and consulting.

Outside of work, Phil lives in Virginia with his wife, Cathleen. Their three adult daughters are scattered across New York, Virginia, and Scotland. He enjoys swimming and is learning to sail – starting small, quite literally, in a very modest boat.

You can connect with Phil on LinkedIn at `https://linkedin.com/in/topness`.

Acknowledgments

This book has been a rewarding journey - one shaped by a decade of working closely with the Microsoft Power Platform and engaging deeply with customers from virtually every industry imaginable. The reality is that I've been profoundly shaped by the countless customers, colleagues, and peers I've met along the way.

To my customers past and present - thank you. Your challenges have become my challenges. Your openness, willingness to innovate, and trust have deeply influenced the insights and practical advice contained within these pages. Every scenario, every strategy, every bit of guidance in this book is inspired by real conversations, real challenges, and real solutions.

My sincere appreciation extends to my colleagues at Microsoft, especially within the Power CAT team, who have provided an environment of continuous learning, innovation, and genuine camaraderie. Your expertise and dedication set an inspiring standard.

Personally, heartfelt thanks to my family for their understanding and patience, particularly when late nights and early morning blurred together in the final phases of writing.

Lastly, to you - the reader - my deepest gratitude. May this book provide clarity, inspire confidence, and offer practical steps as you navigate the complexities of responsible AI adoption, governance maturity, and ultimately achieving sustainable, meaningful success with low-code and AI.

Foreword

I have three daughters, all pursuing careers in science. One of them, Rebecca, is getting her PhD by digging up cores of mud from lake beds and studying them to help predict future climate change.

Why is this relevant to a book on AI transformation?

To analyze her data, Rebecca had to use a software model built by others – so she had to teach herself to code. She learned Python. And Fortran. Fortran! My 25-year-old daughter had to master a programming language that put people on the moon 60 years ago. That's the power of inertia.

You are reading this book because you are, perhaps without realizing it, seeking to break the inertia in your organization. The value in AI comes not from having the technology but from the people using the technology. Inventing with it. Transforming businesses. Transforming lives.

And the best part? You won't do it alone.

This book gives you the guidance to bring others with you. You'll learn how to recruit executives (Chapter 1), which metrics to track and share (Chapter 7), how to use technology to support policy (Chapter 11), and how success can not only transform your organization but position you as an industry leader (Chapter 12).

This isn't theory. You'll find practical tools and technologies, including how innovations like the Model Context Protocol (MCP) and use of autonomous agents can strengthen your governance process (Chapters 10 and 11).

I've worked with Steve for years. He's remarkably calm under pressure, not because he has all the answers but because he has the unique ability to gather data, methods, and insights - and apply them to new, complex situations. He often sees the path forward before the rest of us do. This book helps you see that path, too.

Breaking inertia takes more than vision. It requires your whole organization to move with you - and now is the moment to do it. This moment is rare in history. Many have the tools, but not all will thrive. This book gives you the model, techniques, and mindset to persist, succeed, and lead.

The value in this technology is unlocked by people - by you. I can't wait to see what you and your team create.

—Phil Topness

Introduction

What would it mean for your organization if anyone, at any level, could solve complex problems in days instead of months? Imagine the possibilities if your team could move from ideas to fully realized solutions with a few clicks, powered with Artificial Intelligence (AI). This isn't a distant future - it reflects the capabilities of the Microsoft Power Platform today.

Since 2016, I've been immersed in low-code technologies, partnering with countless organizations across diverse industries to help them navigate their challenges - some straightforward, others far more complex. Through these experiences, I've had the privilege of seeing how organizations across almost every industry and size have embraced the Microsoft Power Platform. It's been inspiring to watch teams empower themselves to solve problems in ways they never thought possible and create real impact across their organizations. I've also seen where they falter - how the absence of vision, governance, or clear leadership turns promising initiatives into missed opportunities.

We are at a pivotal moment, where the boundaries of innovation are rapidly expanding. If we navigate it well, the opportunities are transformative, yet without responsible implementation, there are risks in compromising trust, compliance, and security.

This is where responsible AI (RAI), governance, and trust come into play. These are not just technical concerns; they are foundational to making sure the technology delivers sustainable value while protecting your organization and its stakeholders.

What This Book Will Do for You

Whether you're just starting or looking to refine your existing efforts, this book is designed to guide you through the common challenges organizations face when adopting Power Platform.

This book is not just about the technology – it also emphasizes cultivating a strong leadership vision, fostering a culture of innovation and collaboration, establishing robust governance frameworks and clear accountability, guaranteeing data readiness and security, embedding responsible AI practices throughout your workflows, managing organizational change effectively, and building secure, reliable, and scalable systems. It's a holistic guide to harnessing AI in a way that aligns technology initiatives with your organization's strategic values and goals.

To help you address common challenges, I've structured this book around a maturity model,[1] which is a framework designed to help organizations assess where they are on a particular journey. Whether you are adopting low-code, implementing governance, or fostering innovation, this book provides a structured way to identify strengths and pinpoint areas for improvement.

Think of this book as a road map. Each stage in the maturity model represents a level of organizational capability, starting from the basics and progressing toward advanced integration and excellence. It helps you answer the following key questions:

- Where are we today?
- What is holding us back?
- What do we need to focus on next?

[1] Carnegie Mellon University. (2010). Capability Maturity Model Integration (CMMI) for Development, Version 1.3. Software Engineering Institute.

This book takes every lesson I've learned and every observation I've made and attempts to combine them into an easy-to-read and understandable guide for you. I don't know the specifics of your industry, your organizational structure, or how far along you are with your Power Platform adoption journey. I am confident, however, that the challenges you're facing, or yet to face, will likely echo those I've seen time and again. Whether it's regarding creating a culture of innovation, establishing governance frameworks, or building a foundation of secure, accessible data, this book is designed to help you tackle these challenges head-on.

In this book, the maturity model focuses on three critical areas of capability:

- **Leadership and culture**: Creating a workplace where innovation thrives and decision-making aligns with strategic goals
- **Trust and governance**: Establishing secure, ethical, and responsible AI practices to minimize risk and maximize value
- **Data and security**: Building a scalable, reliable foundation for Power Platform adoption powered by accurate and accessible data

Each part will guide you through these areas, revisiting them at different stages of maturity:

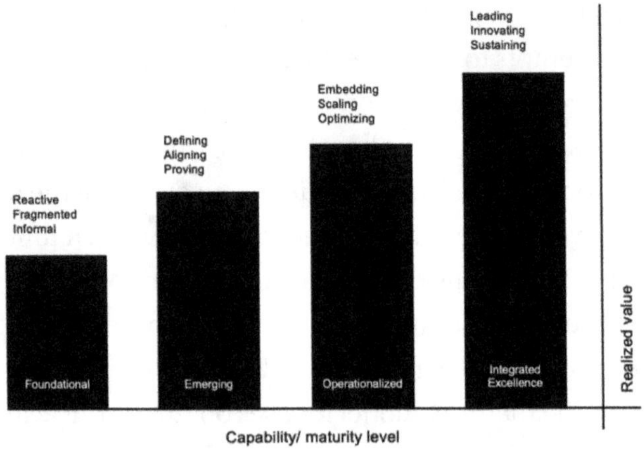

Figure 1. *Maturity models covered in this book*

- **Foundational**: Your organization is just getting started. Efforts are mostly reactive, with basic governance, vision, or administrative practices in place, though they are most likely informal or fragmented. The focus is likely to be on developing awareness, understanding the key concepts, and laying the groundwork for future growth. There will be limited value being realized due to the fragmentation and inefficiency. Digital transformation is minimal and uncoordinated.
- **Emerging**: You're making progress! Roles and processes are being defined, and some early wins are being achieved through structured initiatives. Digital transformation is now gaining momentum, though efforts are likely still localized or siloed.
- **Operationalized:** Processes and practices are established, consistently applied and scaled across the organization. Significant value is realized as systems are optimized and adoption spreads. Digital

transformation is well underway with enhanced operational efficiency, collaboration and strategic alignment driving measurable outcomes.

- **Integrated excellence:** Your organization is fully aligned with strategic goals. Processes are embedded in the culture, continuously improving through feedback and innovation. At this level, technology is seamlessly integrated, creating agility and a competitive advantage.

You can use this framework to assess where your organization currently stands and determine the level of capability you need to achieve. Not every organization needs to aspire to "integrated excellence" in every area - your goals should be aligned with your unique circumstances. Remember, progress isn't always linear, and moving between levels in certain areas may require significant effort. However, think of this book as your compass, helping you navigate your way toward your desired outcomes.

How long does it take to achieve advanced adoption and operational excellence?

Organizations rarely start from the same position or have the same level of sponsorship, support, team structure, or vision, so it's very difficult to state exactly how long things will take. I've seen many organizations achieve operationalized levels of capability within 18–24 months when they've had the sponsorship, resources, and budget well aligned with a strong vision. Every small step can yield outsized results, creating momentum and early wins to build upon.[2]

Below is an illustration showing typically how long an organization takes to achieve each level of capability:

---

[2] Harvard Business Review. (2019). Driving Organizational Change Without Abandoning Tradition.

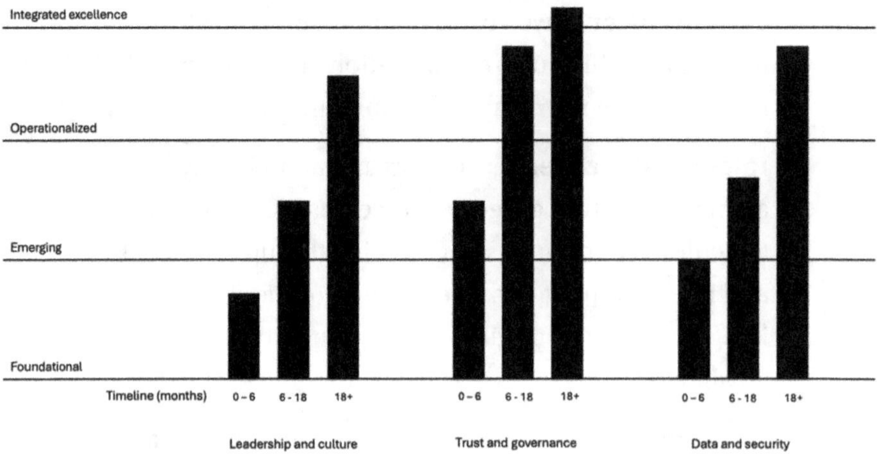

Figure 2. *Maturity progression across three dimensions of capability*

Adopting an AI-Powered Platform

It's incredible how quickly AI has transformed Power Platform. At the time of writing, we've reached a point where an end user can describe a challenge and the platform will identify the user personas, design (and even create) a data schema, and then build the necessary apps and flows.[3] It's remarkable! AI is no longer a tool reserved for data scientists and pro-developers; it's an enabler for anyone with a problem to solve or an idea to explore.

This rapid transformation brings a new set of challenges. Many organizations – especially those already mature in their low-code journeys – are now grappling with questions on how to integrate AI responsibly and effectively. The core challenges remain familiar: identifying where AI offers the most value, deploying it securely, and guaranteeing compliant usage across the enterprise.

[3] Microsoft. (n.d). *Use the Plan Designer.*

However, a major difference between AI and low-code is that AI is pervasive – it organically finds its way into organizations and employees' day-to-day workflow. Employees may already be using AI tools without formal oversight or governance, creating risks of data leakage and compliance violations.

Adopting an AI-powered platform calls for a shift in how you approach digital transformation.[4] You need a strategic framework that aligns this technology with your organization's goals while addressing governance, training, and cultural shifts. As AI capabilities in Power Platform continue to evolve, your organization needs to remain agile, adapting your strategies to keep pace with technological advancements. Power Platform offers a unique opportunity to integrate AI seamlessly into existing processes and develop new and innovative ones, but success depends entirely on thoughtful planning and execution. Understanding how AI fits into your digital transformation strategy is essential, and this book focuses on where Power Platform intersects with these broader initiatives.

AI in this book

AI is rapidly becoming an integral part of organizational workflows, and Power Platform is just one of the ways it can be leveraged. In this book, when I refer to AI, low-code AI, or Power Platform AI, I'm focusing on AI technologies within Power Platform – tools like Microsoft Copilot Studio, AI Builder, and similar features, as well as broader AI integrations that may complement these capabilities.

However, it's important to recognize that Power Platform is not the sole AI initiative within most organizations. AI adoption spans multiple systems, tools, and departments, each with its own objectives and constraints. For example, while Power Platform drives operational efficiencies and empowers business users to create AI-driven solutions,

[4] McKinsey & Company (2023), Rewired to Outcompete: AI and Digital Transformation.

other AI initiatives might focus on customer insights, predictive analytics, or advanced machine learning outside the platform.

This broader scope means the concepts and strategies discussed in this book – particularly around governance, responsible AI, and adoption – are not confined to Power Platform alone. The principles of transparency, accountability, fairness, and security are universal to any AI implementation.[5]

How Power Platform Fits into Broader AI Initiatives

Governance frameworks for AI are typically established at the organizational level, encompassing all AI-related activities across departments, platforms, and tools. Organizations are unlikely to create a separate governance framework exclusively for Power Platform's AI capabilities. Instead, they will adapt or expand existing frameworks to include exceptions or specific guidance relevant to Power Platform's unique use cases.

For instance:

- **Unified policies**: Organizations may have a central AI governance policy that outlines universal requirements, such as fairness in algorithmic decisions, explainability, and compliance.[6] Power Platform initiatives would follow these guidelines while tailoring certain aspects to the platform's strengths, such as low-code democratization.

[5] OECD (2023), AI Principles.

[6] IBM. (2022). AI Governance: Building Trustworthy and Transparent AI Systems.

- **Divergent rules**: In some cases, organizations might establish exceptions for Power Platform. For example, citizen-developed apps using AI may require a simplified approval process compared to enterprise-level AI systems.
- **Integration challenges**: Power Platform solutions will often need to coexist with broader enterprise AI systems, requiring interoperability and consistency in governance practices. This guarantees alignment with overarching organizational objectives, such as data security and ethical AI use.

By understanding how Power Platform fits into your organization's broader AI governance, you can guarantee your initiatives align with existing policies while addressing the platform's specific opportunities and risks.

Concepts covered in this book

The strategies and frameworks you'll encounter in this book are designed to be both specific and adaptable:

- **Specific to Power Platform**: I address challenges unique to low-code and AI within Power Platform, such as governance for citizen developers or tailoring responsible AI frameworks for low-code use cases.
- **Broader organizational context**: These strategies also integrate seamlessly into broader AI governance. Whether it's establishing transparent practices, fostering responsible innovation, or creating scalable systems, the principles apply universally across AI initiatives.

For example:

- **Governance**: While Power Platform might introduce unique governance challenges (e.g., managing AI solutions created by nontechnical users), the underlying principles of responsible AI are shared across all AI implementations.
- **Training and culture**: Building AI literacy among Power Platform users complements enterprise-wide training initiatives, creating a consistent understanding of AI's risks and opportunities.
- **Data security and privacy**: The same rules that govern data use for enterprise AI should apply to Power Platform, guaranteeing consistency in how sensitive information is handled.

Why this matters

Understanding how Power Platform fits into the bigger picture helps you make informed decisions. It makes sure that your initiatives align with your organization's overarching goals and governance frameworks and avoids duplicating effort by integrating Power Platform into broader policies.

By addressing both the specific and universal, this book aims to give you the tools you need to navigate Power Platform adoption while supporting a unified vision for responsible AI within your organization.

Ready to begin?

Power Platform is an enabler of transformation. With the right vision, governance, and strategy, your organization can harness these tools to create value, innovate responsibly, and gain a competitive edge.

This book is your guide to navigating this journey, tailored to Power Platform's unique strengths, while aligning with organizational objectives. Together, we'll chart a path to meaningful, sustainable success. Let's get started.

PART I

Leadership and Culture

What happens when innovation outpaces leadership? Or when technology sits unused - not for lack of potential but because teams feel unprepared or uncertain about its value or, in the case of AI, even threatened? I've seen this play out in organizations I've worked with - hesitation, innovation stalling, and technology that could drive real change doesn't get used.

At the heart of any successful Power Platform adoption are leadership and culture. While technology offers transformative capabilities, these technologies can only thrive in an environment where they are trusted, embraced, and used strategically.

Without strong leadership and a supportive culture, even the most advanced tools risk failing to deliver meaningful value.

Leadership sets the vision that motivates teams to think beyond immediate challenges and focus on the bigger picture. Effective leaders champion low-code and AI's potential, articulate its strategic value, and position Power Platform as central to digital transformation. Leaders also ensure that adoption is responsible and ethical, aligning initiatives with organizational values to build trust.

Culture, on the other hand, shapes how an organization adapts to change. An innovation-friendly culture empowers employees to explore the potential of Power Platform, share ideas openly, and address concerns openly. It fosters collaboration, reduces resistance to change, and embeds principles like fairness, transparency, and inclusivity into everyday practices.

However, leadership and culture often pose the greatest challenges. Resistance can stem from fear (replacing jobs, unfair judgement, bias, lack of explainability), ethical uncertainties, or a lack of understanding about the technology's role in the organization.

It's understandable, but the key is to address them head-on with transparency, trust, and clear guidance.

Getting leadership and culture right is no small task, but it's where successful adoption starts. In this part, we'll dive deep into how leadership and culture drive low-code and AI success across four maturity levels:

- **Foundational**: *Establishing the groundwork* through sponsorship, vision, and awareness
- **Emerging**: *Building momentum* by nurturing a low-code and AI-friendly culture and fostering adoption
- **Operationalized**: *Scaling success* through leadership accountability, strategic integration, and a culture of continuous responsibility
- **Integrated excellence**: *Leading with purpose* by ensuring long-term sustainability and influencing global conversations on responsible AI

Through practical strategies, real-world insights, and actionable guidance, this part of the book will help you align leadership and culture to unlock the full potential of Power Platform and responsible AI.

CHAPTER 1

Foundational: Leadership and Culture

Reactive | Fragmented | Experimental

Leadership and culture often set the tone for how new technologies are adopted. At the foundational level, these elements are in their infancy, and organizations are just beginning to grapple with the potential of Power Platform and AI.

Organizations might acknowledge the importance of low-code and AI as a tool but lack the awareness, integration, and investment needed to address their impact effectively. Adoption efforts are often fragmented, with limited vision, resources, and executive buy-in.

As organizations take their first steps with low-code and AI technologies, it is crucial to recognize their current level of adoption. The illustration below provides an overview of leadership and cultural traits typically seen at the foundation stage. These traits highlight the gaps in awareness, integration, and alignment that must be addressed before meaningful progress can be achieved.

S. Jeffery, *The Low-Code AI Maturity Model*, https://doi.org/10.1007/979-8-8688-1730-4_1

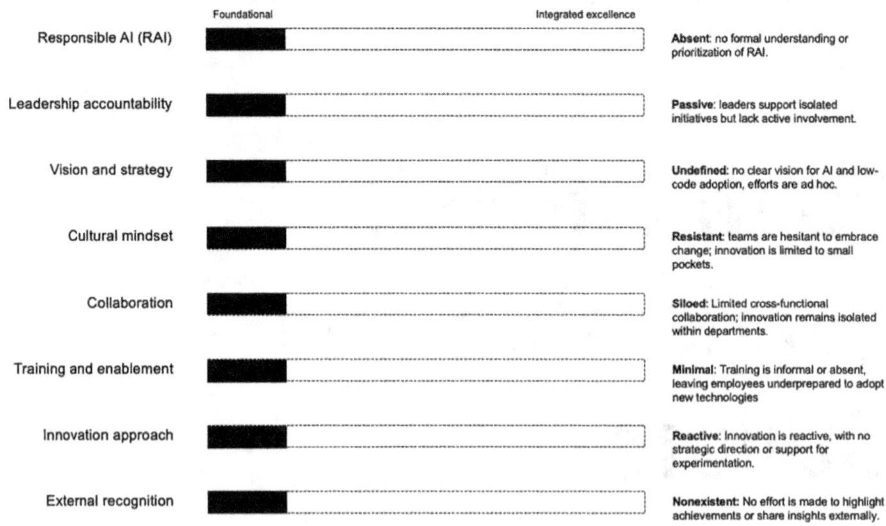

Figure 1-1. *Run state of an organization at the foundational stage*

Before we go further, let me ask you this: *Do the attributes of the foundational stage reflect your organization? Are efforts siloed or lacking strategic direction?* Many organizations face these same hurdles at the foundational level.

Key challenges at this level include the absence of executive sponsorship, limited access to software capabilities and tools, a lack of strategic vision and minimal awareness of responsible AI principles at the organizational level. Together, these issues create a cycle where low-code and AI are seen as experimental rather than transformational.

Let's dive a little deeper into these challenges:

- **Lack of executive sponsorship**: Without a senior leader championing Power Platform, initiatives will struggle to gain traction. Leadership buy-in is essential to elevate low-code from a grassroots effort to a strategic priority.

Executive sponsors also play a role in embedding responsible AI practice, ensuring ethical considerations are part of the decision-making process from the very start.

- **Limited licensing**: Many organizations may not have sufficient licenses or access to advanced features to experiment and prove the value of low-code and AI. Teams will struggle to demonstrate tangible benefits, making further investment difficult to justify. Additionally, limiting use may hinder efforts to explore the transformational capabilities of low-code and AI, reinforcing the perception that low-code and AI are experimental rather than strategic.
- **Absence of a strategic vision**: Leadership isn't involved enough and hasn't articulated why low-code and AI matters or how it aligns with the broader organizational goals. Efforts are fragmented, with no road map or understanding of what success looks like. A vision that incorporates responsible AI principles can help bridge this gap, positioning low-code and AI as a trusted driver of innovation that is aligned with organizational value.
- **Minimal awareness of using AI responsibly (responsible AI/RAI)**: Leadership and teams often lack a shared understanding of what RAI entails or why it matters. Culturally, RAI may even be viewed as a hindrance that it slows down development or limits creativity, or there may be a "not applicable" mindset where there is engrained belief that RAI doesn't apply to simple use cases.

Here's a thought experiment: *How often does your team discuss the ethical implications of AI?* If these discussions feel absent or surface-level, there's an opportunity to deepen awareness and integrate RAI into everyday practices.

Every organization starts somewhere. Acknowledging these gaps is the first step toward transformation. Before diving into actionable strategies, it's critical to clearly understand why these conversations matter in the first place. Let's begin by exploring why responsible AI (RAI) matters - both strategically and ethically - to every organization adopting AI.

Why Responsible AI (RAI) Matters

Artificial Intelligence has the potential to shape our world in very profound ways, revolutionizing industries, improving lives and healthcare, and solving challenges once thought insurmountable.[1] However, successful AI adoption demands fundamentally on gaining the trust and confidence of employees, who often approach AI with deep skepticism. Without addressing these concerns, even the most innovative AI initiative risks internal rejection or failure.

Historically, advancements in technology have often outpaced the frameworks necessary to govern their use.[2] AI, with its ability to influence decisions and automate tasks, demands that we proactively establish guidelines to ensure its benefits are widely and equitably distributed while minimizing risks and harm.

Large organizations like Microsoft, Amazon, Google, and IBM have recognized the critical importance of responsible AI and are leading the way in shaping global best practices. Microsoft, for example, has

[1] McKinsey & Company. (2023). *The State of AI Adoption in 2023.*

[2] Bostrom, N. (2014). *Superintelligence: Paths, Dangers, Strategies. Oxford University Press.*

embedded its commitment to responsible AI into its core principles, emphasizing transparency, fairness, inclusivity, reliability, and security.[3] Through initiatives like the AI Ethics and Effects in Engineering and Research Committee and collaborations with policymakers and academics, Microsoft has demonstrated a clear commitment to building AI systems that customers can trust.

These efforts underline a shared understanding: AI's transformative potential must be balanced with safeguards that protect individuals and society. The stakes are too high for anything less. AI influences hiring decisions, loan approvals, and healthcare diagnoses, so a lack of oversight in these areas can lead to discriminatory practices, loss of trust, and irreversible harm.

Leaders play a pivotal role in mitigating these risks by prioritizing transparency in how AI solutions are selected, deployed, and monitored. Clear communication regarding AI use cases and impacts significantly reduces uncertainty and fear. Additionally, organizations must proactively invest in AI literacy, ensuring employees understand how AI enhances their roles rather than replacing them.

The responsibility doesn't lie solely with technology providers. It extends to every organization that adopts AI, to every developer creating AI-powered tools, and to every leader steering their organization's adoption strategy. Responsible AI is about avoiding harm and actively working to do good. It is about ensuring AI aligns with human values and trust, equity, and accountability at every level.

By proactively addressing skepticism through transparent leadership and robust employee education, organizations can turn initial resistance into meaningful engagement, unlocking AI's full potential.

---

[3] Microsoft. (n.d.). *Responsible AI Standards Overview*.

As we look ahead, the importance of this collective responsibility only grows. By prioritizing responsible AI today, you are both addressing the ethical challenges of the present and laying the foundations for a future where AI serves as a force for progress and innovation.

The Role of an Executive Sponsor

Securing an executive sponsor is more than a tick-box exercise. Too often, I've seen organizations appoint sponsors who lack clarity about their responsibilities, leaving initiatives without the advocacy or needed support. Many sponsors are chosen based on their proximity to a product owner rather than their ability to drive strategic alignment.

An executive sponsor is not simply a symbolic figurehead - they must be positioned to lead strategically. While they should champion low-code and responsible AI initiatives, their influence must extend beyond the platform itself. The ideal sponsor has enterprise-level oversight, with the ability to align low-code adoption to broader business and technology priorities. This role is best filled by someone with cross-functional authority - able to connect the dots between IT, business units, governance, and innovation strategy.

The illustration below summarizes the typical roles, responsibilities, and the impact of each stakeholder in securing and driving Power Platform adoption responsibly.

Role	Key responsibilities	Reports to	Impact on transformation effort
Executive sponsor	• Advocate for the initiative at the leadership level • Secure funding (funding, licenses, personnel) • Resolve escalations and mediate resistance • Communicate successes	C-Suite (e.g., CTO, CIO, COO)	Provides visibility and strategic alignment
Product owner	• Coordinate execution of the initiative • Translate strategy into actionable steps • Manage timelines and team deliverables	Executive sponsor	Ensures day-to-day progress and alignment with goals
RAI champion	• Promote responsible AI principles within their department • Provide feedback from users • Support teams in adopting Power Platform	Department head or CoE lead	Drives cultural alignment and responsible AI compliance at a team/ department level
Governance board	• Set policies and standards for ethical use • Review initiatives for RAI compliance • Address risk and ethical concerns	Executive sponsor	Ensures initiatives adhere to governance and ethical best practices.
Center of Excellence	• Provide technical oversight and support • Implement guardrails and governance • Enable teams through training and resources.	CoE lead, or IT Leadership	Drives platform readiness and technical compliance

Figure 1-2. *Typical roles and responsibilities in Power Platform and AI adoption*

What about your organization? Do you feel leadership is supporting these efforts actively, or are there gaps in visibility and alignment?

Let's dive deeper into the characteristics of an effective sponsor and how you can secure one.

Tips for Finding the Right Executive Sponsor

1. **Influence and authority**: You should look for leaders within your organization with decision-making power and strong organizational influence. Typically, this is a C-suite executive, VP, or director. If your executive sponsor is responsible only for Power Platform, you haven't reached high enough – try going two levels up in the organization!

2. **Alignment with innovation**: Seek individuals who have previously championed innovation or large-scale change initiatives. In many organizations, AI efforts and low-code platforms are led by different teams. Start by identifying leaders driving each area, and assess whether they align under a common executive. If not, consider which leader is best positioned to unify these efforts and act as a bridge between business transformation, IT strategy, and AI governance.

3. **Cross-functional reach and influence**: An ideal sponsor is someone who can connect multiple departments, ensuring there is alignment across the organization.

4. **Passion and openness**: Identify someone who demonstrates a genuine passion for low-code and AI and its potential to impact the organization.

Based on these criteria, *who in your organization* comes to mind as a potential executive sponsor? Does this potential sponsor have experience or interest in championing ethical and responsible practices?

Jot down one or two names, and think about how you might approach them. Once you've identified a potential sponsor, the next step is crafting a compelling approach to secure their commitment.

Approaching an Executive Sponsor

When you've identified the sponsor, you should be very clear on what you are asking from them. Pitch this as an opportunity for them to influence how low-code and AI aligns with organizational strategic goals.

Imagine yourself either composing an email or having a conversation with your target sponsor. The process of securing and engaging an executive sponsor can be broken into four key steps, as shown below.

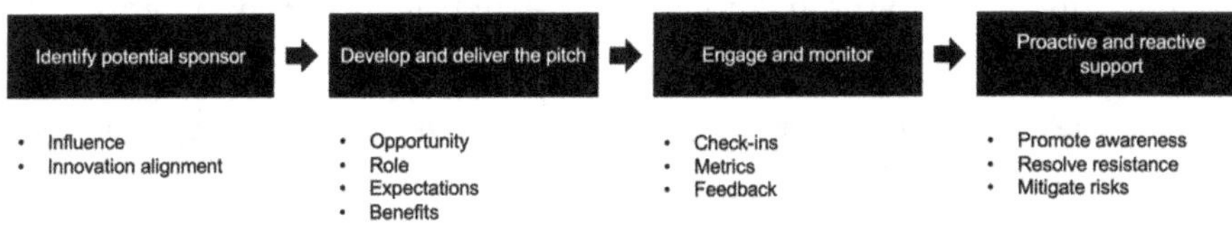

Figure 1-3. *Four steps of securing and engaging a sponsor*

Developing the Pitch

Start with the Opportunity

This frames the conversation positively and immediately shows the value of the initiative. It sets the tone, focusing on tangible benefits and strategic alignment. This step grabs their attention by presenting a clear reason why this initiative matters to the organization.

Example:

> *"We've identified a great opportunity to use low-code and AI to automate our invoice processing, which could cut manual effort by 60% and reduce processing time from five days to one. With the right leadership support, we can align this with our broader goals around operational efficiency and intelligent automation at scale."*

Key points to cover

Explain why their sponsorship matters: Position this as a strategic opportunity rather than a request for operational support. Without visible executive backing, initiatives often struggle to gain the cross-functional visibility, prioritization, and resourcing required for long-term success.

Acknowledge their seniority with confidence: It's likely that the person you're speaking to is more senior than you. That's expected - and appropriate. Focus on the fact that you're inviting them into a leadership role that only they can fulfill. You're not instructing them; you're equipping them with the information they need to lead effectively.

Consider this: *If you were pitching this to your sponsor today, how would you start?* Write down your opening sentence, and refine it based on the examples here.

Be Clear About Expectations

Clarity around expectations avoids misunderstandings and ensures the sponsor knows exactly what they are committing to. It also provides you with ways to track their involvement and hold them accountable for driving progress.

Example:
"We're asking you to support us in four ways:

1. *To advocate for this initiative during leadership discussions to align teams and resources.*
2. *Help us secure funding and licenses to expand the initiative.*
3. *Celebrate and communicate early successes to build momentum and ensure longer-term engagement.*
4. *Help us resolve roadblocks, mediate when there is resistance to change, and reassure teams about ethical concerns of AI."*

Key points to cover

Specify actions: Break down what you need from them into clear and actionable steps. They are busy; you need them to process what you are asking from them quickly to keep their attention.

Set milestones: Include measurable indicators of their involvement, such as promoting early success or supporting budget applications.

Be realistic: Ensure the expectations are achievable and aligned with their role and capacity.

Highlight the Benefits

This step is the "what's in it for me and the organization" - framing their sponsorship as an opportunity rather than a burden. Reinforce the value of their involvement and how it aligns with their leadership goals.

Example:

> *"Your sponsorship will ensure this initiative is successful, driving measurable improvements in efficiency and innovation whilst adhering to responsible AI principles... We see your leadership skills and experience as being vital to advance digital transformation that aligns with our organizational goals."*

Key points to cover

Demonstrate organizational impact: Highlight how the initiative will drive measurable success for the organization. Include KPIs that you intend on tracking for credibility. The sponsor needs to know you've planned this well.

Emphasize long-term gains: Connect the initiative to sustainable growth, improved efficiency, or competitive advantage.

Something to consider: How might this initiative align with your sponsor's existing priorities? Tailor your pitch to address those specifics.

How to Make Use of Your Sponsor

Assuming everything has gone well and that you now have a sponsor, we're going to move on to review some of the ways that you can check on progress and ensure they are accountable. An executive sponsor who will really fight your corner is like finding gold - you need to monitor their engagement as it can fade over time.

Monitoring Their Engagement

- **Schedule check-ins**: Establish regular (e.g., quarterly) review with the sponsor to discuss progress, challenges, and next steps.
- **Use metrics to guide conversations and to arm them with metrics that demonstrate success**: Track adoption metrics (active users, number of apps, number of agents). Keep discussions grounded with data.
- **Provide feedback**: Share insights on what's working well and where additional support is needed.

Monitoring their engagement requires both regular check-ins and data-driven feedback. You also require a thoughtful approach on how you involve the sponsor in driving the initiative. Sponsors can be pivotal in addressing challenges and celebrating wins, but their engagement should be both strategic and adaptive. The table below illustrates how to leverage the sponsor both reactively and proactively to ensure their involvement remains impactful and aligned with the organization's goals.

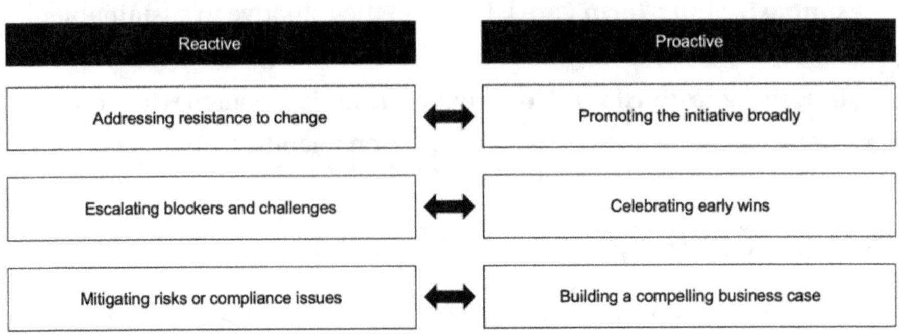

Figure 1-4. *Examples of using your sponsor reactively and proactively*

Navigating Challenges with Sponsors

Even with a strong pitch and careful selection, challenges may arise in securing or maintaining an engaged sponsor. Circumstances can change, priorities can shift, or the alignment might not be as strong as anticipated. Recognizing these challenges early allows you to address them constructively.

Signs to watch for:

1. The sponsor is less involved than expected.
2. The sponsor doesn't actively communicate about the initiative.
3. Support for allocating resources or resolving obstacles slows down.

If you spot any of the signs above, be proactive! Here are a few solutions:

Revisit the relationship: Schedule conversations to discuss capacity, clarify expectations, or realign on priorities.

Expand the network: Look for opportunities to involve co-sponsors or other advocates who can help balance the workload and increase visibility. I worked with an organization that had two sponsors for Power Platform, which worked well - each brought unique skills and perspectives, enabling faster and more successful adoption of the platform.

Focus on their role: If they're unable to engage fully, consider narrowing their responsibilities to areas where they can have the most impact, such as advocacy. Understand their priorities; have they shifted? If you can realign the digital transformation to your sponsor's new priorities, you can regain or even strengthen their sponsorship.

Wrapping Up

Securing an executive sponsor is one of the most critical steps in the foundational stage of adoption. The involvement is pivotal for success, but it requires careful planning, alignment, and sustained engagement. If your sponsor isn't as involved as you'd like, revisit their role, reassess their capacity, or consider other potential champions. Let's move on to another common challenge organizations face at this stage - *building a compelling business case.*

Creating a Compelling Business Case

Building a business case is your chance to connect existing successes with strategic goals, demonstrating how further investment can unlock value.[4] Think of it as a bridge between where you are now and where your organization could be. Let's break this down into three actionable steps:

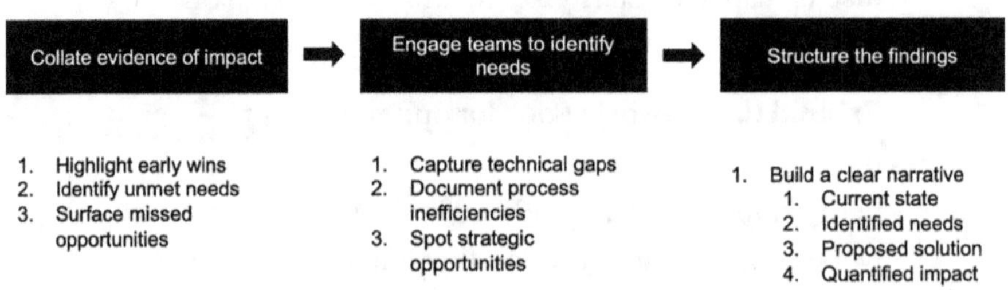

Figure 1-5. *The three steps to creating a business case for expanded licensing*

[4] Forbes (2011), The Seven Keys to Successful Strategic Planning.

Collating Evidence of Impact

Collate evidence of impact: You should be looking to gather information to demonstrate

- **Early wins**: What value has been derived from the current level of investment
- **Unmet needs**: How restricted access to licenses is constraining departments and business units' ability to solve problems and innovate
- **Missed opportunities**: Provide examples of abandoned projects or poorly performing solutions that would have been architected differently if sufficient licensing was available

What unmet needs or missed opportunities have you observed that you could highlight in your case? Can you identify an example of an early win in your organization? What made it impactful?

Identifying Early Wins

There is a mixed approach to collecting the data required to produce a compelling impact report. For solutions that are in use, there is the admin center in Power Platform. This is a useful way of identifying assets (e.g., apps, flows, agents) and ordering them by their usage (labeled as "sessions" in Power Platform Admin Center).

Defining use case characteristics

The type of organizations that I work with typically have thousands, or tens of thousands, of apps, flows, agents, etc., so filtering the top "n" is usually a good place to start. I usually advise defining the "*characteristics of something that is adding value*."

Here are a few examples of use case characteristics:

Use case definition	Personal productivity	Broadly-used	Enterprise
Audience	Personal use only	Used by team, department or project	Used by department, business unit or enterprise
Criticality	Not critical to business operations	May or may not be critical to business operations	May or may not be critical to business operations
User count	Shared with < 5 users	Shared with > 15 users	Shared with > 50 users
Usage	Frequent, or infrequent use	Frequent, or infrequent use	Frequent, or infrequent use

Figure 1-6. *Three common use case definitions for Power Platform solutions*

You may find this sufficient. I rarely have seen other categorizations of use cases; however, there might be subtle differences in between use case characteristics, so try to avoid overfitting and define use case characteristics that are more aligned to how Power Platform is used within your organization.

Building Your Inventory

With clear use cases in mind, the next step is to take stock of what you have. There are several approaches to building this inventory, depending on the tools, resources, and level of customization your organization requires. Here, we'll introduce three primary methods:

- Using Power Platform Admin Center[5]
- Leveraging the Center of Excellence (CoE) Starter Kit[6]

[5] Microsoft. (n.d.). *Power Platform Admin Center Documentation.*

[6] Microsoft. (n.d.). *Center of Excellence Starter Kit Overview.*

- Applying the Business Value Toolkit[7]
- Building a custom solution

Each approach comes with its own benefits, challenges, and levels of effort. Let's look at each in detail.

Power Platform Admin Center

Power Platform Admin Center is a centralized portal that among many things is a good place to manage apps, environments, flows, pages, copilots, and users. It provides quick access to metadata that can help you identify active and impactful solutions.

Steps to Build an Inventory

- **Export app data**

  Use the Apps section to export metadata, such as app names, creators, modified dates, users count, sessions, etc.

- Sort apps based on sessions (usage).
- **Engage app owners**

  Develop a process to communicate with app owners of high-usage apps.

  Use email campaigns or automated surveys (e.g., adaptive cards in Teams) to gather insights on value, time, savings, and impact.

[7] Microsoft. (n.d.). *Power Platform Business Value Toolkit.*

- **Iterative review**

  From my experience, the top 15–20% often have 80% of the value for an organization - so start with the top "n" apps based on usage.

Here are some considerations for using Power Platform Admin Center:

Pros	Cons
Quick and easy to start with built-in tools.	Limited metadata depth for detailed insights.
Minimal setup and no additional cost.	Manual effort required for analyzing and engaging app owners.
Helps prioritize most-used assets quickly.	Relies heavily on session count, which might not indicate strategic importance.

Figure 1-7. *Pros and cons of Power Platform Admin Center for inventory analysis*

CoE Starter Kit: Reference Tooling Providing Governance and Insights

The Center of Excellence (CoE) Starter Kit is a ready-made solution provided by Microsoft to help organizations manage, govern, and monitor their Power Platform adoption. Built and maintained by Microsoft's Power CAT team, it comes with prebuilt dashboards, inventory tracking tool, and automation. As a member of the Power CAT team and a contributor to the kit, both in applications and documentation, I can state that the kit was built to address challenges from organizations; with over 10,000 customers installing it, the kit has grown in popularity to the point where it is almost a checklist item in any successful deployment of Power Platform.

Steps to Build Inventory

- Install the CoE Starter Kit.
- **Inventory tracking**

  Enable and run the cloud flows that populate your inventory.

  Use the CoE's inventory reports to view detailed data about apps, flows, users, etc.
- **Identify high-value apps**

  Use the dashboard to filter for broadly shared or frequently used apps.
- **Automate engagement**

  Develop flows to send surveys to app owners, and gather qualitative data on value and impact.

Considerations for using the CoE Starter Kit

Pros	Cons
Comprehensive out-of-the-box solution for inventory tracking.	Requires set up and ongoing maintenance.
Automates reporting and reduces manual efforts.	May need additional customization to align with unique goals.
Provides rich telemetry and dashboards.	Requires organizational buy-in to implement fully.

Figure 1-8. *Pros and cons of using the Center of Excellence Starter Kit*

Business Value Toolkit

The Business Value Toolkit is well-suited for the purpose of establishing value of specific solutions.

It provides both citizen developers and administrators with strategies of assessing the value of solutions that have already been developed or proposals and ideas.

This toolkit does not explicitly build up an inventory; its utility is in assessing value of a specific application. It uses generative AI to capture the "story behind the app" and analyze it against the goals of your organization. Using this approach, it is far easier to demonstrate that a solution or application is driving impact against something that your organization already views as an objective. The Business Value Toolkit also generates success stories that can be used to form the early wins section of your business case.

Considerations for the Business Value Toolkit:

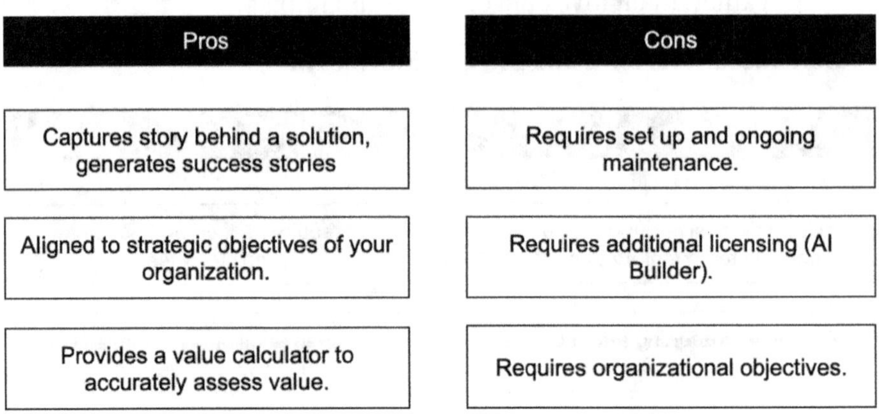

Figure 1-9. *Pros and cons of using the Business Value Toolkit*

Custom Solution

The final option is to develop your own solution. This offers the most flexibility as you can meet your exact requirements. There are trade-offs in this approach as outlined below:

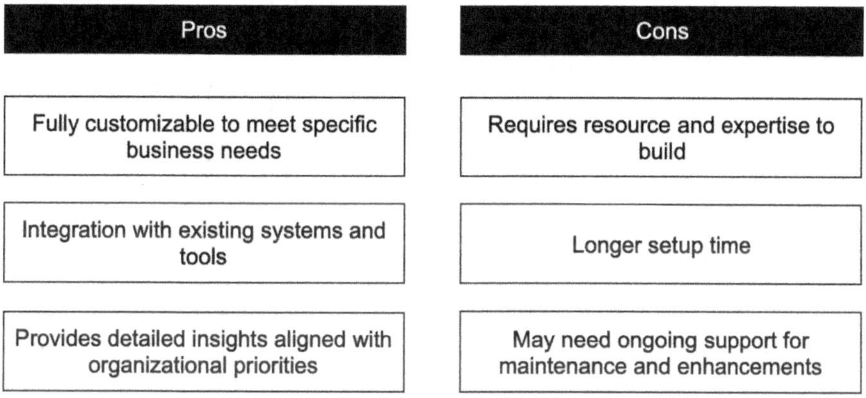

Figure 1-10. *Pros and cons of developing a custom solution*

Steps to Build a Custom Solution

Power Platform has well over 1000 connectors that enable apps, flows, and agents to connect to data sources and services. There are administrative connectors for Power Platform, which can be used in conjunction with cloud flows to gather metadata. Below are a few example data sources that contain useful metadata:

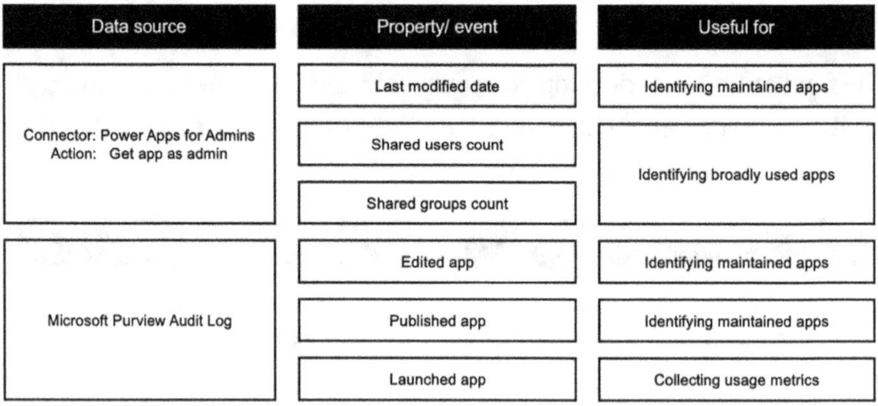

***Figure 1-11.** Useful metadata and data sources*

Use these data sources (and other related metadata) to identify apps in your organization that are unused. Using properties such as last modified date and published can indicate an app that is maintained; using shared users and groups count establishes how broadly used the app is.

Develop a cloud flow to iterate across your environments and the apps within them; save the metadata you need into a Dataverse table.

Why Dataverse and not SharePoint?

A quick segue into why I recommend using Dataverse.[8] In this use case, there are three very good reasons:

- **Corpus of data**: If you have thousands of apps, flows, agents, connection references, environments, etc., you will most likely reach a point where SharePoint becomes untenable as a viable data storage solution.
- **Relational data**: If you are trying to implement any integrity on data, using SharePoint lists is not a great idea. SharePoint isn't a relational database; it's a collaborative

[8] Microsoft. (n.d.). Dataverse vs SharePoint: Choosing the right data storage option for Power Platform.

platform. It often requires additional calls and has query limitations[9] that mean careful considerations into the formulas you can use in Power Apps.

- **Portability**: If you plan on implementing ALM and pipelines to deploy your inventory solution to a production environment, the Dataverse will be the best way to manage your schema.

Limited Licensing

Limited licensing remains one of the most significant roadblocks for organizations at the foundational level, constraining the ability to unlock Power Platform's full potential.

When I first encountered this challenge, I saw how frustrating this was - a small group of employees had access to licenses for exploration, leaving everyone else watching from the sidelines. There was potential inside the organization, but it was stifled, making it very difficult to demonstrate value or scale solutions effectively.

Here are a few symptoms of limited licensing. As you read, ask yourself: *Which of these resonates most with your current situation?*

- **Patchy adoption across teams**: When only a small group of employees has access to licenses, it leads to silos that limit collaboration.
- **Missed opportunities for innovation**: Restricting teams from experimenting with advanced AI capabilities, it reduces their ability to identify and solve departmental and organizational problems. Innovation becomes centralized rather than democratized.

[9] https://learn.microsoft.com/en-us/power-apps/maker/canvas-apps/delegation-overview

- **Inability to prove value**: Limiting access makes it difficult to generate enough success stories or identify metrics to justify additional investment. Without having measurable results, leadership may view licensing as a cost rather than an enabler.
- **Minimal awareness of responsible AI**: Teams might overlook responsible AI principles due to constrained access, limiting the organization's ability to develop frameworks early on – this ultimately introduces levels of risk to the initiative.
- **Suboptimal solutions**: Without access to premium features, teams often rely on the capabilities available to them, leading to poorly architected, hard to maintain solutions. Over time, these solutions can introduce a negative perception that low-code and AI solutions are unreliable or unsuitable for critical business needs.

These challenges often create a frustrating paradox: *How do you demonstrate value when you don't have enough access to explore it?* The key is to work with what you have, optimize it, prove small wins – then communicate to the right people.

One experience I recall was an organization discovered that a single app was saving hundreds of hours and increasing compliance. It wasn't just the time saved or improved compliance that helped demonstrate ROI but the success of an app built by a true citizen developer that opened the door to more ambitious projects. Communicating wins like this one helps prove the impact employees can have on your organization when empowered with the correct tools.

There may be other organizational barriers like budgeting constraints, complex chargeback models, or competing priorities. Understanding these challenges is the first step toward overcoming them.

Limited licensing, suboptimal solution design, the lack of alignment around responsible AI, and organizational barriers are all symptoms at the foundational stage, but they are far from insurmountable.

With the correct strategies, you can transform these challenges into opportunities that build momentum, demonstrate value, and pave the way for scalable and impactful adoption.

Maximizing Value: Reassigning Licenses

What can you do to ensure that you are getting the most from the licenses that you have? Have you evaluated the usage of all your current licenses? If not, what might be the first step in auditing them? You can't ask for more investment if you can't prove that you are maximizing value from the licenses you do have.

Here are a few approaches you can take to ensure that your organization is getting the most value from the current level of investment:

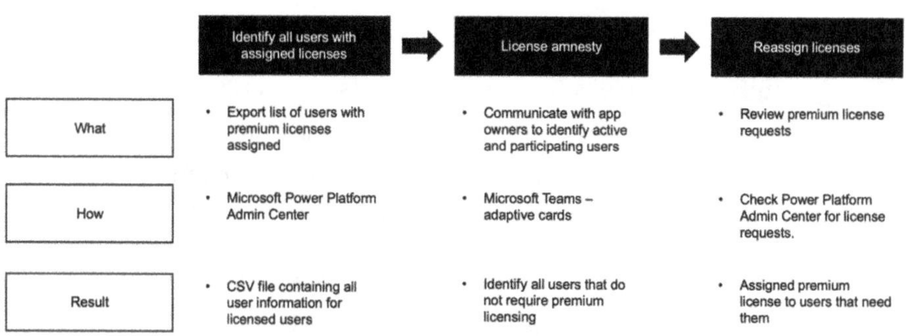

Figure 1-12. *Process for optimizing license assignment*

Step 1: Conduct a Usage Audit

Begin by identifying all users with assigned licenses. This step ensures you have a clear picture of who holds licenses and how they are being used.

What to do: Export a list of all users with premium licenses assigned. Use tools like the Microsoft Power Platform Admin Center to extract this data.

Result: A detailed CSV file containing user information, including the licenses they hold and how often they use the platform.

Step 2: Analyze Usage Patterns

Once you've gathered the data, categorize users based on their level of engagement and the business-criticality of their usage. Look for patterns such as

- **Inactive or minimally active users**: Identify licensed users who rarely or never use the platform. This could involve comparing license assignment data against actual usage patterns. This could be surfaced in the Microsoft Purview Audit log - where you could search for events such as "Launched app," "Created app," "Published app," "Created flow," and "Edited flow" to see if the user is active.
- **High-value users**: These individuals are using the platform to deliver measurable outcomes, such as automating key workflows or building solutions that impact critical business functions.
- **Overprovisioned licenses**: In some cases, users might hold licenses with features they don't fully utilize, which could be reassigned or adjusted.

Step 3: Execute a License Amnesty Program

Implement a license amnesty initiative to engage users and encourage transparency about their license needs. This process includes

- Communicating with app owners and participants using adaptive cards or internal communications to identify active users and their requirements

- Creating a safe space for users to declare underutilized licenses without penalty, enabling them to voluntarily relinquish unused or unnecessary licenses

Step 4: Reassign and Optimize Licenses
After the amnesty period, reassign licenses based on need:

- **Prioritize high-value users and initiatives**: Ensure that premium licenses are allocated to projects or individuals delivering the most significant impact.
- **Review ongoing requests**: Use Power Platform Admin Center to manage license reassignment and track any new requests.
- **Monitor results**: Set up periodic reviews to ensure that licenses continue to align with organizational priorities and user needs.

Step 5: Establish a Feedback Loop
Maximizing license value should be a continuous system for regularly collecting feedback from users and monitoring license usage metrics to help you adapt as organizational needs evolve.

Uncovering Early Wins

An inventory of your apps and flows is a goldmine of insights. By analyzing patterns of usage and criticality, you can identify early wins that showcase the value of Power Platform and build momentum for broader adoption.

Start by looking at the frequency of usage and the role of each app or flow plays in supporting key processes. High-frequency apps indicate widespread adoption, while high criticality apps are essential to business operations, even if they are used less frequently. By combining these dimensions, you can uncover which assets provide the greatest impact.

Below is an illustration of how you can classify your inventory of assets to highlight early, high-value opportunities:

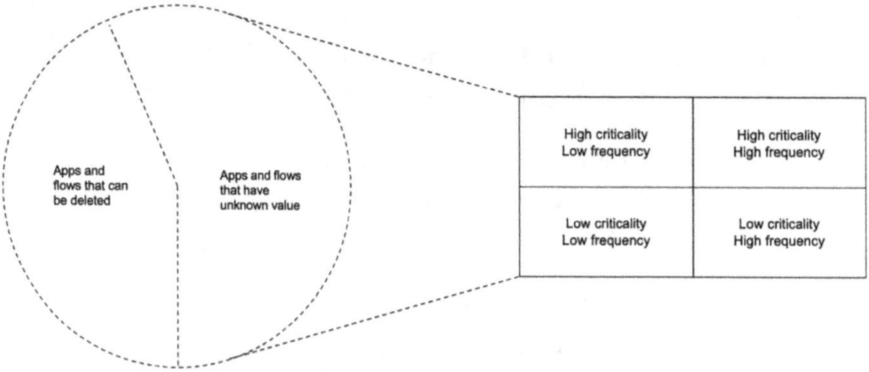

Figure 1-13. *Classifying asset inventory to highlight high-value opportunities*

Some examples of high criticality apps and flows:

Frequency/criticality	Example
High frequency/high criticality	Daily operations dashboard
Low frequency/high criticality	Year-end financial reporting workflow
High frequency/low criticality	Office supply request app
Low frequency/low criticality	Team building event sign-up

We'll discuss more about measuring and communicating value later in this book, but for the purposes of gathering data that is in-depth enough to build a business case, I would suggest that for each broadly used app, you would ask questions like:

KPI	Question
Time-saving	How much time has this app saved your team per week/ month?
Productivity gains	Has this app enabled your team to handle more tasks or requests? Does this app address any critical pain points or bottlenecks within your department?
Innovation and growth	Has this app enabled your team to explore new opportunities or tackle challenges that were previously difficult to address?
Cost reduction	Has this app reduced manual processes?
Revenue contribution	Has this app directly contributed to revenue generation? If so, how much?
Organizational goals	How does this app align with the organization's broader goals?
Scaling challenges	Are there technical limitations or resource constraints that could hinder the app's performance?
Future goals	Would adding AI capabilities enhance this app (e.g., sentiment analysis, generative AI)?

Unmet Needs

Unmet needs highlight the gap between the current capabilities and the desired outcomes that departments and business units are trying to achieve. These unmet needs often center around resource constraints, scalability issues, or the inability to maximize Power Platform's potential due to insufficient investment in product licensing

You should aim to document the technical limitations of solutions built with features available under existing licenses, as well as the missed opportunities teams and departments face – both in terms of process inefficiencies and strategic potential.[10]

To understand the appetite and plans for low-code and AI, you'll need to engage in a structured outreach effort. Start by having conversations with leaders within areas of your organization that are already using Power Platform. These conversations can provide valuable insights into successes and challenges. Additionally, identify areas within your business where there is a high level of manual process or repetition.

Here's how to scope and structure these conversations effectively:

Identifying technical gaps and limitations of unlicensed solutions

Start by asking the right questions

To uncover gaps, leaders and teams should reflect on their current usage and identify areas for improvement. Consider the following:

- What specific features would enable your team to deliver higher-impact solutions?
- Are there performance challenges due to scalability issues?
- Are there tasks or processes that you avoid automating due to platform limitations or lack of resources?

Bridging these gaps into focus

Once you've asked these questions, the answers will often reveal the barriers holding your organization back. For example:

> "Our department needs access to Dataverse for scalable data solutions, as SharePoint lists are causing performance issues during peak usage."

[10] Forrester Research (2022). *The economic impact of low-code platforms.*

> "We require AI Builder capabilities to automate invoice processing but are limited by the current license model."

Identifying these limitations enables you to prioritize where additional licensing or platform enhancements will have the greatest impact. These insights provide the foundation for building a stronger case for investment and ensuring that resources align with the most pressing needs.

Missed opportunities

In every organization, gaps in platform utilization can translate into missed opportunities - both operational and strategic. These missed opportunities often stem from limited licensing or resource constraints that restrict the platform's full potential.

Identifying process inefficiencies

Many missed opportunities arise from inefficiencies in manual workflows or workarounds created due to platform limitations. Identifying these inefficiencies is a critical step toward maximizing the value of your investment.

Ask yourself and your teams these questions to uncover inefficiencies:

- What manual processes still exist that could be automated with better platform access?
- How much effort is placed on workarounds due to inability to use premium features?
- Are there integrations with other systems that you cannot achieve with current tool? How is this restricting progress?

Examples of process inefficiencies:

> "Our team spends 10+ hours a week reconciling data manually, which could be automated with the use of AI Builder Invoice processing."

> "Performance issues caused by M365 data sources are leading to frustration, reducing user adoption by 20%."

Connecting inefficiencies to missed opportunities

These inefficiencies represent not only lost time but also reduced adoption and morale. Addressing these gaps allows teams to focus on more meaningful work and helps fosters a culture of innovation across the organization.

Identifying missed strategic opportunities

Operational inefficiencies are only a part of the picture. At a higher level, strategic opportunities are often delayed or abandoned due to resource constraints, limiting the ability to align technology with organizational goals.

Use these questions to identify missed strategic opportunities:

- What projects or initiatives have been delayed or abandoned due to resource constraints?
- Are there strategic goals that Power Platform could support with better investment?
- How would scaling the capabilities in Power Platform help you meet organizational objectives?

Examples of missed strategic opportunities:

The impact of these missed opportunities can often be profound:

> "We had to postpone automating compliance workflow because we lack licensing for AI Builder."

> "The inability to scale our apps across regions has delayed digital transformation goals in customer service."

From strategy to execution

Highlighting these gaps is critical for demonstrating the broader value of Power Platform. By surfacing these constraints, leaders can make informed decisions about where to invest to drive both short-term wins and long-term growth.

Structuring the findings

After conducting your research and gathering insights, you now have a clear picture of Power Platform's current state. You know which apps are most popular, where refactoring to premium features could unlock additional value, and you've gathered input from leaders across the organization. The next step is to create a compelling business case that takes your key stakeholders on a journey of understanding and action.

To build this narrative effectively, focus on these three steps:

- **Impress them with the current state of Power Platform**: Start by showcasing the progress made so far, even within the constraints of limited licensing. Highlight the great work your colleagues have already achieved and the impact it has had. This sets a positive tone and builds credibility for your case.
- **Share the unmet needs of the business**: Transition into the challenges. Share the inefficiencies that persist, the missed opportunities for process improvement, and the strategic goals that are at risk due to current limitations. These insights should make the case for change feel urgent and grounded.
- **Provide the solution**: Conclude with a clear and actionable request. Specify the additional licensing or resources needed, and articulate the transformative impact they will have on the organization. Use both tangible metrics, such as time saved or costs/errors reduced, and intangible benefits, like improved employee satisfaction or alignment with strategic goals, to strengthen your case.

By structuring your findings in this way, you guide your stakeholders through a logical progression – from celebrating achievements to addressing challenges and ultimately presenting a solution that moves the organization forward.

Below is an example structure based on what we've covered:

- *Context and current state*

Here, you briefly describe overall usage, highlighting successes that demonstrate the potential of Power Platform, e.g.: "The finance department has built over ten apps that automate key processes, saving 500 hours annually."

State the challenges, e.g.: "Despite this success, the lack of investment limits their ability to scale."

- *Identified needs*

List the specific unmet needs, grouped by category, e.g.: technical gaps (missing premium features or integrations). Example: "The finance team requires Dataverse access for broader organizational visibility, robust security control and better integration with existing systems."

Include process inefficiencies that could be automated. Example: "Due to limited licensing, vendor approvals are still handled manually, delaying operations."

Highlight delayed or unrealized projects and goals that impact progress toward strategic goals. Example: "The inability to access predictive AI capabilities has limited our ability to forecast inventory needs effectively."

Include responsible AI-specific gaps where there are governance or ethical concerns about current use. Example: "We lack processes to ensure training data for custom-models is diverse and not reinforcing existing biases or introducing new ones."

- *Highlight the impact of unmet needs*

Quantify the negative impact where possible:

"Manual processes are costing the team 200 hours annually."

"The inability to scale apps has delayed the company's strategic goal of streamlining operations."

- *Provide some recommendations for addressing needs:*

"Investing in AI Builder and Dataverse licensing would enable the automation of repetitive tasks like invoice processing, freeing up resources to work on more strategic work."

"Building an HR Helpdesk agent would free HR specialists time by 70%, allowing them to work on strategic goals."

Here is an example business case using some of the statements above. It's all fictional but covers all the points we raised:

To: [Key stakeholder]
From: [Your name, role]
Subject: Proposal for expanding Power Platform and AI licensing

Executive summary

Power Platform has proven to be a transformative tool for innovation and efficiency within Contoso. However, limited licensing has created barriers to scaling its benefits across teams. This request outlines the challenges posed, the missed opportunities from underinvestment and the potential for expanded licensing to unlock measurable value aligned with our strategic goals.

Current state and challenges

Elements of the business have made substantial progress using Power Platform, developing over 25 apps that streamline processes in HR, Marketing and Operations. These solutions save approximately 1,000 hours annually. However, current licensing limitations are holding back broader adoption and further innovation.

Challenges:

1. Technical limitations:
   Teams are relying on SharePoint lists for data storage due to limited access to Dataverse. This is leading to performance issues and user frustration.
   Lack of AI Builder access restricts the ability to automate repetitive tasks such as document processing.

2. Missed opportunities:
   HR postponed launching a new HR helpdesk that would free up time from five HR specialists, saving the department over $25,000 annually due to limited licensing.
   The Marketing team delayed implementing AI-driven campaign analytics due to lack of advanced feature availability.

3. Process inefficiencies:
   Vendor approval workflows remain manual, taking three days per request instead of one.

———————————————————————

Opportunities and value of expanded licensing

Investing in additional licenses would enable us to overcome these challenges and realize significant operational and strategic benefits.

Opportunities

Efficiency gains:

Automating HR onboarding process with agents could save 300 hours annually.

Streamlining vendor approvals could reduce processing time by 36%.

Scalability:

Migrating/refactoring key applications to store data in Dataverse would enhance performance and security. It would enable larger data volumes, improve user adoption and satisfaction.

Strategic alignment:

Aligns with organizational priorities of improving operational efficiency, empowering innovation and driving digital transformation.

Proposed solution

To address the challenges, and unlock Power Platforms potential, we propose the following:

Expand licensing:

Prioritize high-impact teams like HR and Marketing for additional licenses

Provide training:

Conduct workshops to train users on premium features and ensure effective adoption.

Include responsible AI (RAI) training to embed ethical practices into workflows.

Monitoring and evaluation:

Track usage and outcomes through quarterly reviews to ensure licenses are being used effectively and delivering ROI.

Recommendations and next steps:

1. Approve the expansion of Power Platform licensing.
2. Assign an executive sponsor to oversee rollout and adoption.
3. Begin implementation with high-impact teams in Q1.
4. Conduct a six-month review to assess outcomes and refine strategies as needed.

Supporting data:

Survey results from HR and Marketing teams showing unmet needs.

Current platform usage metrics and success stories.

Please let me know if you need any further details.

[Your name]

———————————————————————

Wrap-Up

We've covered how to tackle licensing challenges head-on and how to optimize what you have, build evidence to change, and make a compelling case for investment. The strategies are field-tested approaches that I've seen unlock value in organizations just like yours.

Licensing challenges don't have to limit ambition. With a clear process, some creativity, and a strong business case, you can transform these roadblocks into opportunities, demonstrating to leadership that low-code and AI are ready for big bets.

Now we've addressed licensing, it's time to set sights on your vision for Power Platform. Let's explore how to align strategy, value, and success. Before moving on however, take a moment to list one action you can implement immediately based on what you've learned here. Small steps lead to significant progress.

Absence of a Strategic Vision

At the foundational level of low-code and AI adoption, the absence of a strategic vision is one of the most common and significant barriers to success that I experience. If an organization hasn't charted a successful course, it can't be successful.[11]

In this section, we'll look at some of the common challenges that an absence of a low-code strategic vision manifests. We'll then explore some strategies for creating a compelling strategy for your organization.

[11] Gartner (2023), AI Strategy: Building a value-driving AI strategy for your business.

Symptoms of an Absent Strategic Vision

Fragmented efforts

- Teams and departments operate in silos.
- Experimentation with low-code isn't often aligned with strategic objectives.
- Solutions are reactive, often overlooking ethical considerations or responsible AI practices, and address only narrow problems rather than a broader set of business capabilities.

Lack of strategic alignment

- Low-code and AI work is not tied to the organization's overarching objectives.
- Stakeholders don't see Power Platform's long-term value or relevance.

Difficulty justifying investment

- Impossible to build a strong case for funding or resources.
- Leadership views low-code and AI as experimental rather than a strategic enabler.

Low team engagement

- Employees struggle to understand where Power Platform fits into their roles or the organization.
- Resistance to change increases without shared purpose of direction.

If any of that sounds familiar, you are far from alone. However, with a strategic vision and a sense of direction, you can (over time) resolve most, if not all, of the problems I just mentioned.

The Elements of a Good Vision

Go ahead and search for vision and strategy templates. There are a LOT of examples for you to take inspiration from! We're not about to invent a new approach to something that has been done <insert number of search results here> times before.

A strong strategic vision serves as your foundation for achieving organizational goals. In the context of low-code and AI adoption, a good strategy aligns this technology with business priorities and should inspire action and sustained progress.

Below are the key elements of a good strategy:

A clear and inspiring vision

A compelling vision outlines the desired future state and inspires stakeholders to act.

Example: *"Achieve operational excellence by enabling teams to create AI-powered, scalable, impactful low-code solutions."*

Well-defined strategic objectives

Follow the SMART framework.[12] SMART objectives – Specific, Measurable, Achievable, Relevant, and Time-bound – are critical because they provide clear, actionable, and trackable goals.

Objectives should not just align with business priorities but also integrate ethical practices to build trust an AI-powered solutions.

Objectives can be segmented by timelines focusing on short-, medium-, and long-term goals or by complexity, emphasizing incremental advancements in capability:

---

[12] George T Doran (1981), *There's a S.M.A.R.T way to write managements goals and objectives.*

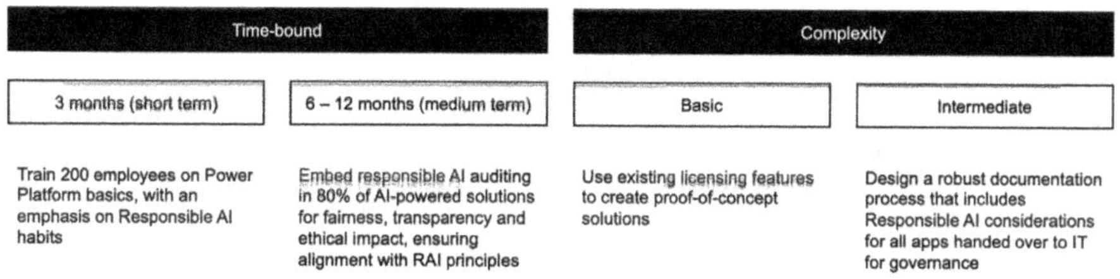

Figure 1-14. *Illustration of time-bound and complexity-based objectives*

Approaches for Creating a Strategic Vision

Creating a compelling strategic vision is as much about engaging the right people as it is about defining the right goals. It requires collaboration, creativity, and a practical framework to guide the process.

In this section, you'll learn about the various approaches that I've used to successfully help customers define their low-code visions.

Engaging with the Right Audience

The first step is identifying who needs to be in the room to shape the vision. Having a diverse audience reflects the different perspectives within your organization and helps align with broader priorities. Typically, the following personas attend:

Executive sponsor: They set the direction, ensure alignment with strategic goals, and hold others accountable.

Department heads: They represent operational needs and challenges specific to their area of responsibility.

Key influencers: Who are enthusiastic about innovation and can help champion the vision.

End users: Who provide practical insights and help ground the vision in day-to-day realities.

Create the Right Setting

The environment plays a significant role in fostering creativity and focus. Poorly structured sessions can lead to frustration and uninspiring results, while the right setting can unlock transformative ideas.

Here are some things to consider:

Collaborative atmosphere: Use a distraction-free space to encourage dialogue.

Structured agenda: Plan sessions with a clear flow, balancing creative brainstorming with practical discussions.

Set expectations and boundaries: Ensure all participants can contribute; set rules around taking/making calls and checking email.

Break often: Ensure you make space for breaks. I usually break for five minutes or offer a break every hour to ensure focus. Break even more often if you must run this workshop virtually – it is very hard to remain focused for lengthy periods when participating remotely.

Facilitation tools: Leverage tools like whiteboards (virtual or physical), sticky notes, and flip charts to capture ideas in real time.

These factors matter. Putting thought into the setting helps foster engagement and reduce hierarchy-driven discussions and ensures everyone feels empowered to contribute.

Use Creative Approaches to Spark Ideas

Defining what the future looks like requires breaking away from routine thinking. Asking the right questions and using creative techniques can inspire innovative ideas. Below are two powerful workshop techniques that combine creativity with focus:

- **Future-state visualization**: Asking participants what they would expect to see more of or less of when Power Platform is successfully adopted. This is a great way to spark creative conversations.
- **Reverse brainstorming**: Start with obstacles and then brainstorm approaches to overcome them.

Each technique approaches vision creation in slightly different ways. I've created the following illustration to help better visualize the stages each technique uses:

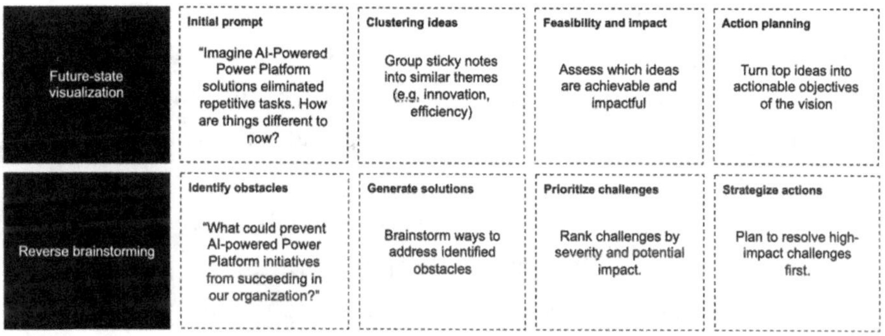

***Figure 1-15.** Different techniques to drive a vision and strategy workshop*

Let's explore each more deeply.

Future-State Visualization

How do you inspire a room of people to envision beyond incremental improvements, to a transformational future?

Future-state visualization is a powerful technique that helps teams articulate what success looks like when new tools, such as Power Platform, are integrated into their workflows. By shifting the focus away from current limitations and instead imagining an ideal future, participants are encouraged to think expansively and explore opportunities they might not otherwise consider.

At its core, this approach is about creating a shared vision. It bridges individual aspirations with organizational goals, helping participants align around a unified direction.

Future-state visualization taps into creativity while grounding ideas in achievable outcomes. This approach allows leaders to uncover latent opportunities, surface hidden challenges, and inspire a sense of ownership among their teams.

To tap into creativity, it's important to guide participants through structured prompts, encourage diverse perspectives, and integrate ethical considerations. By imagining not just the "what" but also the "how" of success, you are creating a road map that is both ambitious and sustainable.[13]

How do you move from abstract ideas to a collaborative and engaging exercise? The success of this approach depends on guiding participants thoughtfully through the process. By setting the right tone, creating the right environment and asking open-ended questions will bring the vision into focus. Here is how to get started.

[13] The Strategy Institute (n.d.), 2024: *How to align your team around a shared vision.*

Running Effective Ideation Sessions for Future-State Visualization

- **Ask open-ended questions to generate ideas**

  Frame prompts in ways that invite broad, imaginative responses. Consider the capabilities of an AI-powered Power Platform. Frame questions that inspire thinking toward a future state where AI is fully integrated into your organization. I've used prompts like

  "What would your ideal day look like with fully adopted AI-powered Power Platform solutions?"

  "How could AI-powered Power Platform tools transform how your team collaborates?"

  "What barriers that frustrate you now would no longer exist?

  Encourage participants to think beyond their immediate roles or constraints, enabling them to focus on possibilities rather than limitations.

- **Encourage multiple perspectives**

  Diverse ideas come from individuals viewing the same problem through different lenses:

  Prompt participants to think not just about their own responsibilities and processes but how other teams or departments could benefit.

  Ask participants to consider how different stakeholders - executives, front-line employees, or customers - might experience the future state differently.

- **Incorporate responsible AI prompts**

  Integrate ethical considerations into the visualization or ideation exercise to help ensure that the ideas generated align with responsible AI principles. Since responsible AI will be unfamiliar to many, use prompts that raise awareness of real-world risks and support thoughtful design.

  Example prompts:

  "Could this solution unintentionally exclude certain users or communities?"

  "What would happen if this AI model made an error – who would be impacted, and how would we know?"

  These types of questions encourage thinking beyond functionality and consider fairness, accountability, and safety from the start.

1. **Documenting ideas to maximize impact**

   To ensure ideas generated during the session are actionable and don't get lost, it's important to document them in a structured way. Here's how to do this effectively:

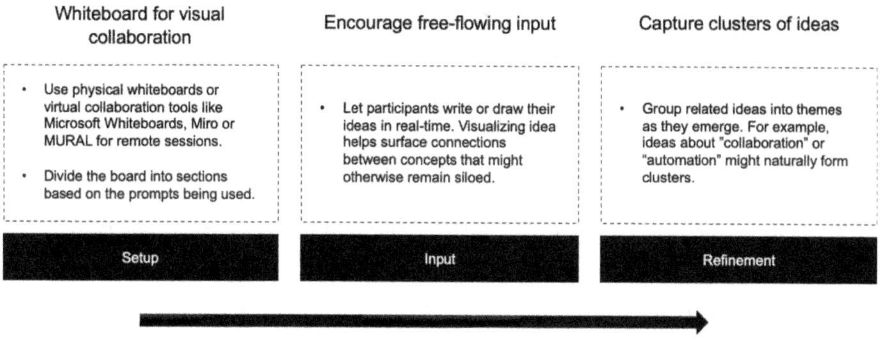

Figure 1-16. *How to effectively document ideas to maximize impact*

- **Encouraging participation and refining ideas**

  After documenting, revisit the ideas as a group to refine and prioritize the outcomes. Begin with collaborative review and clustering session, walking through the documented ideas to consolidate any overlapping or similar suggestions. Encourage prioritization by using techniques like dot-voting to identify ideas with the highest impact and feasibility. This ensures the most valuable concepts resonate with the group.

Finally, focus on the top ideas by expanding on key themes – encourage participants to discuss potential barriers, outline next steps, and explore how ideas can be translated into actionable outcomes.

Turning Grouped Ideas into a Vision

Now you've completed the brainstorming and clustering phase of future-state visualization, the next step is to translate these ideas into a compelling vision.

A vision alone is not enough to inspire or drive action – to operationalize the ideas, convert the clustered themes into a list of prioritized objectives categorized by timeline: short-term, medium-term, and long-term.

Steps to create prioritized objectives:

For each theme, list specific, actionable objectives that align with the vision.

Example for the theme of automation:

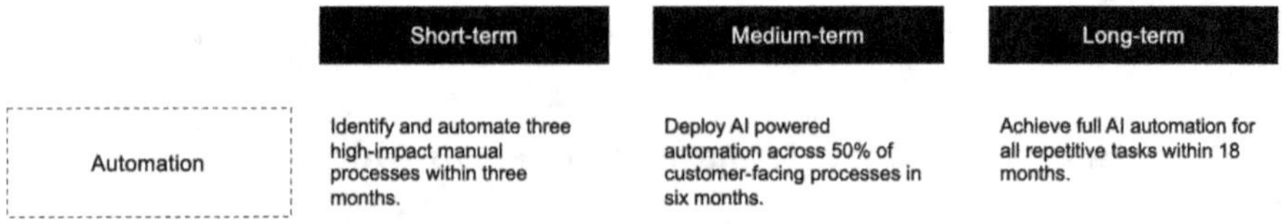

Figure 1-17. *Grouped objectives prioritized by timeline*

Assess feasibility and impact

Step 2 of turning ideas into actionable objectives is to evaluate each proposed objective against two key dimensions:

- Feasibility
- Value measurement

A well-structured evaluation will prioritize objectives that align with your organization's current capabilities and deliver measurable benefits.

Operational and technical feasibility

Understanding whether an objective can realistically be implemented involves assessing both **operational feasibility** (how well it fits within current processes and resources) and **technical feasibility** (whether the technology can deliver the expected outcome).

Operational feasibility

Focuses on whether your organization has the capacity to execute the objective:

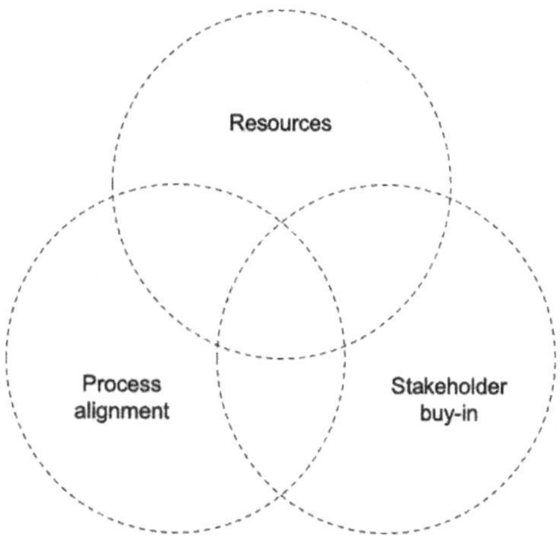

Figure 1-18. *Venn diagram illustrating operational feasibility*

Let's explore each view:

Resources: Does the organization have the necessary budget, time, and personnel?

Example: "Do we have the support staff to support the solution, or will we need additional support?"

Stakeholders buy-in: Are decision-makers and end users on board?

Example: "Will teams resist adopting the new system due to unfamiliarity or work output disruption?"

Process alignment: Does the objective align with current work processes, or will it require significant reengineering?

Example: "Will automating invoice processing integrate smoothly with our existing finance operations?"

Technical feasibility

Technical feasibility focuses on the readiness of the technology:

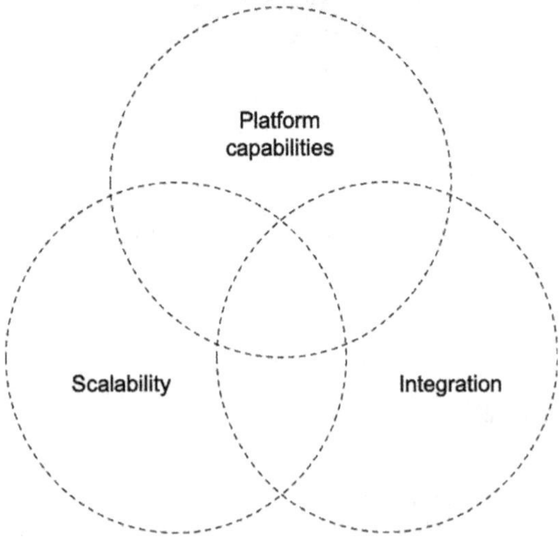

Figure 1-19. *Venn diagram illustrating technical feasibility*

Platform capabilities: Can Power Platform and integrated AI tools support the desired functionality?

Example: "Can we utilize AI Builder capabilities with our current licensing?"

Scalability: Can the solution scale across teams or departments without performance issues?

Example: "Will AI Builder be able to process thousands of invoices daily without latency?"

Integration: Will the solution work with existing systems or require additional development?

Example: "Can this automation integrate seamlessly with our ERP system?"

Value measurement

After determining feasibility, you should assess the potential impact of the objective. This includes considering both the tangible and intangible benefits it will deliver.

Tangible value

Tangible value represents measurable outcomes that directly contribute to organizational efficiency or financial performance, for example:

- **Time savings**: Estimate the hours saved by automating manual processes.
- **Cost reduction**: Quantify the reduction in operational costs.
- **Revenue impact**: Identify potential gains from enhanced productivity or new capabilities.

Intangible value

Intangible value is typically a benefit that might not have immediate financial implications but could contribute to long-term success. Here are some examples:

- **Employee satisfaction**: Will the objective reduce frustration or improve workflows?
- **Customer experience**: Does the objective enhance how customers interact with your organization?
- **Brand and reputation**: Does the objective position the organization as a leader in AI innovation or responsible AI?

Here is an example of what the objectives could look like in a low-code vision:

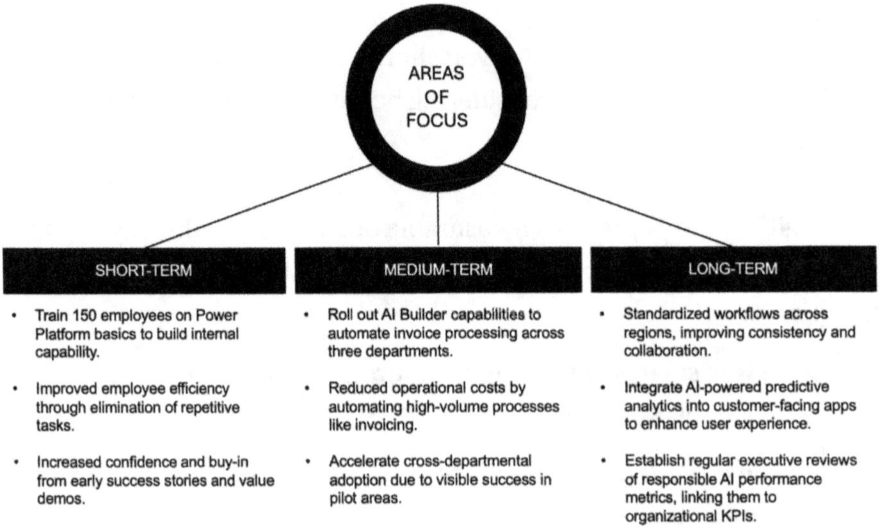

***Figure 1-20.** Illustration of objectives prioritized over timelines*

Reverse Brainstorming

Reverse brainstorming flips traditional problem-solving on its head by first identifying obstacles or ways to make the situation worse. This "reverse" approach can often uncover hidden challenges and lead to more creative solutions.[14] Here is an overview of this approach:

Stage one: Define the problem, e.g., what's the opposite of the desired outcome?

The goal is to clearly articulate a problem or opportunity you're addressing.

Example: Instead of asking "How can we improve Power Platform adoption?" You might ask: "What would prevent Power Platform adoption from succeeding?"

[14] MindTools (n.d.), 2021, Reverse brainstorming.

Stage two: Generate reverse ideas - What could go wrong or sabotage success?

Generating reverse ideas means deliberately thinking about how to cause the problem instead of solving it. Essentially, you identify the actions, behaviors, or circumstances that could lead to failure or make the problem worse. It flips traditional brainstorming into a more counterintuitive but highly creative exercise. Here is an example:

If the problem is "What would prevent Power Platform adoption from succeeding?"

> **Reverse idea**: "Ensure no one understands AI features by avoiding training programs."
>
> **Reverse idea**: "Avoid getting buy-in from leadership by not involving them early."
>
> **Reverse idea**: "Ensure high-value applications perform poorly and frustrate users by designing confusing interfaces."

Stage three: Flip and uncover solutions - Instead of "how to make the problem worse," you rephrase into "how to solve the problem."

Often, the exaggerated or negative aspects of the reverse brainstorming process reveal hidden gaps, overlooked risks, or untapped opportunities. The goal is to reformulate these negative statements into constructive solutions that are both actionable and aligned with your objectives.

How to approach it:

Analyze reverse ideas

Begin by revisiting the list of reverse ideas generated in the previous stage. Discuss each one as a group, and identify the underlying issues or risks they highlight.

This stage involves flipping the reverse ideas into actionable solutions

What does it mean?

- Analyze each reverse idea for insights

Reformulate the negative statements into proactive strategies that address risks or challenges identified in previous stage.

- Reformulate into solutions

Transform these insights into proactive strategies.

Here are some examples:

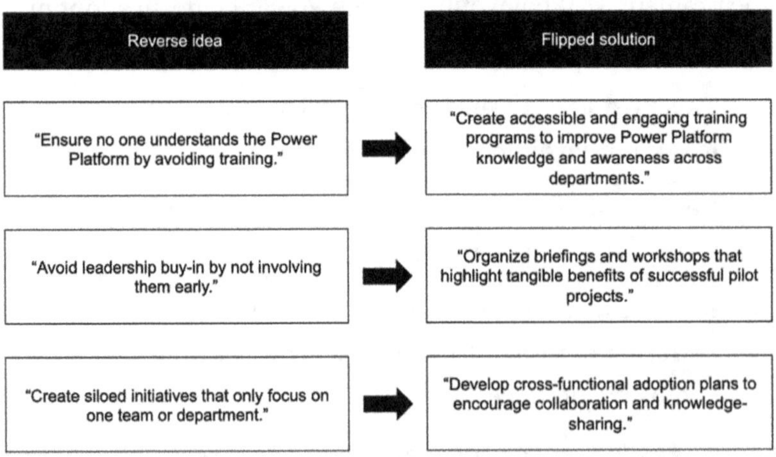

Figure 1-21. *Examples of reverse ideas used in reverse brainstorming*

Why it matters?

This approach provides actionable insights that are directly tied to specific risks. Teams get a clear path forward that is rooted in addressing real-world challenges, not just theoretical solutions.

Stage four: Prioritize solutions by risk and impact – In this final stage, participants evaluate the flipped actions to prioritize them into timelines based on

- **Risk**: The potential consequences of not acting
- **Feasibility**: The resources, time, or expertise required to implement

Here is an example of what it might look like when risk and impact prioritization is complete:

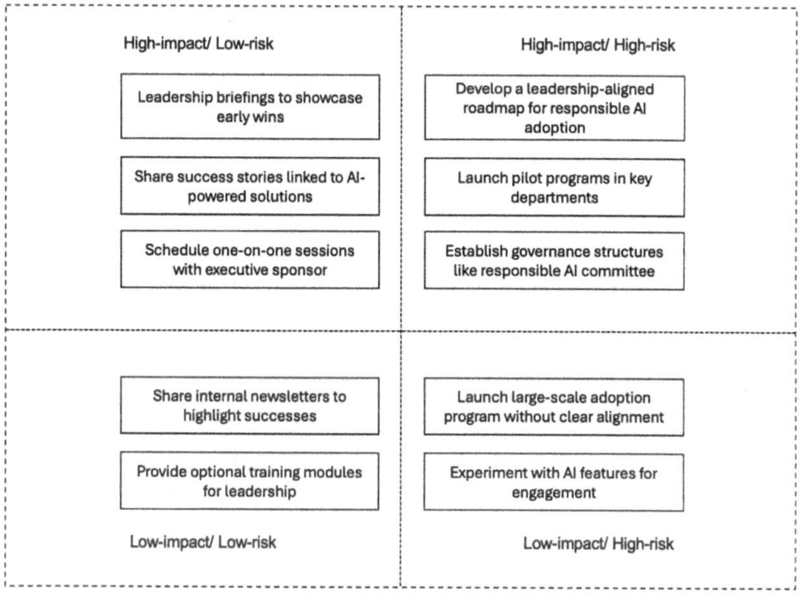

Figure 1-22. *Completed risk and impact prioritization*

Each action is mapped into a timeline:

- **Short-term (1–3 months)**: Actions that can deliver quick wins, like pilot programs or immediate process improvements
- **Medium-term (6–12 months)**: Priorities and projects that require some planning and alignment but can show meaningful impact
- **Long-term (12+ months)**: Strategic objectives such as embedding organizational-wide responsible AI frameworks into governance processes

Summary of Vision Workshop Techniques

The table below compares the two techniques that we've covered:

Technique	Best for	Key outcome
Future-state visualization	Broad, creative thinking	Ideal vision and desired outcomes
Reverse brainstorming	Identifying hidden challenges	Clear, actionable solutions

Creating Your Vision Statement

Both techniques that we've covered are great to agree on prioritized objectives, grouped by timelines. The one piece of a vision that pulls together the objectives from being a collection of abstract statements to an inspiring declaration that encapsulates your aspirations is the vision statement. This is often the hardest part - you thought getting a room full of diverse people to agree on actions was hard - wait until you've been through this a few times!

Here is an approach I've used with some success:

Reflect on objectives and themes

Start by reviewing the objectives and themes you've developed. Identify the central ideas that resonate most strongly across all the objectives. Ask participants:

- What are the most critical goals we're aiming to achieve?
- How does Power Platform and AI align with the broader organizational priorities?
- What outcomes will have the most transformative impact on our people or processes?

Define the desired future state

Painting a picture of where you want your organization to be once these objectives are realized is far easier now you've defined them. Here, you are looking to focus on the big-impact objectives, inspiring all stakeholders. To inspire participants, consider these prompts:

- "What does success look like in three years?"
- "How will our organization be different once Power Platform is fully integrated?"
- "How will this involve teams, improve processes, or transform our industry?"

Draft statement

Using the themes, and future state, draft a concise and engaging statement. The key is to keep it actionable and inspiring while avoiding jargon or overly technical language. A good vision statement often starts with phrases like

- "To become..."
- "To lead in..."
- "To empower teams to..."

Example: "To lead the industry by empowering teams with low-code and AI tools, streamlining operations, driving innovation, and embedding ethical practices that build trust and long-term value."

Test for clarity and inspiration

Review the draft vision statement with key stakeholders and the workshop participants. Ask:

- Does this clearly reflect our strategic goals?
- Is it both ambitious and realistic?
- Does it inspire and engage both leadership and teams?

Once refined, share the vision statement widely across your organization. It should be included in presentations, communications, and training materials to ensure alignment and buy-in at every level.

Here's an example of how your vision might look:

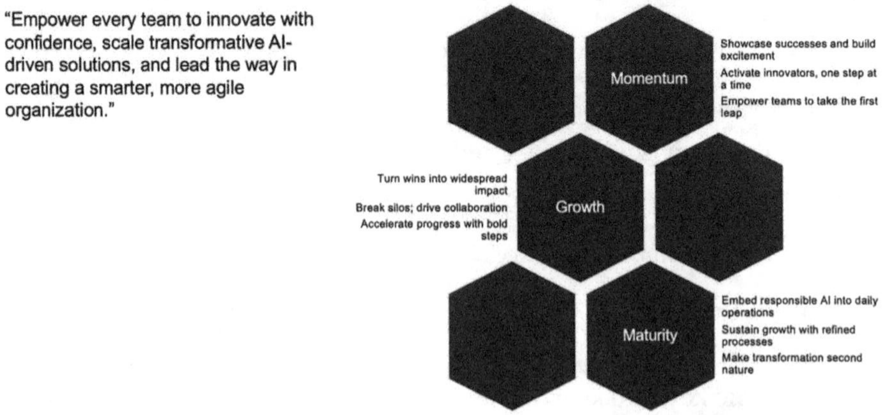

***Figure 1-23.** Example vision with momentum, growth, and maturity*

Next steps after creating a vision

Well done! Creating a vision is a significant milestone. Its success however depends on what happens next. To ensure the vision translates into actions, you need a plan to communicate the vision within your organization.

Communicating the Vision

A vision has no impact if it's not widely understood and embraced. Communicating your vision is critical to establishing trust in the organization. However, getting communication right requires research. Broadly speaking, you are looking to tailor your messaging based on the audience. You have a variety of different stakeholders in your organization who want to hear about the vision but have very different perspectives as far as the details go.

To reach these various audiences, consider:

- How is your message relevant to your audience's role?
- What level of detail are they interested in?
- How do they like to receive information?

Here's how to do it:

Leadership alignment: Ensure that your executive sponsor aligns with senior leaders to champion the vision.

Tailored messaging:

- **Executives, directors and senior leaders**: Focus on strategic alignment and outcomes.
- **Teams**: Highlight how the vision impacts day-to-day work and empowers them.
- **End users**: Explain the practical benefits and encourage engagement.

Multiple channels: Use all channels available - emails, presentations, community calls, town halls, internal newsletters, intranet announcements - to reinforce your message.

Storytelling: Share early wins or relatable scenarios to make the vision tangible.

Wrapping Up

After reading this section, you should now be able to identify how the absence of a clear vision manifests within organizations. You've explored a framework for establishing a compelling vision, learned how to run a workshop that helps define it, and gained insight into tailoring communications for different stakeholder groups.

With your vision in place and actively communicated, the next step is to extend that foundation of trust to the solutions being created. As enthusiasm grows around low-code and AI tools, so too does the importance of using them responsibly. Responsible AI introduces a new dimension – one many organizations have not yet fully considered. Without basic awareness of responsible AI principles, teams may unintentionally create solutions that undermine trust, invite compliance risk, or make biased decisions. Before governance structures can take hold, organizations must first focus on awareness – cultivating curiosity, prompting conversations, and laying the groundwork for a more informed and responsible culture.

Minimal Awareness of Responsible AI

What happens when you don't know the risks of the tools you're using?

For many organizations starting their journey with low-code and AI, responsible AI (RAI) is either misunderstood or overlooked entirely. The absence of RAI awareness and process creates ripple effects that go beyond technical issues, impacting trust, compliance, and organizational success.[15] Where employees and leaders are unaware of RAI principles,[16] they often underestimate the broader implications of AI misuse, unintentionally exposing their organizations and stakeholders to significant risks.

AI systems, by design, often automate or augment decision-making processes. When teams deploy these systems without understanding the need for transparency or accountability, errors or biases can go unaddressed. Without awareness of RAI, these systems are deployed with good intentions, but poor oversight can result in outcomes that undermine organizational credibility.

At the foundational level, you probably haven't established enterprise-wide governance for RAI. Building awareness and encouraging good habits should be the focus. Raising awareness at this level will start with small,

[15] World Economic Forum (n.d.), 2023: *The future of work starts with trust.*

[16] Microsoft - Microsoft Responsible AI Standard.

grassroots initiatives that cultivate curiosity, foster discussions, and lay the groundwork for broader cultural change.

Remember that at this stage, without any training, your peers are often excited by how quickly they can streamline existing tasks. They are excited by the potential and are looking for new ways to use their skills. There are a few, simple things that you can do to help them.

In this section, we'll look at how you can nurture the early phases of responsible AI.

Instilling RAI Habits Through Positive Reinforcement

Developing RAI habits begins by fostering good practices through encouragement and recognition. Research suggests that positive reinforcement - such as rewarding ethical decision-making and celebrating success - can be more effective than relying solely on rules or warnings.[17] By focusing on highlighting achievements and building confidence, organizations can reduce resistance and foster engagement in adopting RAI principles.

This section introduces a practical technique - *new app detection* - designed to instill RAI habits early in the development process.

New app detection: Automating early engagement

The new app detection process proactively identifies recently created apps, agents, and flows and provides developers with relevant resources and guidance.

Here's how it works:

- **Daily monitoring**: The process runs automatically each day, scanning the organization's environments for newly created assets.

[17] Skinner, B.F. (1953). *Science and Human Behavior.*

- **Identification and outreach**: Once the asset is detected, the app maker receives a personalized communication, such as an email, or adaptive card. This message includes essential information on governance policies, training opportunities, and support channels.
- **Encouraging responsible habits**: By reaching out within 24 hours of app creation, this process embeds responsible AI practices early in the development life cycle, helping app makers make informed decisions.

This is a simple process, but very effective. It helps ensure that governance and RAI isn't an afterthought but is an integral part of app development. The visual below illustrates how the detection process works:

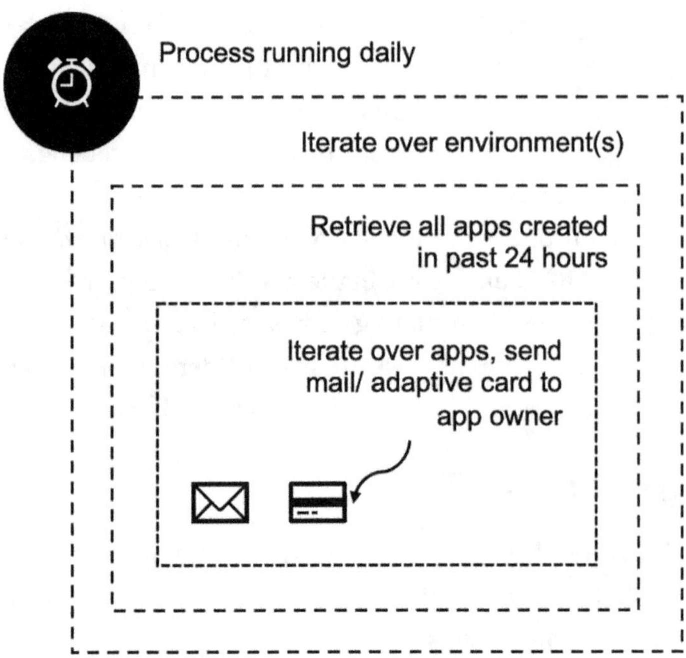

***Figure 1-24.** Process for identifying apps created within the past 24 hours*

Establishing Healthy RAI Habits: A Grassroots Approach

Building responsible AI (RAI) awareness is about embedding the principles into daily practices. At the foundational stage, where formal governance structures might not yet be in place, the focus should be on cultivating grassroots habits that lay the groundwork for broader cultural change. By encouraging teams to adopt simple, repeatable behaviors, you create a ripple effect that fosters trust and accountability.[18]

Where do you start? Establishing healthy RAI habits doesn't have to be complex. Instead, focus on creating small, actionable routines that teams can integrate into their development processes, sparking reflection and thoughtful decision-making.

Let's explore practical ways to make this happen.

The W.A.I.T. Framework

Sometimes, the pressure to deliver quickly or the fast-paced nature of low-code development can sometimes overshadow the need for careful reflection. However, ensuring low-code solutions powered by AI are developed responsibly requires deliberate thought at every stage. To encourage this, I developed two practical frameworks: W.A.I.T. and C.H.A.T. that focus on two critical aspects - decision-making (W.A.I.T.) and communication (C.H.A.T.).

The W.A.I.T. framework encourages teams to ask critical questions before starting development, focusing on who the solution impacts, the risks involved, the integrity of any data, and how outcomes can be explained and trusted.

[18] Duhigg, C. (2012). *The Power of Habit: Why we do what we do in life and business.*

By fostering this mindset early in the process, organizations can avoid unintended consequences and create AI-powered solutions that are ethical, transparent, and aligned with organizational values.

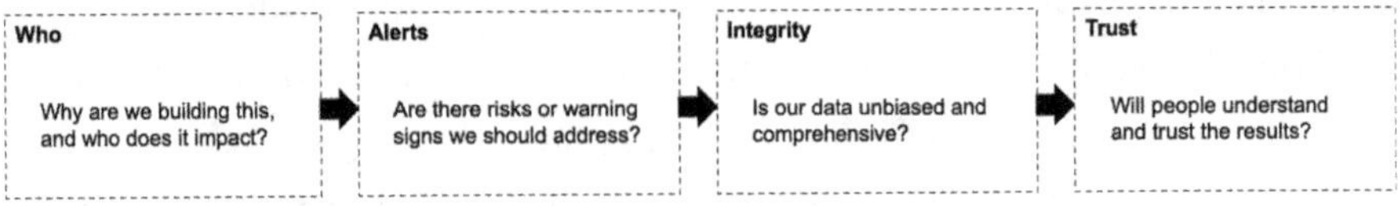

Figure 1-25. *The W.A.I.T. framework*

> **Who**: Who will benefit from this, and could anyone be adversely affected?
>
> **Alerts**: Are there risks or warning signs you should address?
>
> **Integrity**: Is the data you're using representative, and have you evaluated for fairness?
>
> **Trust**: Can you explain the AI output to end users or stakeholders?

Here is an example scenario that illustrates how the W.A.I.T. framework can catch potential early risk during the development phase:

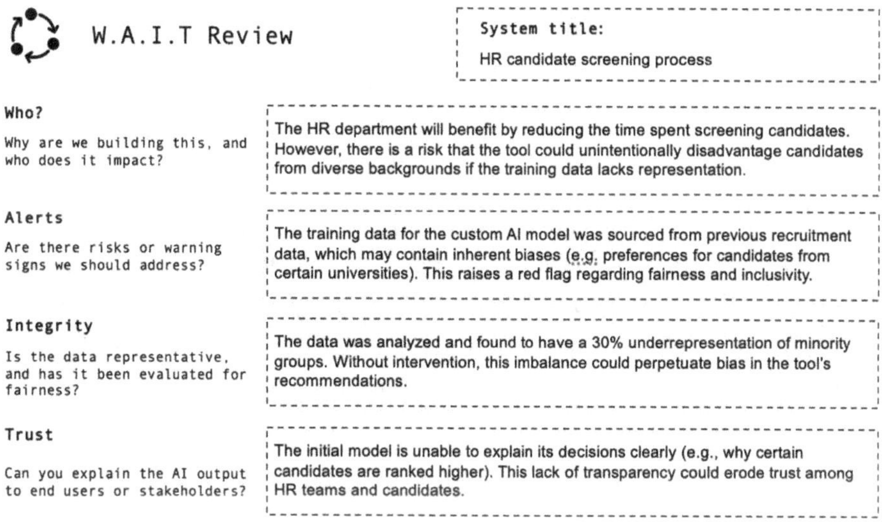

W.A.I.T Review

System title:

HR candidate screening process

Who?

Why are we building this, and who does it impact?

The HR department will benefit by reducing the time spent screening candidates. However, there is a risk that the tool could unintentionally disadvantage candidates from diverse backgrounds if the training data lacks representation.

Alerts

Are there risks or warning signs we should address?

The training data for the custom AI model was sourced from previous recruitment data, which may contain inherent biases (e.g. preferences for candidates from certain universities). This raises a red flag regarding fairness and inclusivity.

Integrity

Is the data representative, and has it been evaluated for fairness?

The data was analyzed and found to have a 30% underrepresentation of minority groups. Without intervention, this imbalance could perpetuate bias in the tool's recommendations.

Trust

Can you explain the AI output to end users or stakeholders?

The initial model is unable to explain its decisions clearly (e.g., why certain candidates are ranked higher). This lack of transparency could erode trust among HR teams and candidates.

Figure 1-26. *A completed W.A.I.T. review form for a new HR candidate screening agent*

Using the example about, the development team identified two key risks early:

- Bias in the training data
- Lack of explainability in AI outputs

The team paused development to address these challenges by

- Expanding training data to include more diverse candidate profiles
- Integrating an explainability feature into the tool to provide HR teams with clear justifications for candidate rankings

Result: By catching these risks during the development stage, the organization avoided potential reputational damage and created a recruitment tool that aligns with RAI principles, ensuring fairness and transparency.

The C.H.A.T. Framework

The C.H.A.T. framework offers a structured way to embed RAI principles into ongoing projects by facilitating meaningful conversations among stakeholders.

Unlike W.A.I.T., which is more suited for early-stage assessments, C.H.A.T. helps teams evaluate AI-powered projects in development, encouraging thoughtful consideration of risks, impacts, and safeguards.

It is designed to promote collaborative thinking and ensure ethical AI practices remain at the forefront throughout the life cycle of a project.

Overview of C.H.A.T.

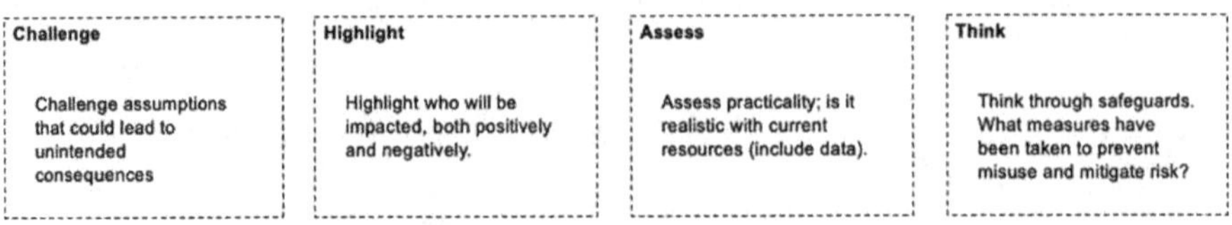

Figure 1-27. *C.H.A.T. framework*

C.H.A.T. stands for

> **Challenge** assumptions that could lead to unintended consequences.
>
> **Highlight** who will be impacted - both positively and negatively.
>
> **Assess** practicality - Is it realistic with current resources (include data)?
>
> **Think** through safeguards. What measures have been taken to prevent misuse and mitigate risk.

This framework fosters a culture of ethical awareness and cross-functional collaboration, ensuring that all perspectives are considered, from developers and stakeholders to end users.

Let's apply the C.H.A.T. framework to a scenario. Imagine your organization is developing an AI-powered customer support agent designed to handle routine queries and provide quick resolutions. The goal is to reduce response times and enhance customer satisfaction (note the tangible and intangible value here).

Applying the C.H.A.T. framework – your organization has a RAI policy whereby C.H.A.T. assessments must happen at key milestones in the project life cycle.

Challenge

What assumptions are we making, and could they lead to unintended consequences?

The team assumes that automating customer support will improve satisfaction without negatively impacting more complex queries. However, there's a risk that the AI agent or app might misinterpret nuanced issues or fail to escalate them to a human representative promptly.

Highlight

Who will be impacted both positively and negatively?

Positive

- Customers with straightforward queries will benefit from quicker resolutions.
- Support teams will have reduced workloads.

Negative

- Customers with complex problems might face delays if the AI agent cannot escalate issues effectively.
- Support teams could become disengaged if their roles are overly simplified.

Assess

Is it feasible to mitigate risks with current resources?

- The team has access to resources to train the AI agent on basic queries but lacks enough annotated data for edge cases.
- Implementing a robust escalation process is realistic but would require time and input from customer support representatives to design and test workflows.

Think

What safeguards can we implement to mitigate risks and ensure a successful rollout?

- Introduce a confidence threshold: If the AI agent is uncertain about a query, it should escalate to a human agent immediately.
- Provide an opt-out option for customers to request human support directly.
- Regularly audit the AI agent's responses using real customer feedback to improve its accuracy and effectiveness

Outcome of C.H.A.T.

By engaging in a structured C.H.A.T. session, the team identified potential risks and implemented safeguards early in the process. This ensured that customers received faster responses without sacrificing quality for complex issues; the support team retained meaningful work, ensuring continued engagement and motivation; and the organization maintained its commitment to RAI by proactively addressing risks.

The W.A.I.T. and C.H.A.T. frameworks encourage reflection and open conversations about AI-powered solutions. With the right level of encouragement (look to your executive sponsor for support), you can start to raise awareness about RAI without it feeling overwhelming. As you progress, you will look to formalize RAI into governance policy and have more rigor, but for now, you are doing the right thing by getting your maker community to reflect prior to development (W.A.I.T.) and to review while in development (C.H.A.T.).

Reinforcing Habits with Small Incentives

Habits are more likely to stick when there's positive reinforcement. Recognizing and rewarding teams and individuals that demonstrate good RAI practices can help your organization build momentum and encourage others to follow. By celebrating success and offering targeted incentives, organizations signal that RAI is a valued part of their culture and not just a requirement.

Incentives can take many forms, depending on the organization's culture, resources, and goals. Here are a few ways that I've seen to incentivize impactful initiatives:

Public recognition

- **What it looks like**: Share success stories in newsletters, town halls, or leadership updates. Recognize individuals or teams that exemplify responsible AI practices and ethical decision-making.
- **Why it works**: Public acknowledgement creates a ripple effect. Employees feel valued, and others are encouraged to follow their example.

Tangible rewards

- **What it looks like**: Offer tangible incentives like certificates, badges, or small awards. For example:
    - Digital badges: Create a badge system that rewards adherence to RAI principles, such as "AI Ethics Ambassador" or "Innovating with integrity."

  - Gifts or perks: Provide small but meaningful perks, such as gift cards, team lunches, or branded items recognizing RAI efforts.

- **Why it works**: Tangible rewards serve as a visible reminder of the value placed on responsible practices, reinforcing the behavior.

Career advancement opportunities

- **What it looks like**: Provide pathways for career growth linked to demonstrated RAI commitments. This could include mentoring opportunities, leadership roles with AI councils, or participation in high-profile projects.
- **Why it works**: Tying incentives to career development aligns personal goals with organizational priorities, making RAI principles a key part of an employee's success story.

Internal and external showcases

- **What it looks like**: Highlight teams that have successfully implemented RAI in internal newsletters or company blogs. For significant achievements, consider submitting their projects to external conferences or award programs.
- **Why it works**: Being featured externally as a thought leader strengthens the organization's reputation while motivating employees to strive for excellence.[19]

While individual incentives are critical to building good RAI habits, creating a broader sense of shared accountability and collaboration is the next step in embedding RAI into organizational culture. The foundation of

[19] Freeman, R.E. (1984). *Strategic management: A stakeholder approach.*

any sustainable RAI program lies in fostering a community of advocates who champion these principles across teams, departments, and even externally.

This next section explores how to establish and nurture a network of RAI champions - equipping them with the tools, authority, and support needed to lead the charge for responsible AI practices within your organization. Let's learn how to create a community that drives collective success.

Setting Up a Community or Network of RAI Champions

At the foundational level, organizations often lack clarity about how to handle ethical concerns or uncertainties. Providing a centralized point of contact - a person, team, or dedicated network - can ensure that RAI-related questions are addressed promptly and consistently.[20]

One key to success is starting with a clear and inspiring purpose for your community. A well-defined purpose helps to attract members as well as ensuring alignment with the organization's goals.

This section will guide you through creating that purpose, setting measurable objectives and fostering engagement. By the end, you'll have a road map for building a thriving RAI community that drives meaningful impact.

Community purpose

To thrive, a community or network should have a clear and compelling purpose. With sponsorship to support events, consistent communication, and engaging content, the community can grow in membership and influence. Equally important is setting measurable targets to track its impact. For example, how is the community driving awareness, fostering collaboration, or addressing RAI challenges? Defining these objectives ensures alignment with organizational goals and keeps the community focused.

[20] Kotter, J.P. (1996). *Leading Change.*

In my experience, using techniques like the "future-state visualization" (we covered this in the earlier section) workshop can help focus the community's mission. This method enables leaders to craft a vision and articulate the value of the community to potential members, sponsors, and stakeholders.

The example below highlights the essential components that help define and guide a successful community: a clear vision to inspire, a well-defined purpose to unify efforts, and tangible goals that provide direction and focus. Together, these elements ensure that members, sponsors, and leaders understand not only why the community exists but also what it strives to achieve.

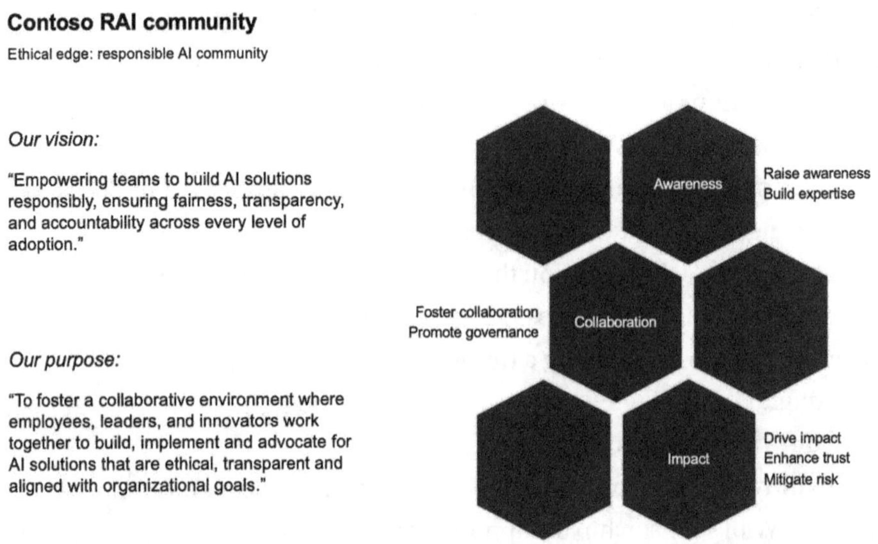

Figure 1-28. *Example community vision, purpose, and objectives*

Defining a vision and purpose is just a part of the journey. To measure the community's success and drive continuous improvement, it's essential to establish clear metrics. These metrics will allow you to track progress, demonstrate value to stakeholders and sponsors, and refine your approach over time.

Success metrics

Defining the goals of the community means you have something to measure the impact it's having.

Here are some examples that you could publish alongside the community vision and purpose:

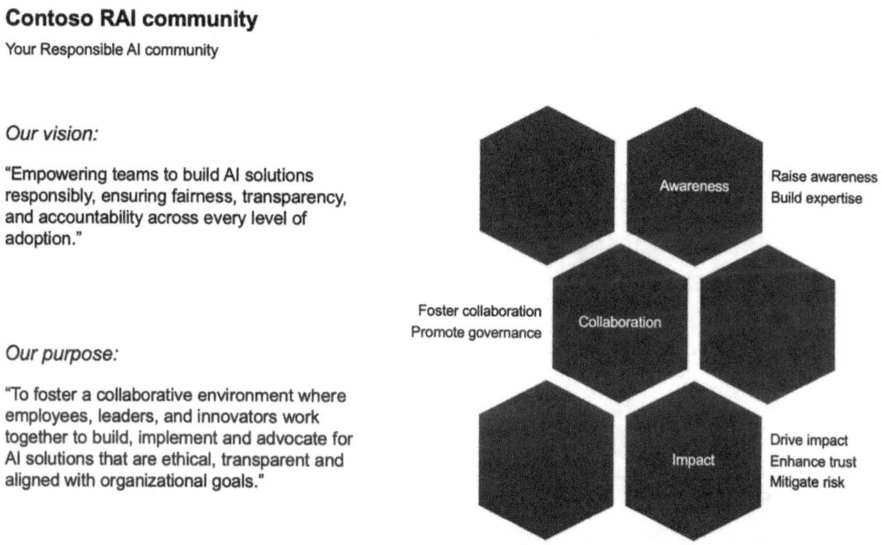

Figure 1-29. *Example community goals and targets*

Unlike broader organizational objectives, community goals and targets are often more focused on fostering engagement, demonstrating progress, and addressing specific challenges within the scope of the community's purpose.

Here's a step-by-step approach to creating effective goals and targets:

Step 1: Align Goals with the Community's Vision and Purpose

Begin by revisiting the vision and purpose of your RAI community. Goals should directly contribute to the overarching mission. For example, if the vision emphasizes collaboration, set goals that encourage cross-departmental initiatives or knowledge-sharing events.

Example:

- **Vision**: "To foster a collaborative environment where employees, leaders, and innovators build ethical AI solutions"
- **Goal**: "Host three cross-functional workshops annually to address ethical AI challenges and share best practices."

Step 2: Use SMART[21] Criteria for Target Setting

When defining targets, ensure they are

- **Specific**: Clearly outlining what the target aims to achieve.
- **Measurable**: Include quantifiable metrics to track progress.
- **Achievable**: Ensure they are realistic given the available resources.
- **Relevant**: Tie them back to the community's goals and organizational priorities.
- **Time-bound**: Establish a clear timeframe for achieving each target.

Example target:

"Train 50% of employees on AI principles within one year."

Step 3: Incorporate Feedback and Metrics

Engage the community to co-create goals where possible to ensure buy-in and relevance.

---

[21] Doran, G.T. (1981). There's a SMART Way to Write Management's Goals and Objectives. Management Review, 70, 35–36.

Use existing metrics or historical data to inform realistic targets: for example, if past engagement rates for training programs were 60%, aim for a modest increase rather than an overly ambitious 100%.

Example: "Increase RAI training completion rates from 60% to 75% within the next 12 months."

Step 4: Focus on Both Tangible and Intangible Metrics

While measurable metrics like training completion rates are important, don't overlook intangible aspects such as community sentiment, member satisfaction, or the quality of discussions in forums and events. Include periodic surveys or feedback loops to capture these insights - we'll talk about using feedback more later.

Step 5: Set Iterative Goals

Building flexibility into your targets allows you to accommodate evolving community needs and organizational priorities. Iterative goals can be updated as the community grows or as the RAI landscape changes.

Example: "Pilot a new RAI review process in two departments and refine based on feedback before scaling organization-wide."

Setting clear goals provides both a road map and opportunities for member contribution. However, the journey doesn't end there. You need to ensure progress and maintain momentum, and to do that, you need a way to measure the community's health and effectiveness over time. Metrics are the vital signs of your community.

Let's explore how to establish and leverage metrics to ensure your community remains a vibrant, purposeful, and impactful place.

Community health

To ensure your RAI community is thriving and delivering measurable impact, it's essential to define and track key performance indicators (KPIs).[22] These metrics provide a clear picture of the community's health, highlighting areas of strength and uncover opportunities for improvement.

[22] Kaplan, R.S., & Norton, D.P. (1996). *The Balanced Scorecard.*

KPIs should align to the roles and responsibilities within the community (we'll talk more about these shortly), ensuring that each contributor – whether an executive sponsor, community lead, or champion – has clear objectives and measurable outcomes. Below is a table offering examples of role-specific KPIs to inspire your approach:

Role	Key performance indicator
Executive sponsor	• Number of executive-level forums advocating RAI principles • Budget allocated and utilized for RAI-related initiatives • Feedback from leadership on alignment with organizational goals
Community lead	• Growth in community membership (e.g., % increase per quarter) • Engagement metrics (e.g., active participants in discussions) • Number of successfully completed community-driven initiatives
Content and events planning	• Number of events held (e.g., training sessions, community calls, newsletters written) • Attendance rates and feedback scores for events • Frequency of newsletters or internal communications
RAI champions	• Number of teams adopting RAI practices • Number of RAI-related consultations or support sessions provided • Success stories shared within the community (e.g., improvements made through RAI integration)
Knowledge contributors	• Number of resources contributed (e.g., training guides, FAQs, templates) • Contribution to resolving technical or ethical challenges (measured by member feedback or surveys)

Outlining responsible AI community roles

A community can thrive when members understand the roles and responsibilities. By defining key roles, you ensure that the network remains active, effective, and aligned with organizational goals. Here are the primary roles and their contributors:

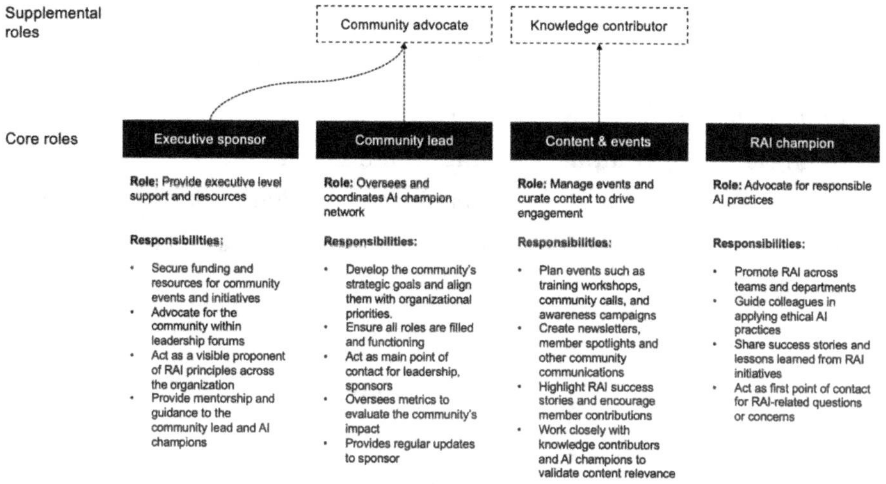

Figure 1-30. *Typical community roles and responsibilities*

Core roles

Executive sponsor

The executive sponsor is the linchpin of a successful community, offering strategic support and elevating the community's voice within the organization. Their involvement reinforces the importance of RAI initiatives and ensures alignment with organizational goals and priorities.

Key responsibilities

- **Securing resources**: Whether its funding for community events or time allocated for champions to participate, sponsors ensure the necessary resources are in place.

- **Advocating at the top level**: By bringing the community's achievements and challenges to leadership meetings, the sponsor ensures buy-in at all levels of the organization.
- **Mentorship and guidance**: They provide critical coaching to the community lead and other members, fostering accountability and helping navigate challenges.

Community lead

Acting as the operational hub, the community lead oversees the community's strategy, day-to-day activities, and alignment with long-term goals. They are responsible for ensuring that the community remains dynamic and focused.

Key responsibilities

Developing and executing the community's strategy

Monitoring engagement metrics and ensuring content resonates with members

Acting as the point of contact for the executive sponsor

Content and events manager

This role is focused on planning and delivering high-value content to keep the community engaged and informed. From newsletters to training sessions, they create opportunities for members to learn, connect, and contribute.

Key responsibilities

- Managing events calendar and promoting attendance
- Sourcing and curating content that aligns with community goals
- Ensuring accessibility and relevance of resources

RAI champions

Champions act as advocates for responsible AI within their teams, raising awareness and bringing grassroots energy to the community.

Key responsibilities

- Sharing success stories
- Providing support to colleagues with RAI-related challenges
- Acting as a bridge between the community and individual teams

Supplemental roles

In addition to the core roles, supplemental roles such as **Community Advocates** and **Knowledge Contributors** add depth to the community by promoting engagement and sharing expertise. These roles can evolve as the community grows and matures.

Reflections on the Foundational Stage

In the foundational stage, we explored how leadership and culture lay the groundwork for responsible AI and Power Platform adoption. From securing executive sponsorship to fostering awareness and forming early habits, this stage sets the tone for everything that follows.

This is where organizations begin to understand the importance of aligning technology with ethical practices and recognize that leadership is supporting the initiative and shaping its trajectory.

The foundational stage introduces a wide range of responsibilities: creating a vision, building awareness, ensuring alignment with organizational goals, securing appropriate licensing, and addressing early challenges. It can feel overwhelming, and that is natural. Remember though that these steps are the essential building blocks of longer-term success. Each effort contributes to a broader cultural shift that ensures responsible AI and Power Platform adoption are both sustainable.

By focusing on leadership accountability and nurturing a culture of collaboration and responsibility, organizations prepare themselves for the next phase.

Emerging maturity builds on this foundation, taking the initial momentum and turning it into meaningful progress. *Let's explore what it means to move from foundational groundwork to a stage where low-code and AI adoption begin to thrive.*

Foundational Stage Progression Checklist

This checklist helps you assess whether your organization has laid the foundational groundwork for successful and responsible low-code and AI adoption. Use it to identify where you stand and what gaps need addressing before progressing further.

Leadership support and sponsorship	Licensing readiness
☐ You have identified an executive sponsor who advocates for Power Platform and AI.	☐ AI license usage audit has been conducted to identify underutilized or misassigned licenses.
☐ The executive sponsor understands their role and is actively involved in strategy and communications.	☐ A reallocation process is in place to optimize license value (e.g., license amnesty initiative).
☐ Sponsorship responsibilities are clearly defined and revisited quarterly.	☐ Early wins using current licensing levels have been documented and communicated.
☐ Engagement of the sponsor is monitored through regular check-ins and data-informed discussions.	☐ The organization has a structured process for building a business case to expand licensing.
☐ There is a plan for managing disengaged or underperforming sponsors (e.g., role realignment or adding co-sponsors).	☐ Licensing limitations, missed opportunities, and unmet needs are documented and regularly reviewed.

Strategic vision and alignment	Responsible AI awareness
☐ Leadership has articulated a clear strategic vision for Power Platform and AI.	☐ A basic understanding of Responsible AI principles has been introduced.
☐ Strategic objectives are defined using SMART criteria and aligned to business outcomes.	☐ There is a proactive process for reaching new makers (e.g., daily app detection + outreach).
☐ Vision workshops (e.g., future-state visualization or reverse brainstorming) have been conducted.	☐ The W.A.I.T framework is being used by makers before development to encourage ethical reflection.
☐ The vision is communicated effectively across leadership, teams and end users.	☐ The C.H.A.T framework is used mid-development to assess and mitigate AI-related risks.
☐ The vision includes short-, medium-, and long-term goals and is tied to ethical and responsible adoption.	☐ Responsible AI is presented as an enabler, not an inhibitor, through positive enforcement.

Community and culture

- [ ] A grassroots community or informal network of RAI champions is being established.
- [ ] Community vision, goals, and purpose have been defined and shared.
- [ ] Roles within the RAI community (e.g., sponsor, lead, content manager, champions) are clearly outlined.
- [ ] Metrics are in place to measure community health and effectiveness.
- [ ] Teams or individuals demonstrating good RAI practices are recognized and rewarded (e.g., shoutouts, digital badges, showcases).

CHAPTER 2

Emerging: Leadership and Culture

Exploratory | Structured | Evolving

What happens when curiosity replaces fear? At the emerging level, organizations begin to shift from hesitant experimentation to a more structured approach to low-code and AI.

The transition from foundational to emerging represents a significant cultural and leadership shift. In this stage, organizations begin aligning their leadership vision with actionable strategies and establishing collaboration across teams. The illustration below maps this progression, showing how leadership becomes more supportive, and culture starts embracing innovation:

S. Jeffery, *The Low-Code AI Maturity Model*, https://doi.org/10.1007/979-8-8688-1730-4_2

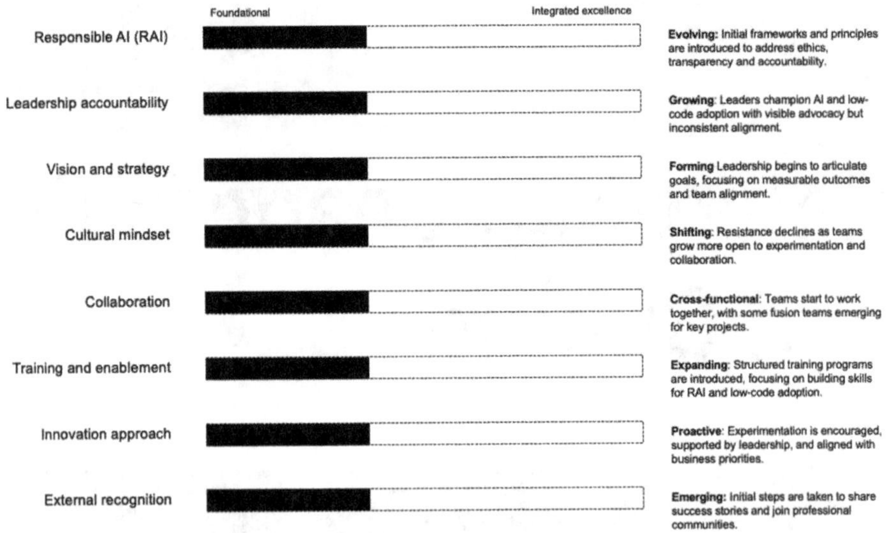

Figure 2-1. *Organizational run state at the emerging stage*

Leadership is increasingly aware of Power Platform's potential, and cultural dynamics start to shift as teams become more open to exploring and collaborating using AI technologies. However, this phase is also prone to growing pains: inconsistent adoption, lingering resistance, and a need for stronger alignment between leadership vision and execution.

This is also where responsible AI (RAI) reaches a critical turning point. What started as basic awareness and habit forming now matures into the first steps of operationalizing RAI. Organizations begin to align AI initiatives with ethical, transparent, and accountable practices.[1] Governance frameworks are introduced, addressing concerns around bias, data privacy, and accountability.

This stage is transitional. It's when low-code and AI moves from being a niche tool to a strategic enabler, and leadership plays a crucial role in this evolution.

[1] NIST. (2023). *AI Risk Management Framework.*

Fixed vs. Growth Mindsets in Low-Code and AI

One of the biggest barriers to shifting culture is the persistence of fixed mindsets – limiting beliefs that prevent individuals and teams from realizing their potential. Shifting to a growth mindset unlocks creativity, collaboration, and innovation.[2]

Here are some common fixed beliefs and actionable strategies toward a growth-oriented approach:

Fixed mindset	Growth mindset
"Only IT teams should hand AI and low-code projects."	• Encourage cross-functional collaboration to build enterprise-grade systems. Commonly referred to as "fusion teams" where subject matter experts and developers collaborate on an app or agent. • Create citizen developer initiatives to enable all teams to contribute to projects.
"I don't have a technical background, so I can't contribute to low-code or AI initiatives."	• Create beginner-friendly learning paths and mentorship programs to build confidence. • Encourage community participation when seeking assistance. • Run frequent "drop-in" clinics for any employee who may have challenges in Power Platform.
"Low-code tools such as Power Platform are only for beginners, and not suitable for real applications."	• Highlight real-world success stories where Power Platform was used to build enterprise-grade solutions. • Provide hands-on workshops to showcase capabilities.

[2] Dweck, C.S. (2006). *Mindset: The New Psychology of Success.*

Shifting to a growth mindset is the first step in transforming how your organization approaches Power Platform. But mindset alone isn't going to be enough – it's the foundation upon which new habits, practices, and opportunities are built.

To drive adoption and innovation, you need to nurture a culture that empowers experimentation, encourages collaboration, and provides the tools for success.

This requires deliberate actions from leadership to create an environment where Power Platform and AI are actively embraced. Let's look at how to cultivate a low-code and AI-friendly culture that transforms individual mindsets into collective progress.

Nurturing a Low-Code and AI-Friendly Culture

Nurturing a low-code and AI-friendly culture requires a fundamental shift in mindset. I recall back to 2017 when Satya Nadella released his book *Hit Refresh.*[3]

As a Microsoft employee, I received a special annotated version, and the message resonated deeply. The book focused on (among other things) the importance of embracing a growth mindset to drive innovation and cultural change – exactly what organizations need today. Reflecting on Microsoft's journey, I've seen firsthand how deliberate actions can transform culture. What made this shift possible at an organization as large as Microsoft? And what, at the emerging state of maturity, is required to drive engagement, adoption, and sustained growth within yours?

[3] Nadella, S. (2017). *Hit Refresh: The Quest to Rediscover Microsoft's Soul and Imagine a Better Future for Everyone.*

- **Training and skill development**: Programs focused on skill development, leadership growth, and adapting to change taught me to embrace new challenges. I learned to see change as an opportunity to learn something new, or to help others grow, rather than something to fear. Training is the cornerstone of any cultural transformation. At the emerging stage, organizations should
    - **Develop tailored training programs**: Create learning paths specific to roles, such as developer, IT administrator, and AI champions.
    - **Leverage hands-on workshops**: Provide opportunities to employees to solve real-world problems using Power Platform.
    - **Recognize and reward progress**: Celebrate milestones such as completing certifications, contributing to the community, sharing knowledge, or launching successful solutions.

Actionable tip Establish a mentorship program where experienced Power Platform users guide newer users.

- **Encouraged risk taking**: Experimentation was supported by a safe environment, free from judgement. This made Microsoft feel more like a startup, where proposing and testing ideas led to significant innovation and productivity gains, both for Microsoft and our customers. I received (and do still to this day) passionate encouragement from my manager (thanks Phil) to try something new. Encouraging experimentation is essential to normalizing innovation. At this stage:

- **Create a psychologically safe environment**: Leaders must demonstrate that failure is an acceptable part of learning. They must recast it as an opportunity to learn and try something new.
- **Reward initiative**: Publicly acknowledge teams or individuals who try new approaches, even if they don't succeed on the first attempt.
- **Launch innovation challenges***:* Organize hackathons or contests where employees can pitch and build solutions, reinforcing the idea that innovation is a shared responsibility.

Actionable tip Use internal communication channels to share examples of risk-taking and experimentation that led to successful outcomes.

- **Access to technology**: I was provided with access to the technology that would allow this experimentation. As this was available to everyone in the company, innovation and collaboration increased, and so did the output of good work. Without the right tools, even the most motivated employee will struggle to innovate. Organizations must
    - **Ensure platform availability**: Provide widespread access to tools like Power Platform and AI Builder, reducing barriers to entry.

- **Empower teams with resources**: Invest in premium licensing, personal developer and sandbox environments, and technical support to encourage experimentation.
- **Facilitate integrations**: Ensure that Power Platform is seamlessly connected to existing enterprise systems, enabling broader use cases.

Actionable tip Regularly audit platform adoption and usage to identify gaps and proactively address them.

The culture change at Microsoft was intentional, guided by leadership, and communicated clearly (and frequently). It empowered individuals to solve problems creatively - first their own and then those of customers and teams.

The same intention is required to nurture a low-code and AI-friendly culture in your organization.

How do you know if your efforts are truly making an impact? Cultivating the right environment is just the beginning - measuring the cultural shift keeps you on the right track and allows you to refine your approach. Knowing which shifts are visible and where additional focus might be needed ensures that the culture you're nurturing continues to grow and thrive.

Measuring the Cultural Shift

Transforming the culture of your company is a gradual process that requires close attention, persistence, and intention. To ensure you're on the right path, it's crucial to establish measurable indicators that track progress.

Key Indicators of Cultural Change

While every organization is unique, there are universal signals that demonstrate cultural transformation is underway. Let's look at some of the key signs:

- **Increased collaboration**: Teams from diverse departments and roles (business and IT, for example) are regularly collaborating on projects. This shows that silos are breaking down, and a shared understanding of Power Platform is taking root.
- **Shared innovation**: Employees are no longer hesitating to share their ideas for improvement. You may notice an increase in proposals during team meetings or through digital platforms like community hubs or intranet forums.[4]
- **Widespread engagement**: Training sessions, workshops, and community events see a growing number of participants, demonstrating enthusiasm for learning and adapting.
- **Observable behaviors**: Employees demonstrate a proactive approach to adopting new tools and experimenting with solutions.

Developing a Measurement Framework

To move on from anecdotal evidence, you need to develop a structured approach to measure cultural progress. Here's a framework you can follow:

[4] Schein, E.H. (2010). Organizational Culture and Leadership.

- **Establish baselines**: Start by identifying where your organization currently stands. Conduct surveys, hold focus groups, or collect data on metrics such as training attendance, collaborative projects, and feedback loops.
- **Set clear targets**: Define what success looks like. For example:
    - A 20% increase in cross-departmental collaboration on Power Platform projects.
    - A specific percentage of employees completing training programs within six months.
    - Achieving a defined number of successful community-driven initiatives.
- **Leverage tangible and intangible metrics**:
    - **Tangible**: Track participation in programs, app creation rates, and resolved use cases.
    - **Intangible**: Conduct employee interviews to gauge changes in attitudes and perceptions of Power Platform.
- **Use technology for monitoring**: Tools such as Power BI dashboards can aggregate data on training participation and app usage providing actionable insights immediately.
- **Regular check-ins**: Schedule periodic reviews to evaluate progress. Involve key stakeholders to align results with broader organizational goals, and adjust strategies where needed.

Example metrics

Here are some practical examples of metrics to help measure your cultural shift:

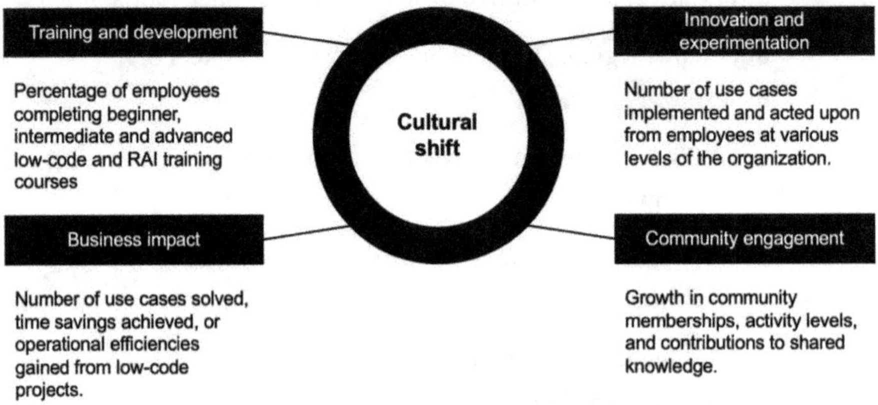

Figure 2-2. *Example metrics tracking cultural shift toward low-code and AI trust*

Using metrics to drive accountability

Tracking progress also acts as a tool to create accountability and ensure that efforts remain aligned with the overarching goal of cultural transformation. For example, rewarding top contributors, teams, or individuals who actively embrace the cultural change or using results to inform leadership decisions about resource allocation or focus areas for further improvement.

Measuring cultural progress provides clarity on the organization's overall direction; it can also reveal disparities. Not all business units or teams may embrace low-code and AI at the same pace, leading to uneven adoption.

Addressing this imbalance is critical to ensure the benefits of transformation are felt across the organization. Let's explore how to identify and bridge these gaps effectively.

Uneven Adoption Across Business Units

Is uneven adoption a feature or a bug? It's not always accidental - sometimes, it's strategic. Certain business units may adopt low-code and AI faster because their goals align more directly with its capabilities, while others are left on the sidelines, either by design or neglect.

There are, however, also a number of deep-seated reasons for uneven adoption which in this section we'll break down:

- **Strategic prioritization**: Leadership may prioritize adoption in areas of the business where its value is easiest to measure, or most impactful, like finance, Human Resources, or customer support. Departments like legal or operations may be deprioritized because their contributions are less quantifiable.

  This can result in these "low-priority" units feeling excluded, which can lead to disengagement.

- **Varying levels of technical capability**: Some business units have a natural affinity for technology, such as IT, data science, or customer-facing teams that already use digital tools. Others, like procurement or facilities, may lack the same technical foundation.

While varying levels of technical capability can influence how quickly different business units adopt Power Platform, it's only one view. The pace of adoption is sometimes shaped by a combination of factors, including the nature of work, the availability of data, or strategic priorities of the organization.

Let's look at why typical business units adopt low-code and AI faster – and why others might need a more tailored approach:

Domain	They adopt quickly because
Customer-facing teams (sales, marketing, customer support)	• Clear and measurable value (increased leads, improved customer satisfaction) • Frequent exposure to digital tools like CRMs or analytics platforms • Competitive pressure to enhance customer satisfaction
IT and data teams	• High technical literacy and familiarity with emerging technology • Direct involvement in implementation • Pressure to improve internal workflows, such as monitoring and automation
Finance and risk management teams	• Need for precision and efficiency in forecasting compliance and fraud detection • Access to structured data that can easily be integrated/analyzed

Lagging adoption:

Domain	Why they may lag
Human Resources	– Concerns about fairness, bias, and ethical implications of AI solutions in hiring or performance evaluations. – Reluctance to move away from people-driven processes. – Limited exposure to low-code and AI compared to other business functions.
Operations and supply chain	– Complexity and variability in processes can make adoption challenging.
Procurement	– Processes often rely on manual negotiations and vendor relationships, which can be harder to automate entirely. – Less urgency to adopt compared to customer-facing functions.

Tailoring Adoption Strategies

Understanding why certain areas adopt faster gives you valuable insights, but it also highlights the importance of supporting the business units that are slower to embrace the technology. These business units may require a more customized approach to training (as an example).

Take HR teams - they may need bespoke workshops to demonstrate how generative AI and low-code can *streamline* their operations, rather than *replace* decision-making.

Similarly, operational teams may view low-code platforms as overly simplistic or unsuitable for solving complex challenges. By designing a pilot project that addresses a critical operational bottleneck, such as automating repetitive workflows or improving resource allocation, you provide a real-world example of the platform's capability to drive meaningful change. This pilot approach not only builds confidence but creates an internal success story that can inspire other teams and departments to follow suit.[5]

Cultural differences within your organization also play a significant role in adoption rates. Risk-averse teams may require a structured introduction that emphasizes safeguards, while teams that thrive on collaboration may benefit from cross-departmental projects that illustrate how AI enhances teamwork and communication.

This culture map helps identify areas where targeted interventions, such as training or cross-functional collaboration, can support more equitable AI and low-code adoption across departments.

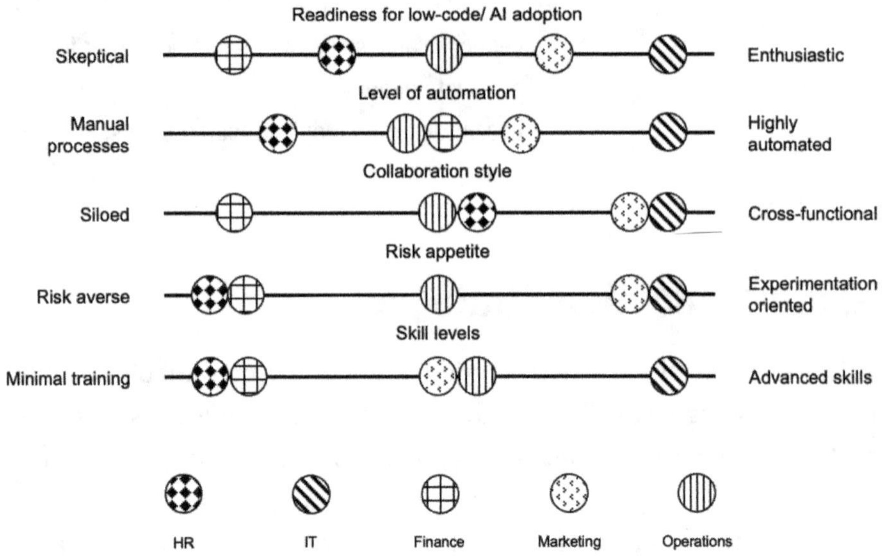

Figure 2-3. *Cultural variability in AI and Power Platform adoption across business units*

[5] Brown, T. (2009). *Change by design: how design thinking creates new alternatives for business and society.*

Some teams may hesitate to adopt low-code and AI-powered tools, perceiving them as unnecessary or misaligned with their focus on intuition or creativity. However, presenting a tailored use case - such as demonstrating how AI Builder and sentiment analysis can enhance decision-making through data insights, like analyzing marketing campaign performance - can bridge the gap between skepticism and enthusiasm. The success of an initial pilot project often leads to broader adoption, helping to shift perceptions and showcase AI-powered solutions as valuable and complementary tools.

By investing time to understand each team's unique needs and providing targeted support, you not only build trust but also lay the foundation for sustained adoption.

When hesitant teams begin to see the tangible benefits of Power Platform and AI-powered solutions, they often transition from resistant participants to active advocates.[6]

As organizations move through the emerging stage of their low-code and AI journey, the foundations of leadership, culture, and adoption strategies begin to take root. With these elements in place, the focus naturally shifts to embedding responsible practices that not only guide development but also build trust across stakeholders.

Responsible AI is no longer an optional add-on; it becomes a defining characteristic of how organizations innovate and operate. The challenge now is not just adopting responsible AI principles but elevating them to become a natural and visible part of the organization's DNA. This next section will explore strategies, structures, and actions needed to deepen your commitment to responsible AI and ensure it aligns with your organization's broader goals.

---

[6] Rogers, E.M. (2003). *Diffusion of innovations (5th ed.).*

Elevating Responsible AI

At this stage, you've laid the foundations for responsible AI (RAI). Awareness campaigns have sparked interest, and you may have a handful of enthusiastic champions within your network. But how do you move from isolated efforts to something more structured and strategic?

The path forward requires a clear framework - an approach that will move from enthusiasm to enterprise-wide action and integrate RAI principles into the fabric of your organization. There are already several leading frameworks that provide guidance, offering tested methodologies and practical insights.

In this section, we'll explore three frameworks:

- The National Institute of Standards and Technology (NIST) AI Risk Management Framework, emphasizing trust and accountability
- The Responsible AI Institute's responsible generative AI best practices, guiding the ethical and scalable use of generative AI capabilities
- Microsoft's blueprint for governing AI, offering practical strategies rooted in real-world enterprise experience

Each framework stands up on its own and brings something unique. Collectively, they provide a robust toolkit for integrating RAI principles into your Power Platform and AI initiatives. Let's take a closer look at what each framework has to offer, why its impactful, and how it can support your organization as it transitions to the emerging phase of responsible AI adoption.

Diving Deeper into Responsible AI Frameworks

Let's examine the three established frameworks that shape how organizations approach responsible AI. Rather than exploring each in full detail, we'll focus on how their guidance intersects - especially when it comes to leadership behavior, team alignment, and culture change at the emerging stage of responsible AI adoption. By synthesizing these frameworks, we can provide a practical, unified view that equips you to apply their collective insights within your organization. The goal is not to educate you on the frameworks themselves but to make them useful - to help you elevate responsible AI from abstract principles to actionable habits.

NIST AI Risk Management Framework (AI RMF)

The NIST AI RMF[7] provides a comprehensive guide for identifying, assessing, and managing risks associated with AI systems. It emphasizes trustworthiness through principles such as accountability, reliability, and transparency. The framework's core functions - Govern, Map, Measure, and Manage - offer organizations a life cycle approach to building and maintaining AI systems responsibly.

Key elements

- **Structured risk assessment**: The emphasis on "Map" and "Measure" ensures organizations systematically identify and evaluate risks, rather than reacting to issues after deployment.
- **Governance focus**: The "Govern" function ties directly into leadership, requiring clear oversight and accountability structures, which are critical to building trust in Power Platform solutions.

[7] NIST. (2023). *AI Risk Management Framework.*

- **Adaptability**: The framework is designed to be flexible and applicable across industries, making it ideal for organizations integrating Power Platform.

How this can apply to Power Platform:

- Leadership can adopt the "Govern" function by establishing governance committees for Power Platform projects.
- Culturally, teams can align on risk-aware practices, ensuring trust and accountability in citizen developer solutions.

Responsible AI Institute's Responsible Generative AI Best Practices

This framework[8] emphasizes ethical, scalable adoption of generative AI, dividing guidance into categories like strategy, workforce, capacity, practice, and proactivity. It addresses organizational challenges such as bias, security, and transparency while providing actionable steps for governance and workforce enablement.

Key elements

- **Workforce-first approach**: By prioritizing training and capacity-building, this framework directly supports cultural-transformation – helping employees embrace AI as an ally, not a threat.
- **Governance integration**: The framework ties strategic planning with on-the-ground practices, ensuring alignment from leadership to citizen developers.

[8] Responsible AI Institute. (2023). *Best practices for Responsible Generative AI Adoption.*

- **Proactivity**: Encouraging organizations to anticipate challenges and build resilience aligns well with the dynamic nature of AI.

How this can apply to Power Platform

- Leadership promoting accessible training on tools like AI Builder, Microsoft Copilot Studio, and Power Automate
- Culturally normalizing AI experimentation while embedding ethical checks into workflows

Microsoft's Blueprint for Governing AI

Microsoft's responsible AI framework[9] focuses on embedding ethical AI principles into corporate strategy, governance, and operational practices. Its five-point plan includes developing safety frameworks and ensuring transparency.

Key elements

- **Real-world relevance**: Rooted in Microsoft's own experiences, this framework offers practical guidance for organization's scaling AI initiatives.
- **Transparency and communication**: The focus on transparency ties closely to building a culture of trust - critical when scaling citizen development initiatives in Power Platform.

[9] Microsoft. (2022). *Responsible AI Standard: v2.*

How this can apply to Power Platform

- Empower champions in departments to advocate for responsible practices.
- Normalize open discussions about ethical concerns in Power Platform projects, building trust in both the technology and leadership's vision.

Let's finally look at the overlap between the three frameworks:

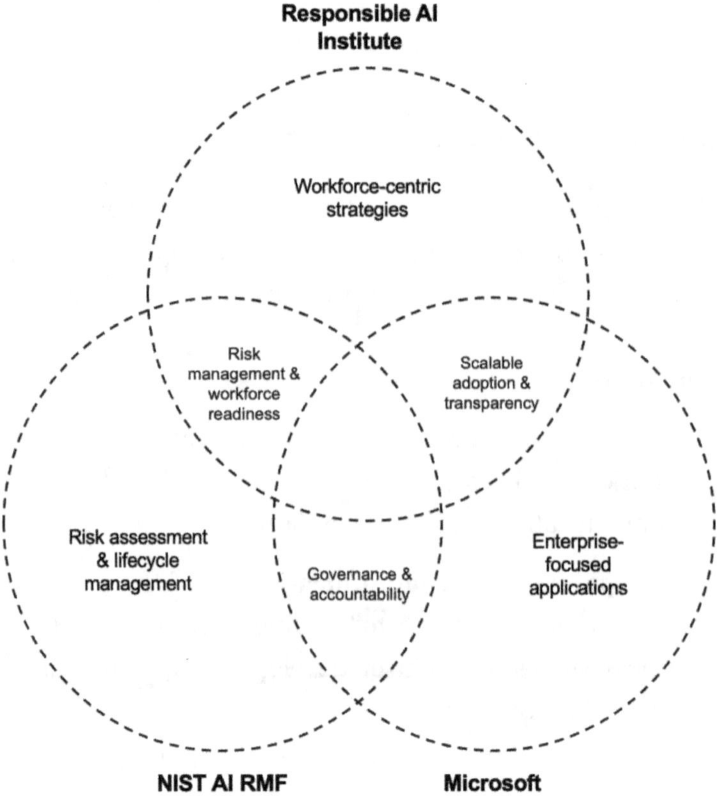

Figure 2-4. *Venn diagram showing overlap between three frameworks*

- **Ethics and trustworthiness**:
    - All three build trust through transparency, fairness, and accountability. This is foundational for scaling AI responsibly.
- **Governance and accountability**:
    - NIST AI RMF provides a structured life cycle approach; Microsoft focuses on operational governance and enterprise alignment; the Responsible AI Institute highlights workforce enablement for consistent accountability.
- **Scalability and practicality**:
    - Microsoft and Responsible AI Institute frameworks align on making RAI principles scalable across organizations. NIST adds rigor with its risk assessment framework.
- **Cultural readiness**:
    - Workforce training and capacity-building are a strong focus in the Responsible AI Institute framework. This is supported by governance (NIST) and practical tools (Microsoft), creating a culture where RAI is embedded in everyday workflows.
- **Strategic integration**:
    - NIST's mapping and measuring approach aligns with the strategic focus of Microsoft's blueprint, which create a feedback loop for continuous improvement.

Adopting the best practices from all three frameworks creates a balance that integrates ethics, risk management, and operational governance. It's an adaptable approach for organizations, scale, and use cases. It also creates a strong link between leadership, culture, and workforce readiness, ensuring RAI principles are both scalable and practical.

Tools and Templates

The three frameworks that we've just explored provide a wealth of deeper reading, and resources, from assessing risks to creating governance frameworks and introducing cultural readiness.

Here are templates and tools from Microsoft, NIST, and the Responsible AI Institute to assist in implementing responsible AI practices:

Organization	Resource	Description
Microsoft	Responsible AI Impact Assessment Template	A template to systematically assess the impact of AI systems, focusing on their purpose, features, and societal impact
	Responsible AI Standard v2	Microsoft's internal guidelines for designing, building, and testing AI systems responsibly
	Responsible AI Toolbox	A suite of tools providing interfaces and libraries to understand and assess AI system, promoting responsible development and monitoring
NIST	AI Risk Management Framework (AI RMF) 1.0	A framework offering guidance and managing AI risks and promoting responsible development and monitoring

(continued)

Organization	Resource	Description
	AI RMF Playbook	A companion resource suggesting ways to navigate and use the AI RMF to incorporate trustworthiness in AI system design, development, deployment, and use
Responsible AI Institute	AI Policy Template	A template to help organizations build comprehensive frameworks guiding AI development, procurement, supply, and use

What Responsible AI Means in Practice (for Power Platform)

You've seen how frameworks like the NIST AI RMF, Microsoft's Responsible AI standard, and the Responsible AI Institute offer strong foundations for building ethical and trustworthy AI systems. But for most organizations using Power Platform, especially in the earlier stages, responsible AI can still feel abstract - more a list of good intentions than practical actions.

Responsible AI isn't just a set of rules. It's a **commitment to intentional design, transparent decision-making, and ongoing accountability** - especially as we begin to rely on AI agents and copilots to interact with customers, make recommendations, and automate judgement calls. Below, we introduce the six core principles that form the backbone of most responsible AI frameworks. You'll see how these principles apply specifically in a Power Platform context and why each is essential for avoiding risk and enabling trust.

Why Generative AI Requires Special Attention

Traditional low-code solutions were largely deterministic - if "X" happens, do "y." But with generative AI, the platform introduces nondeterministic behaviors: dynamic text generation, humanlike conversation, recommendations, and decisions that aren't easily traced to fixed logic. This increases the risk of

- Hallucinations (AI generating convincing but false responses)
- Prompt injection or manipulation
- Ambiguous accountability when AI actions are delegated

These risks are new to many makers and business stakeholders; your governance strategy must extend beyond technical safeguards. It must include education, review workflows, and cultural awareness of AI's potential impact.

Rather than adopting a single framework in full, the remainder of this chapter aligns and combines key elements from NIST, Microsoft, and the Responsible AI Institute to form a practical, Power Platform-specific approach. This blended approach will support you in building trust, reducing harm and operationalizing responsible AI in a way that matches how your organization is adopting low-code tools today.

Scaling Responsible AI Across Your Organization

How do organizations scale RAI efforts across departments and teams without overwhelming them or creating resistance? What role do leadership and culture play in ensuring success at this stage?

Scaling RAI across an organization is complex because AI initiatives span both low-code and pro-code environments. It's important to recognize that AI governance, risk management, and ethical principles apply universally across development approaches, but the ways they are implemented might differ significantly. Power Platform democratizes development, enabling citizen developers to rapidly innovate. Meanwhile, pro-code environments usually require more rigorous governance to handle the complexities of advanced systems.

At this stage, the focus in not purely about new frameworks and tools (although this plays a large part in enablement); it is also about ensuring the RAI principles become embedded in both the mindset of your users and day-to-day operations. Leaders need to focus on both equally and must guide the organization in addressing challenges, such as uneven adoption, resistance to both the technology, and the formal processes, and aligning low-code and pro-code initiatives under a unified vision.

For low-code initiatives, integrating RAI into the Center of Excellence (CoE) becomes critical. The CoE serves as the central hub for governance, training, and collaboration, ensuing that Power Platform is deployed responsibly.

Let's explore some of the key roles involved in RAI.

Defining the Key Roles in Responsible AI

Integrating RAI into a CoE requires a clear definition of roles, accountability, and integration into existing organizational structures. Here are the key roles, their responsibilities, and advice for selecting the right individuals:

RAI Lead (or RAI Program Manager)

Responsibilities:

- Developing and articulating the organization's RAI strategy and vision across both low and pro-code
- Aligning RAI efforts with overall business objectives and enterprise goals

- Overseeing the execution of RAI initiatives across departments
- Collaborating with the CoE to ensure RAI principles are integrated into low-code governance structures, such as app approval, environment requests, and monitoring

Who's right for the role?

- Senior leaders with expertise in governance, strategy, and a good understanding of AI technology
- Someone with the ability to manage cross-department collaboration and communicate effectively with executives

Who they typically report to:

- The Chief Data Officer (CDO) or Chief Technology Officer (CTO), depending on the organization's structure

Ethics Officer (or Ethics Advisor)

Responsibilities:

- Ensure compliance with RAI principles, ethical guidelines, and regulatory requirements.
- Lead discussions on fairness, inclusivity, and potential societal impacts of AI systems.
- Collaborate with governance boards to address ethical risks proactively.

Who's right for the role?

- A professional with expertise in ethical decision-making, legal frameworks, or compliance
- Strong interpersonal skills to navigate sensitive discussions and provide balanced recommendations

Who they typically report to:

- The RAI lead or directly to the Chief Ethics Officer (if present)

Risk Manager
Responsibilities:

- Identify, assess, and mitigate risks in AI projects (including operational and compliance risks)
- Integrate risk management (e.g., NIST AI RMF) into organizational processes

Who's right for the role?

- An experienced risk manager or quality with an understanding of AI technologies

Who they typically report to:

- The RAI lead

RAI Champions (Departmental Advocates)
Responsibilities:

- Act as an advocate for RAI within their departments.
- Provide feedback to governance teams on how RAI principles affect day-to-day operations.
- Share success stories and help address challenges related to RAI adoption.

- Ensure that citizen developers understand and follow RAI principles, leveraging CoE-provided resources and templates.
- Work closely with the CoE to address questions or concerns specific to low-code.

Who's right for the role?

- Passionate employees with strong knowledge of their department's workflows and a willingness to drive change

Who they typically report to:

- Department heads, while connecting to the RAI lead for alignment on broader RAI initiatives.

Establishing Governance and Lightweight Monitoring

Getting the balance right: As you start to formalize governance (we'll go deep on this topic in Part 2), leaders must avoid overly rigid governance structures. This may stifle innovation and impact adoption should structures need to change. Instead, you should advocate and aim to implement lightweight monitoring systems that encourage creativity while ensuring ethical compliance.

It's important to communicate governance as an enabler of trust and innovation rather than a barrier. Position governance to protect teams in a way that enhances their impact and productivity to drive a positive cultural response.

Explicit leadership actions: Here are some actions that you can perform now:

- **Establish a RAI committee** within the CoE to manage governance, including representatives from the business, IT, and ethics teams.

- **Empower AI champions** to self-monitor low-code projects and report on compliance with RAI principles.

Training and Capacity Building

Training and capacity building are the foundations of driving cultural change, understanding, and embracing RAI principles. These are some of the core aspects outlined in the Responsible AI Institute's governance framework.

At scale, the challenge lies in the diversity of teams with different levels of experience and expertise: from citizen developers to pro developers - maintaining consistency in delivery and depth of knowledge.

For leaders, the task is two-fold: *first,* create inclusive and accessible training, and *second,* create engagement through innovative methods like gamification, practical challenges, and real-world use cases.

Key Considerations for Training and Capacity Building

- **Tailored learning paths**: Develop learning programs that cater for the varying needs of citizen developers, pro developers, and managers/leadership. Different audiences require different depths of knowledge and application.
- **Gamification**: Engage employees with interactive learning experiences that reward participation, encourage competition, and celebrate success.
- **Mindset**: Shift the mindset from training as a one-time event to an ongoing process of learning and growing, ensuring teams stay aligned with evolving RAI practices and technologies.

Feedback Mechanisms

Effective feedback mechanisms are critical to scaling RAI practices. They ensure that teams can surface challenges, share insights, and refine processes. Establishing these systems requires a thoughtful approach to design to make feedback actionable and accessible

Consider feedback channels:

Channel	Purpose
Anonymous reporting channel	These allow employees to voice concerns about ethical issues or operational barriers without fear of repercussions.
Training feedback channel	Capturing feedback from employees who have completed training – Do they understand RAI after completion? Have they formed the habits in their workflow? Do they know the policies of your organization?
Operational feedback	Captures insights from teams actively using AI tools to improve implementation and identify friction points – e.g., app users can report inconsistent results.
Governance feedback	Allows employees to provide input on leadership decisions relating to RAI, such as policies, initiatives, or priorities.

Defining and Measuring Successful RAI

The success of your RAI initiative can be measured by its tangible and cultural impact across the organization. This involves tracking quantifiable metrics - such as the number of RAI compliant solutions and completion rates for training.

Successful RAI impact is reflected by how seamlessly RAI principles are integrated into everyday workflow. Below is a table containing some tangible indicators:

Measure	What success could look like
Training courses completed count	High participation in RAI training programs across departments, indicating growing workforce capacity.
AI champions registered count	A robust network of champions actively promoting RAI principles within their departments.
Power apps, flows, and agents using AI features count	Steady increase in the adoption of AI-powered tools across business units, showing widespread utility.
RAI-reviewed power apps, flows, and agents count	A high proportion of AI-integrated solutions undergoing RAI reviews, ensuring governance and compliance.
Number of anonymous ethical concerns count	Regular submissions, indicating employee trust in feedback channels and vigilance about ethical AI practices.
Ethical issues resolved percentage	A high percentage of submitted issues resolved within a time frame demonstrates responsiveness and accountability.
RAI-reviewed power apps, flows, and agents per team count	Majority of teams adopting and implementing RAI policies in their workflows.
Time to review RAI compliance issues	Reduction in average time for RAI reviews indicates streamlining of processes and organizational efficiency.
RAI committee meetings held	Regular meetings with high attendance, leading to actionable outcomes that influence governance and policies.
Average attendance to RAI calls	Monitors sentiment to RAI overall.

Reflections on the Emerging Stage

As you reflect on the strategies, tools, and approaches outlined in this section, consider how far your organization has progressed in building a foundation of leadership and culture that supports responsible AI and Power Platform adoption. Leadership and culture are now the driving forces behind innovation, collaboration, and accountability. You've set the stage for sustainable and systemic growth.

Scaling responsible AI and adoption enterprise-wide is a journey. In this section, we explored the critical steps for transitioning from foundational efforts to an integrated, enterprise-wide approach. We discussed how to establish governance structures, foster workforce capacity through training programs, and how to create feedback mechanisms that enable continuous improvement. Success in RAI is measured not only by adoption rates but by embedding trust, accountability, and innovation into your organization's DNA.

As we move forward, you'll focus on embedding RAI into your leadership, culture, and day-to-day operations at a more advanced level. At the "operationalized" level, RAI becomes a core part of how decisions are made, projects are executed, and values are demonstrated across the organization.

Emerging Stage Progression Checklist

Use this checklist to assess your progress toward embedding a growth mindset, nurturing cultural transformation, and scaling responsible AI adoption in the emerging stage.

Shift to a growth mindset	Nurture a low-code & AI-friendly culture
☐ Identify and challenge common fixed beliefs within teams.	☐ Develop role-specific training pathways and run hands-on workshops.
☐ Promote fusion team models and citizen development initiatives.	☐ Launch innovation challenges or hackathons to encourage creative problem-solving.
☐ Offer beginner-friendly training and mentorship programs.	☐ Provide broad access to Power Platform, AI Builder and sandbox environments.
☐ Highlight real-world success stories to shift perceptions of low-code tools.	☐ Invest in premium licenses and integrations to remove barriers to adoption.

Elevate Responsible AI	Establish roles and light governance
☐ Introduce a clear framework for RAI	☐ Define key roles: RAI lead, ethics office, risk manager, RAI champions.
☐ Align leadership messaging with ethical AI practices.	☐ Form a cross-functional RAI committee within your Center of Excellence.
☐ Provide access to RAI tools, templates, and policy guides.	☐ Enable lightweight monitoring for citizen-developed solutions.
☐ Begin embedding RAI into day-to-day Power Platform processes and templates.	☐ Position governance as a trust enabler, not a blocker.

Shift to a growth mindset

- [ ] Identify and challenge common fixed beliefs within teams.
- [ ] Promote fusion team models and citizen development initiatives.
- [ ] Offer beginner-friendly training and mentorship programs.
- [ ] Highlight real-world success stories to shift perceptions of low-code tools.

Nurture a low-code & AI-friendly culture

- [ ] Develop role-specific training pathways and run hands-on workshops.
- [ ] Launch innovation challenges or hackathons to encourage creative problem-solving.
- [ ] Provide broad access to Power Platform, AI Builder and sandbox environments.
- [ ] Invest in premium licenses and integrations to remove barriers to adoption.

Build capacity through training

- [ ] Roll out inclusive training paths for citizen and pro developers.
- [ ] Use gamification and practical challenges to boost engagement.
- [ ] Promote continuous learning rather than one-time training events.

Create feedback channels

- [ ] Set up anonymous reporting for ethical concerns.
- [ ] Collect feedback after training and project delivery.
- [ ] Use operational feedback to identify friction in adoption or trust.
- [ ] Enable employees to shape RAI governance through adoption or trust.

Define and track RAI success metrics

- [ ] Monitor training completion rates and champion engagement.
- [ ] Track the number of RAI-reviewed solutions across departments.
- [ ] Measure average time to resolve RAI issues and number of ethical concerns raised.
- [ ] Regularly evaluate attendance and engagement in RAI governance meetings.

CHAPTER 3

Operationalized: Leadership and Culture

Strategic | Scalable | Integrated

Going beyond initial success and creating lasting impact

At this level, responsible AI (RAI) moves beyond frameworks and early, but sustained successes – it becomes embedded into the core of how the organization operates, makes decisions, and delivers value. RAI is no longer seen as a separate initiative or compliance requirement; it is fully integrated into leadership priorities, organizational strategy, and cultural norms.

The illustration below captures the characteristics of an organization at this level, where responsible AI is fully integrated, leadership drives alignment across business units, and cultural adaptability is a core strength.

S. Jeffery, *The Low-Code AI Maturity Model*, https://doi.org/10.1007/979-8-8688-1730-4_3

This visualization highlights how far the organization has progressed and provides a benchmark for what systemic, sustainable adoption looks like:

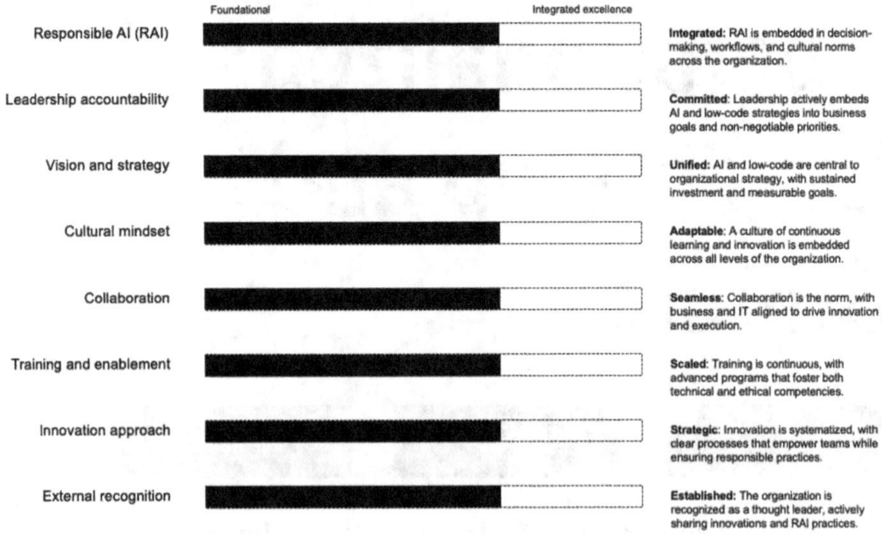

Figure 3-1. *Organizational run state at the operationalized stage*

The Push to Operationalized Adoption from Emerging

In the previous stage (emerging), we briefly covered organizational attitude, and we've touched on it again here; but in my experience, organizations that push for operationalized adoption can

- **Allocate dedicated resources** for scaling initiatives, including champions (both platform and RAI), budgets for high-value initiatives, and advocacy from leadership.
- Establish processes that are **enablers of innovation and not barriers**, ensuring tools are used responsibly without stifling creativity. For example:

- **Self-service enablement**: Empowering employees to solve their own problems using pre-approved resources
- **Sandbox/developer environments for experimentation**: Encouraging risk-free prototyping
- **Flexible security layers**: Adapting policies based on risk, providing more freedom where appropriate
- **Iterative feedback loops**: Ensuring solutions are user-centric and continuously improved

- **Fully integrate adoption metrics and feedback mechanisms** into their strategies to measure progress and drive continuous improvement.

This is where leadership evolves into something more intentional – leaders move from "supporting initiatives" to actively embedding them into their decision-making processes and their teams' day-to-day operations and workflows. For example, a leader may require that all new process improvements be prototyped using low-code tools, ensuring AI and automation are evaluated as part of standard solution planning. Putting it another way, leaders start to "walk-the-walk."

Illustrating the transition: emerging vs. operationalized

Understanding the differences between the emerging and operationalized stages helps highlight how leadership accountability and cultural adaptability evolve as organizations mature. In the following illustrations, we compare the approaches at each stage to help you identify opportunities for growth within your organization. The contrasts reflect the tangible shifts required to embed low-code and RAI practices into the core of your business and ensure sustainable and systemic progress.

Leadership now champions RAI as a strategic enabler, allocating dedicated resources, defining accountability, and measuring its success alongside other business outcomes. Teams comply with RAI principles and now actively plan for them into project life cycles. RAI is a shared responsibility across departments, with cross-functional collaboration between business units, teams, and governance boards.

Leadership is a role model for responsible practices. They set clear expectations and develop metrics that make success measurable.

Operationalized RAI requires embedding ethical AI principles into the core of your organizational leadership and strategy. It involves establishing clear accountability, aligning AI initiatives with business objectives, and fostering a culture that prioritizes AI use responsibly.

Operationalizing Success

Organizations at the operationalized stage define success by turning individual wins into *systemic and sustainable progress*. They go beyond the foundational steps of adoption, embracing an ambition to lead with Power Platform and AI, rather than just keeping the lights on. These organizations achieve systemic progress by ensuring that individual wins are not isolated successes but instead become *repeatable, scalable, and embedded* across the organization. Let's look at how they do that:

- Organizations invest in **robust feedback mechanisms** to learn from successes and failure. These often include
    - **Post-implementation reviews**[1]: Analyzing what worked, what didn't, and identifying opportunities for improvement after deploying Power Platform solutions. These are typically aligned to use cases that are broadly shared and business critical.

[1] Edmondson, A.C. (2018). *The fearless organization: creating psychological safety in the workplace for learning, innovation and growth.*

    - **Sharing successes**: Hosting internal forums or showcases where teams share their experiences, enabling others to replicate successes or avoid pitfalls.

- **Standardizing best practices**: Defining and documenting best practices that can be scaled across teams and regions. Leaders play a critical role in supporting:

    - **Creating playbooks**: Offering step-by-step guidance for similar projects.
    - **Encouraging the use of templates**: Providing reusable frameworks for common applications, reducing development time and ensuring consistency.
    - **Community created "best practice shorts"**: Establish a practice of sharing "best practice" shorts. These are concise, actionable snippets (usually video) shared by citizen developers and teams that demonstrate how responsible AI (RAI) principles have been successfully applied, such as improving transparency in app design, addressing potential biases in AI models, or creating workflows that align with governance policies.

Moving from individual wins to systemic progress is a process that combines strategic planning with a commitment to culture. It can result in organizations positioning themselves as innovators and role models, transforming the potential of Power Platform into sustainable and long-term value.

This stage is also marked by a significant shift: leaders recognize the *importance of external visibility* as an extension of internal successes. As organizations mature, they increasingly seek to position themselves as innovators within their industries.

This drive for industry leadership recognition reflects confidence in their responsible AI and Power Platform initiatives; it's not a branding exercise.

The Role of External Leadership Recognition

Leaders at this stage understand power of storytelling. It's a strategic enabler that amplifies internal achievements, positioning the organization as a leader in responsible AI and Power Platform adoption.[2] Let's look at two of the goals showcasing achievements externally provides:

- **Highlighting innovation success**: Leaders use case studies, conference presentations, or community events to share how their RAI and Power Platform initiatives have solved complex challenges, created efficiency, and delivered measurable outcomes. This serves as a powerful motivator for internal teams while positioning the organization as a trailblazer in the industry.
- **Advancing RAI practices**: By sharing their journey in adopting responsible AI, these organizations demonstrate that innovation can coexist with accountability. Publicly showcasing their commitment to ethical AI practices reinforces trust among customers, partners, and regulators while encouraging other organizations to follow suit.

---

[2] Freeman, R.E. (2010). *Strategic Management: A stakeholder approaches.*

Now that we've established why external recognition matters, the next step is understanding how to achieve it and how to make it meaningful and aligned with the organization's values.

Starting the Process of External Visibility

I've seen many organizations and individuals follow similar paths to what we'll discuss in more detail here, but the journey *always* starts with intention.

Leaders who approach visibility authentically, consistently, and with a focus on impact tend to set their organizations apart. Sharing both what you do and how you do it naturally differentiates your organization as industry leaders in both responsible AI and Power Platform.

Before diving into visibility efforts, clarify why external recognition matters to your organization:

- *Are you showcasing innovation to attract partnerships or clients?*
- *Do you aim to inspire trust in your RAI practices to strengthen stakeholder confidence?*
- *Are you positioning your organization as a thought leader to attract top talent?*

Once you've identified your primary motivation, the following table can help you determine the most effective strategies and channels to align your efforts with your goals.

Motivation	Strategic focus	Suggested channels
Showcasing innovation	Highlight successful projects, efficiency gains, or innovative solutions.	• Industry conferences and events. • Case studies published on your website or shared via blogs. • Webinars.
Inspiring trust in RAI practices	Demonstrate transparency, ethical decision-making, and accountability.	• White paper on RAI principles and their application. • Social media campaigns emphasizing ethical AI. • Partnerships with RAI advocacy groups.
Positioning as a thought leader	Share insights, best practices, and vision to influence industry trends.	• Speaking opportunities. • Publishing articles in industry journals or websites. • Active engagement on LinkedIn or other professional forums.
Attracting top talent	Showcase your organization's culture of innovation, growth, and accountability.	• Employee testimonials shared on social media. • Highlight RAI initiatives in recruitment marketing materials.

With a clearer understanding of your motivation and the corresponding strategies, you can now focus on selecting and targeting the right channels to amplify your message.

Invest Time and Leverage the Right Channels

Here are some tips and actions to help you target the right channels to ensure your message reaches the right audience:

Professional and industry communities

- **Join industry groups**: Look for groups focused on RAI, AI ethics, and low-code technologies. Actively contribute to conversations to build credibility.
- **Leverage online forums and platforms**: Platforms like LinkedIn groups or Microsoft's Power Platform Community are great for sharing insights and connecting with peers.

Conferences and events

- **Starting small**: Look for local events, user groups, webinars, or regional conferences where your organization can gain speaking opportunities and experience. Build confidence and refine your story before targeting larger stages.
- **Pitching your success stories**: Create compelling proposals that highlight the unique value of your hard work and progress. Frame them around the challenges solved, the measurable outcomes, and lessons learned.
- **Networking effectively**: Use events to make meaningful connections with other leaders and potential collaborators.

Social media and content creation

- **Highlight and share your key achievements**: Use LinkedIn, X, or other platforms to share your successes. Create short, impactful posts with visuals or data that is highly engaging.
- **Blogs and articles***:* Publish thought leadership content on platforms like Medium or industry-specific outlets to showcase your expertise.

Foster internal advocacy

- Empower team members to act as ambassadors for your organization:
    - Encourage employees to share posts or speak about their work at events.
    - Provide templates, talking points, or support to amplify their contributions.

Building external recognition ultimately is a demonstration of an organization's maturity in RAI and Power Platform adoption. It reflects the alignment of internal achievements with external visibility and positions the organization as a thought leader in both innovation and ethics.

Let's now pivot to explore how the ability to culturally adapt is required, not just for external recognition but also for creating sustainable and systemic progress.

Developing Actionable Practices

In this section of the book, we'll take a closer look at how to translate this elevated vision into actionable practices.

First, we'll explore **leadership accountability and strategic integr**ation, diving into how leaders embed RAI into their organization's DNA by aligning it with core business goals and decision-making processes. This includes collaborating across teams and ensuring alignment from the top down.

Next, we'll examine **scaling RAI through governance and performance metrics**, focusing on the systems, frameworks, and tools that sustain RAI at scale. This section highlights how measurable indicators not only track success but also build confidence in AI-powered decision-making.

Finally, we'll shift to **building a culture of continuous responsibility**, where we'll discuss how cultural alignment enables teams to innovate responsibly, collaborate effectively, and uphold trust across the organization.

Leadership Accountability and Strategic Integration

Strategic Vision and Cultural Adaptability

What is cultural adaptability?

At its core, cultural adaptability is the ability of an organization to evolve its behaviors, values, and systems in response to new challenges and opportunities.[3]

From my experience working with adaptable organizations, their success often stems from two forces:

- **Strategic vision**: Leadership recognizes that scaling Power Platform adoption and AI capabilities requires clear direction, sustained investment, and alignment across the business. Real value is realized when leadership integrates Power Platform into broader organizational goals.
- **Cultural adaptability**: Organizations that succeed at this level know that technology alone does not drive transformation. They focus on empowering teams and embedding habits like experimentation and RAI; they foster accountability to create a sustainable culture of adoption.

[3] Schein, E.H. (2010). *Organizational Culture and Leadership.*

This section dives into how organizations cultivate cultural adaptability to align their strategic vision with day-to-day practices, creating a sustainable culture focused on innovation and accountability.

How to Achieve Leadership Accountability and Strategic Integration

Leadership accountability:

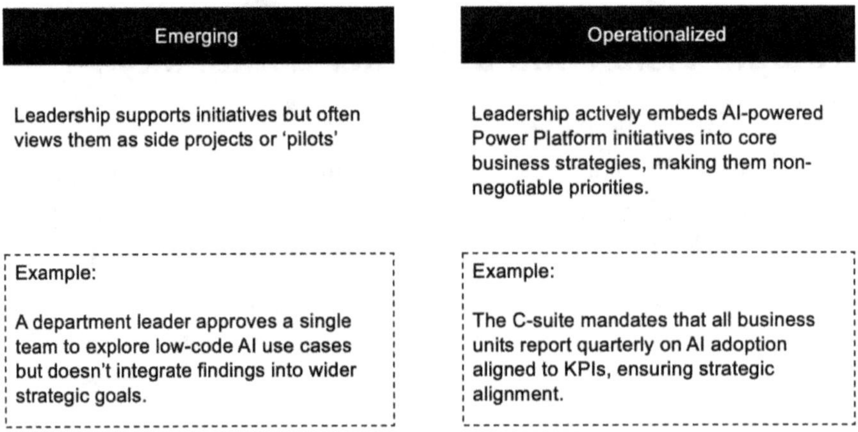

Figure 3-2. *Examples of leadership accountability at emerging and operationalized stage*

Achieving operationalized leadership and integration requires deliberate steps to align technology, culture, and strategy. Here's how you can act:

- **Define accountability mechanisms**
    - **Assign ownership**: Clearly designate leaders responsible for RAI and Power Platform adoption, governance, and integration across departments. Where possible, connect these responsibilities to existing organizational structures – such as digital innovation office, enterprise architecture

function, or an established Power Platform Center of Excellence (CoE). Avoid placing ownership solely within IT or a single department, as this can create silos and limit long-term strategic alignment. Instead, position ownership at a level where leaders can influence cross-functional decision-making and coordinate with broader AI and data governance efforts.

- **Establish metrics**: Define KPIs to measure the impact of RAI and Power Platform initiatives. These should go beyond adoption rates and ethical compliance to include both **efficiency metrics** (like time savings and cost reductions) and **growth-oriented indicators**, such as new revenue streams, customer satisfaction improvements, or accelerated time-to-market. We'll explore KPI design in more detail later in this chapter.
- **Conduct regular reviews**: Schedule periodic leadership reviews to evaluate progress on key metrics – focusing not just on adoption and governance activity but on the outcomes they enable. These reviews should assess how responsible AI and Power Platform initiatives are contributing to business goals such as ROI, efficiency gains, risk mitigation, and innovation capacity. Reframing governance discussions around impact ensures sustained executive engagement and support.

- **Create a strategic RAI and Power Platform adoption plan**
    - **Align with organizational goals**: Tie AI and Power Platform initiatives to high-priority business objectives, such as efficiency gains, customer experience improvements, or revenue growth.
    - **Invest in dedicated resources**: Ensure budget, training, and tools are in place to scale initiatives effectively. This includes funding for RAI training and awareness campaigns (see building a business case).
    - **Build road maps**: Develop clear timelines for rolling out RAI initiatives, from pilot programs to full-scale integration.
- **Foster a culture of experimentation**
    - **Encourage experimentation**: Create safe spaces for teams to explore new ideas, even if they involve elements of risk. Acknowledge that not every project will succeed but each will offer learning opportunities.
    - **Support RAI**: Integrate RAI principles into project approval processes. Encourage teams to use frameworks like W.A.I.T and C.H.A.T for ethical reflection.
    - **Recognize efforts**: Celebrate teams that lead successful AI initiatives, especially those that integrate RAI into their workflows.

Why this matters

This shift to industry leadership recognition ties directly into the core goals of operationalized adoption. By combining innovation successes with public-facing advocacy for responsible AI, organizations establish themselves as both thought leaders and deepen their accountability to ethical principles. This dual focus strengthens internal momentum, drives cultural change, and positions the organization as a benchmark for others to follow.

Cultural Adaptability

Organizations that succeed at this level know that technology alone is not going to drive transformation.

A culture that fosters adaptability enables teams to respond proactively to new opportunities while remaining aligned with ethical principles and strategic objectives.

When that adaptability is missing or uneven, initiatives may face **stagnation, resistance, or slow progress** - but these aren't always signs of failure. Sometimes, they reflect natural friction during transformation, especially as teams shift mindsets or break down long-standing silos.

Some signs that your culture is shifting in the right direction:

- **Navigating change effectively**: Your teams are embracing change as an opportunity rather than a disruption. This kind of mindset reduces friction and accelerates adoption.
- **Fostering collaboration**: By breaking down silos and promoting cross-functional teamwork, your organization is ensuring that ideas and innovations are flowing freely across departments.
- **Ethical practices**: These are now embedded and integrated into daily workflows. They are no longer treated as separate compliance checklists.

If you aren't seeing any of these signs just yet – or if you're encountering uncertainty or pushback – it may be a normal part of the change curve. There are tried and tested strategies that you can use to move forward with confidence:

- **Empower teams with autonomy**

  Adaptability will flourish when teams are trusted to make their own decisions.[4] Leaders should provide clear guardrails – like RAI guidelines and performance metrics but avoid micromanaging.

  Example: A marketing team creates a low-code solution to streamline customer outreach campaigns. Leadership allows the team to design and test different approaches without seeking constant approval – **as long as they adhere to governance policies and responsible AI principles.** For example, the team ensures that customer data is handled in accordance with privacy guidelines and that outreach algorithms are designed to avoid bias or exclusion. This balance of autonomy and ethical oversight enables innovation without compromising trust.

- **Invest in learning and development**

  Continuous learning is the backbone of adaptability. Offering training programs that go beyond introductory technical skills and incorporate topics like ethical decision-making, collaboration in fusion teams, and cross-functional problem-solving.

[4] Pink, D.H. (2009). Drive: *The surprising truth about what motivates us.*

Example: Conduct workshops that simulate real-world scenarios where teams must make decisions involving RAI principles, such as addressing bias in AI models or managing sensitive customer data.

- **Encourage open dialog and feedback**

  Adaptability requires an environment where feedback is welcomed and acted upon. Create channels - such as feedback forms, AMAs (Ask Me Anything), townhalls, or anonymous surveys - where employees feel safe to share their feelings, concerns, or ideas about cultural challenges.

Leaders must focus and actively remove barriers that hinder change. This includes ensuring that policies, processes, and communication channels align with the adaptability goals. For instance, leaders can make adaptability a core value by integrating it into team charters and performance evaluations, where employees are assessed both on outcomes and how they navigate change and embrace innovation.

When leaders actively participate in cultural change, it sends a strong signal that adaptability isn't just an expectation for others but a shared commitment across the organization - for example: I've seen leadership attending training programs, building Power Apps, and sharing their personal challenges in learning low-code.

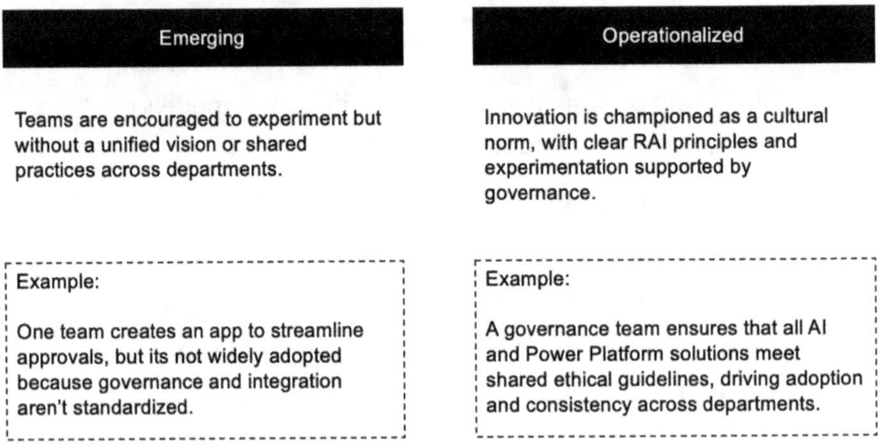

Figure 3-3. *Examples of cultural adaptability at emerging and operationalized stages*

Cultural adaptability ensures that teams are open to change, aligned with organizational values, and prepared to embrace innovation. As organizations mature, they need robust systems and processes (that are measurable) to sustain progress and build trust across stakeholders. This is where governance and performance metrics come into play.

In the next section, we'll look at how organizations can implement governance frameworks and performance metrics to scale RAI and ensure that innovation is both measurable and impactful.

Scaling RAI Through Governance and Performance Metrics

As organizations move beyond the emerging stages of RAI adoption, the challenge shifts from introducing principles to sustaining, scaling, and normalizing them. At this stage, the focus becomes embedding RAI into everyday practices, ensuring it is not only operational but also measured and continuously improved.

This progression involves aligning governance frameworks with practical implementation and developing performance metrics that provide visibility into progress and impact.

The Value of a Thoughtful Foundation

I once worked with a global organization that demonstrated what thoughtful planning can achieve with Power Platform. From the outset, they recognized the importance of embedding compliance into their adoption strategy - not as an afterthought but as a foundation. They invested significant time upfront to establish clear boundaries and metrics to measure the effectiveness of their governance.

What struck me most was their patience and foresight. They understood that getting governance right wasn't a quick fix but a process that would take months. By aligning their efforts with a clear vision and instilling these principles into their citizen developers from the very beginning, they created a culture where compliance was second nature.

Years later, their success continues to be a testament to the decisions they made early on. This approach not only safeguarded their innovation but also enabled them to scale with confidence, proving that the right foundation can set an organization up for lasting impact.

This story underscores the importance of thoughtful, strategic planning for organizations aspiring to operationalize their RAI efforts. Scaling RAI is building the systems, processes, and indicators that ensure ethical AI practices are upheld at scale, across departments, teams, and use cases. It requires leaders to move from talking about RAI to demonstrating it through measurable, transparent outcomes.

The Next Step: Embedding RAI into Strategy

The question is no longer "Why does RAI matter?" but "How do we ensure RAI is *pervasive, accountable and impactful*?". This step up in maturity is akin to implementing robust governance structures (we'll cover this deeply in the following part of this book) where clear communication, comprehensive training, and a coordinated approach that bridges both top-down leadership and bottom-up employee engagement are required.[5]

This section explores how organizations can achieve this level of maturity by focusing on

- **Governance**: Establishing scalable, enforceable systems that align ethical AI practices with broader strategic objectives
- **Performance metrics**: Establishing scalable, enforceable systems that align ethical AI practices with broader business strategies

Scaling RAI is an ongoing journey of refinement and growth. If you get it right, you'll elevate trust, accountability, and innovation, positioning your organization as both a responsible steward of AI and a leader in its effective use. Sounds good right? Let's explore how to make that happen.

The Leadership Mindset for Scaling RAI

Embedding RAI at scale demands a strategic shift led by forward-thinking leadership. At this stage (operationalized), the role of leadership is to ensure that RAI principles are actively woven into decision-making, innovation, and accountability across the organization. This mindset fuels sustained growth, minimizes risks, and aligns AI initiatives with the broader goals of the business.

[5] NIST. (2023). *AI Risk Management Framework.*

Moving from Tactical to Strategic

Leaders move from a tactical approach (focused on solving immediate problems) to a strategic outlook.[6] They view AI adoption as a transformation that touches every facet of the organization. This move manifests in several ways:

- **Proactive governance**: Leaders anticipate risks and establish frameworks to prevent them, rather than responding reactively to incidents.
- **Innovation with guardrails**: They foster a culture of experimentation but within clearly defined ethical boundaries, ensuring AI powered solutions deliver value without compromising trust.
- **Integration into core strategy**: AI initiatives are no longer side projects but are integrated into the organization's long-term strategic goals, driving competitive advantage.

The illustration below breaks down how leaders can implement these principles to foster a culture that prioritizes RAI while driving measurable business outcomes:

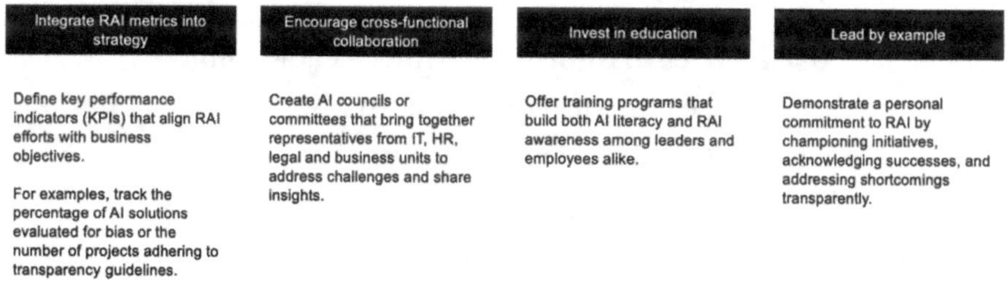

Figure 3-4. *Illustrating steps to integrate RAI into your strategic outlook*

[6] Kotter, J.P. (1996). *Leading Change.*

The illustration outlines the foundational steps leaders can take to scale RAI effectively. To bring them to life, let's dive deeper into what each step entails and how it translates into real-world actions.

- *Integrate RAI metrics into strategy*

  What it means: Embedding RAI metrics into the organization's strategic planning ensures ethical AI use becomes a measurable and integral part of business success.

- *Key actions*

  - **Define measurable KPIs, for example**:
    - Percentage of AI-powered solutions evaluated for transparency and fairness
    - Number of RAI champions trained and actively participating in RAI activities
    - Reduction in time to review AI-powered solutions
  - **Regular reporting**: Create dashboards that provide visibility into these KPIs for leadership and key stakeholders.
  - **Use metrics to inform decisions**: Leadership should use these insights to adjust priorities, allocate resources, or address shortcomings. We'll cover metrics in more detail later in this chapter, but below is an example of what could be tracked to make informed decisions:

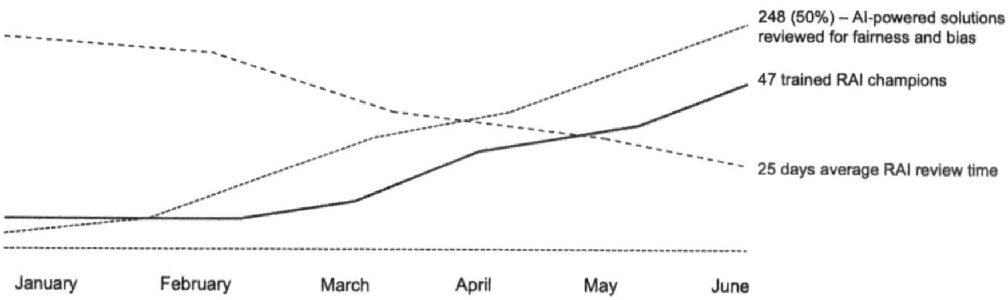

***Figure 3-5.** Example report demonstrating impact of RAI initiatives*

- *Encourage cross-functional collaboration*

  What it means: Building collaboration across departments ensures diverse perspectives and collective accountability for AI governance.

- *Key actions*

  - **Establish an AI council**: Include representatives from IT, HR, Legal, and business units. Rotate members to keep perspectives fresh.
  - **Hold regular meetings**: Use these meetings to share insights, address challenges, and drive alignment on RAI initiatives.
  - **Create shared ownership**: Ensure every department sees RAI as part of their responsibility, not just IT or legal domain.

AI councils work well at aligning and securing buy-in from departments that typically might not work closely together. They ensure that AI adoption aligns with the organization's values, ethical principles, and strategic objectives.

How to Put Together an AI Council

Establishing the AI council is a critical step, as I've just mentioned. Remember, in this book, we're mainly focused on adopting Power Platform and its AI features, but your organization will likely have *broader AI objectives* and other platform initiatives. Aligning with the broader view on AI within your organization will likely speed up standing up a functional AI council.

You might discover that the AI council naturally establishes itself as an extension of your existing Center of Excellence – there are some new faces, but the team is largely the same; we'll discuss that in a moment; however, here are some steps to follow or check on within your own organization:

1. **Define the council's purpose and scope**

   Much like earlier in this book where we briefly looked at creating a community focused on Power Platform and RAI, clarifying the AI council's mission, scope, and goals is crucial. Clearly define the council's mandates to ensure the focus is aligned with the broader RAI objectives.

   Questions for the council to address:

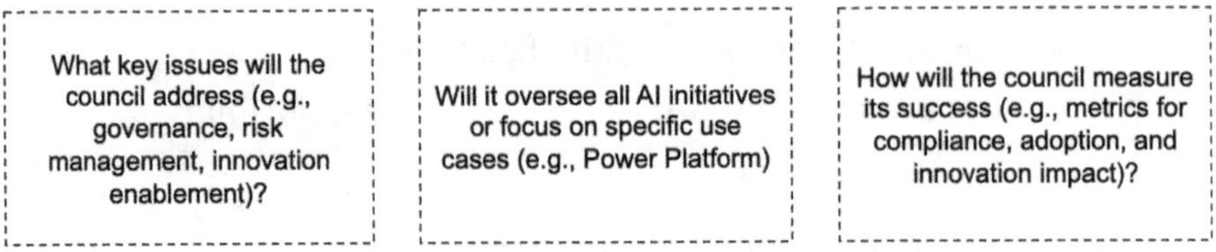

Figure 3-6. *Key questions to establish the AI council's purpose and scope*

2. **Assemble the right team**

   The effectiveness of an AI Council hinges on bringing together the right mix of expertise, perspectives, and leadership authority.

   Core roles to include:

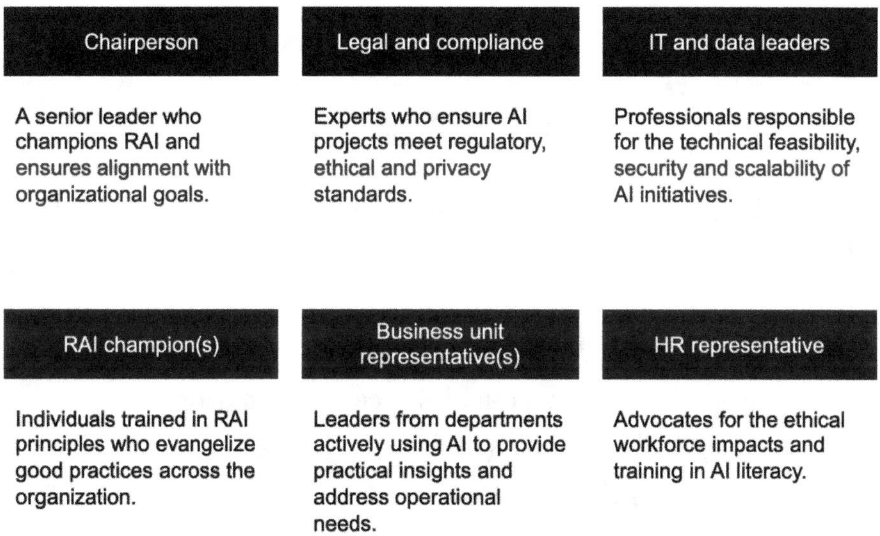

Figure 3-7. *Core roles of the AI council*

3. **Establish roles and responsibilities**

   Clearly define what each member is responsible for to avoid ambiguity and ensure accountability.

Example responsibilities:

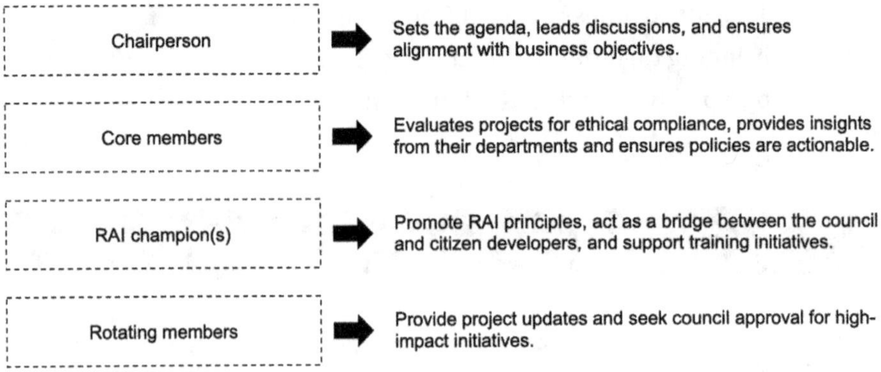

***Figure 3-8.** Examples of common roles and responsibilities in AI councils*

4. **Set operational guidelines**

   To ensure efficiency, establish clear processes for the council's operations.

   Key guidelines to consider:

   - **Meeting frequency**: Hold regular meetings, typically quarterly, but allow for ad hoc sessions if urgent issues arise.
   - **Decision-making process**: Determine how decisions will be made (e.g., majority vote, chairperson approval).
   - **Documentation**: Record meeting minutes, decisions, and action items to ensure accountability and transparency.
   - **Communication**: Develop a process for sharing updates with the broader organization, such as newsletters or town halls.

5. **Create initial policies and templates**

   The council should develop foundational policies and tools to guide RAI initiatives.

   Some examples:

   - **AI governance policy**: A document outlining ethical AI principles, approval processes, and compliance requirements. The Responsible AI Institute website contains a number of superb resources to help you get started - here is a link to their AI policy template: `https://www.responsible.ai/ai-policy-template/`.
   - **Solution evaluation template**: A checklist for assessing Power Platform solutions against RAI standards (e.g., fairness, transparency, risk mitigation).
   - **Training plan**: The council should prepare resources for educating employees on RAI principles and frameworks (e.g., W.A.I.T and C.H.A.T).

6. **Measure and communicate impact**

   To sustain momentum and demonstrate value, regularly evaluate the council's impact, and share results with senior leadership and the wider organization. Some metrics you might consider:

   - Number of projects reviewed and approved over time
   - Percentage of existing Power Platform assets adhering to RAI principles

- Employee engagement with training programs
- Reduction in time to review AI-powered solutions
- Number of identified RAI risks mitigated or avoided
- Examples of RAI incidents prevented or lessons applied from review outcomes

Alongside throughput and adoption metrics, consider how you will communicate value and impact. Publishing updates from your AI council or governance body should include not only how many solutions were reviewed but also why it mattered. Did the council prevent a potentially biased deployment? Did they flag a privacy concern before go-live? Did they help a team redesign a prompt to avoid hallucination or confusion?

These stories reflect trust in action – they show leadership that RAI governance is about ensuring the right things move forward with confidence.

Why this matters?

An AI council is a strategic enabler. It demonstrates that your organization is serious about the ethical and responsible use of AI. It ensures that citizen developers have a place to go if they have questions and that you have policies and processes in place that support innovation.

A high-performing AI council drives the momentum to innovate new products and services and augment existing processes and roles while, at the same time, fostering transparency and creating sustainable adoption.

We've discussed how governance frameworks and performance metrics provide the structural and operational backbone for scaling RAI, but we have yet to explore how organizational culture plays a pivotal role in ensuring these systems are embraced, followed, and continuously improved. Let's look at that now...

Cultural Foundations for Sustained Governance

I've mentioned "shifts" a few times in this chapter already, from strategic thinking to implementing new RAI specific structures and processes. I'm going to mention another one - we need a shift in organizational behavior that reinforces accountability, transparency, and continuous improvement. We'll explore how a strong cultural foundation underpins sustained governance, and I'll share some practical strategies to embed RAI principles into the fabric of your organization.

The role of culture in governance

Governance systems ensure that policies are in place, but culture determines whether people follow, trust, and embrace them. Without cultural buy-in, governance is a tick-box exercise; with a strong culture, it transforms into something that employees champion every day.

I've seen several organizations in both camps, and it might not come as a surprise to you when I say that the latter become more mature, innovative, and gain more strategic value from Power Platform.

Culture helps amplify governance in multiple ways. It

- Shapes behaviors, ensuring citizen developers internalize ethical principles and act within the guidelines
- Builds trust by demonstrating transparency and consistency
- Fosters resilience, enabling organizations to adapt governance practices as technology evolves

At this stage, remember, we're trying to form a "governance habit" rather than a set of rules that need to be obeyed. As your governance culture matures, scaling responsible AI introduces new demands that test and elevate that foundation. The principles you've already built - trust, transparency, and adaptability - remain essential, but responsible AI places greater pressure on how deeply these values are lived out across teams. AI systems amplify both the value and the risk of decisions made at scale, which means governance isn't enough. It must be reinforced by a culture that not only encourages ethical development but actively enables people to make responsible decisions without constant oversight. The following cultural principles - transparency, empowerment, accountability, and adaptability - extend what's already working in successful Power Platform environments and reflect the deeper commitment to scale AI responsibly.

Key cultural principles for RAI success

Organizations scaling RAI must foster a culture built on the following core principles:

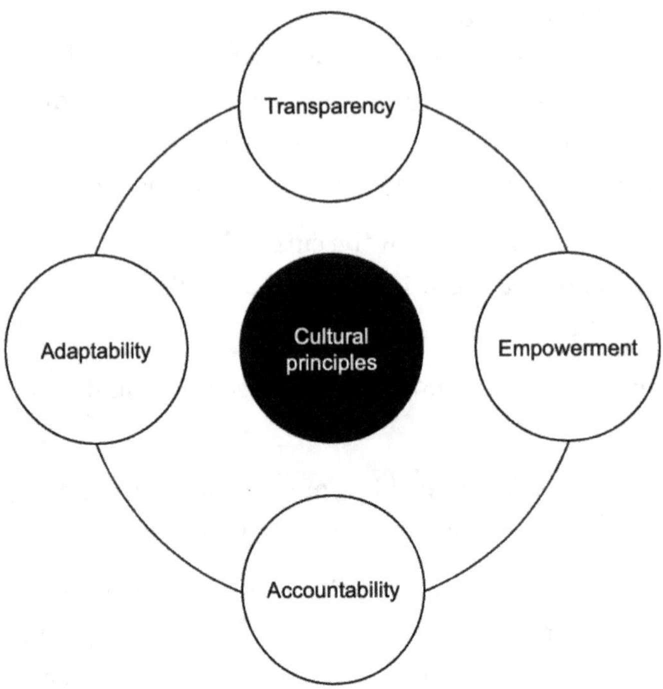

Figure 3-9. *Core principles for scaling RAI*

Transparency

Encouraging open communication about RAI and AI systems, their risks, and their decision-making processes.

Ensure employees feel safe raising concerns or questioning AI outcomes without fear of backlash.

Empowerment

Equipping employees at all levels with the tools, knowledge, and confidence to adopt RAI practices.

Democratize AI while reinforcing governance, so citizen developers and technical teams work within a shared framework.

Accountability

Building trust through mechanisms that encourage ownership of AI outcomes, including the ethical implications of solutions.

Make accountability visible at all levels, from leadership to individual project teams.

Adaptability

Promote continuous education to keep pace with evolving AI technologies, Power Platform features, and governance requirements.

Treat governance challenges as opportunities to learn, adapt, and strengthen systems.

Integrating the Core Principles

Below are some examples of strategies I've seen work well:

- **Leadership "walks the walk"**: Employees follow leaders that actively demonstrate their commitment to RAI by embodying its principles, both in decision-making and communication. Leadership should seek examples of how they've prioritized RAI into their own projects or strategies – this showcases transparency and accountability.

- **Integrate RAI into training**: Making RAI a core part of your organizational training program. Consider the different audiences, from citizen developers to executive leaders, and ensure everyone understands their role in upholding governance. I've delivered a lot of training and "snuck" in RAI into it, either by using group activities that discuss a scenario or by using storytelling – every time, the topic has resonated with the audience: the key is ensuring your training is targeting the right level of user. For example – training citizen developers: they are more interested in learning the good habits they need vs. an executive who is probably more interested in knowing about the regulatory risks that RAI helps mitigate.
- **Recognition and incentives**: Celebrating success is essential for embedding responsible AI practices into culture. Recognition should be intentional, visible, and inclusive – highlighting not just outcomes but the behaviors that make them possible. Recognize individuals, teams, and especially your RAI champions – those who serve on the front lines, handling queries, reviewing solutions, and guiding others. Consider using certificates, awards, internal communications, or public acknowledgements in company town halls to spotlight examples of ethical decision-making, thoughtful risk mitigation, or cross-functional collaboration. These recognitions reinforce desired behaviors, elevate champions as role models, and create a positive feedback loop that strengthens organizational commitment to responsible AI.

- **Storytelling**: Take the successes and share the stories to illustrate RAI in action. For example, highlight how responsible AI practices mitigated risks, improved outcomes, or enhanced trust with stakeholders. Use internal channels such as intranets, town halls, and company all-hands, and use executives or leaders to tell the story.
- **Feedback loops**: Communication can't be one-way. Encourage and regularly gather input from employees and other stakeholders about the effectiveness of governance systems and cultural alignment. Use this feedback to iterate and refine your RAI practices, ensuring they remain relevant and impactful.

Everything above helps build a cultural foundation that can sustain governance. To truly sustain the culture, you need to continually reinforce expectations and policies; here are some examples:

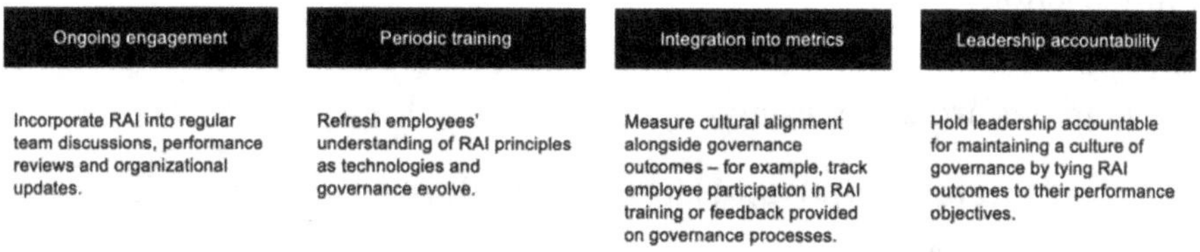

***Figure 3-10.** Example policies that reinforce and sustain RAI*

Let's dig a little deeper...

Reinforcement Through Training and Development

Training can go beyond initial onboarding - in fact it should become a continuous process. AI features in Power Platform will mature, becoming more advanced; regulations will change and update as AI capabilities advance, therefore training needs to stay current.

Offer regular refreshers - such as ongoing training sessions, webinars, workshop, or clinics that focus on RAI topics, such as transparency and data ethics.

Consider role-specific learning - tailor your training programs for different roles:

- **Developers**: Advanced RAI frameworks, fairness in model development
- **Leaders, managers**: Ethical considerations in AI strategy and risk
- **Citizen developers**: How to interpret AI outputs responsibly

Embed RAI into Performance Reviews and Incentives

To sustain a culture, employees must see that RAI is valued, not just in principles but also in outcomes that impact their career growth and compensation.

Performance metrics: Including RAI adherence as part of individual and team performance reviews; for example:

- Developers (pro and citizen) could be evaluated on how well their solutions and projects meet ethical guidelines.
- Teams could be evaluated on the ethical implications of their use cases.
- Individuals could be evaluated on their RAI training completion status.

You could also consider incentivizing contributions, for example, recognize and reward employees who identify ethical risks, improve RAI frameworks, or contribute to solutions that align with RAI principles. You could award the most impactful RAI champion for their contributions.

Actively Encourage Open Feedback and Dialogue

Creating spaces for employees to discuss challenges, share insights, and raise concerns helps the shift toward accountability and ownership of RAI principles. Some ideas for you to consider:

- **AI forums**: Organize monthly or quarterly forums where teams can discuss ethical dilemmas, share success stories, and propose ideas for improvement.
- **Anonymous feedback channels**: Set up mechanisms for employees to flag potential risks or concerns without fear of reprisal. These could include AI audit findings, unintended biases, or ethical questions.
- **Leadership-led discussions**: Regular town halls or AMAs (ask me anything) sessions where leadership addresses RAI-related questions and showcases their commitment to ethical practices.

Example in action: how sustained culture looks

A financial services company integrates RAI into its culture by

- Rolling out an annual RAI bootcamp to all new hires and refresher courses for existing staff
- Setting up cross-departmental AI council to review quarterly metrics on ethical AI practices and addressing flagged concerns
- Including RAI impact metrics in employee and team bonus calculations, such as reducing bias, or increasing AI-powered solution explainability
- Having the top executive regularly update the entire organization on RAI progress, highlighting success stories and lessons learned

Call to action: sustaining RAI at scale

As we conclude this chapter, it's essential to emphasize the scaling RAI is not just a technical or governance challenge – it's a leadership imperative. Embedding ethical AI principles into your organization requires commitment, accountability, and cultural transformation. Here's how you can take action to ensure sustained success:

1. **Align your leadership team**
    - Schedule a leadership workshop to define your organization's vision for RAI.
    - Assign clear ownership for RAI initiatives across departments.
    - Regularly communicate the strategic importance of RAI to your teams.
2. **Establish clear goals and metrics**
    - Identify the key performance indicators (KPIs) to measure the success of your RAI initiatives.
    - Use these metrics to create visibility and accountability at all levels of the organization.
    - Publish RAI progress reports to key stakeholders and employees to maintain transparency and build trust.
3. **Build cross-functional collaboration**
    - Form an AI council to drive ongoing discussions and actions around RAI.
    - Foster collaboration between technical and nontechnical teams to address ethical challenges holistically.

- Encourage open dialogue and feedback to continuously refine your approach.

4. **Invest in training and awareness**

    - Roll out regular training sessions tailored to different roles within your organization.
    - Provide practical tools like the W.A.I.T. and C.H.A.T. frameworks to help employees make responsible decisions.
    - Recognize and reward RAI champions who contribute to the organization's ethical AI journey.

5. **Be a leader in your industry...**

    - Share your organization's RAI journey through case studies, conference presentations, and community events.
    - Position your organization as a thought leader by advocating for ethical AI practices both internally and externally.
    - Inspire others by showcasing how ethical AI can be a source of competitive and cultural strength.

The step-up to operationalized RAI and Power Platform adoption is not without its challenges, but the rewards - greater trust, innovation, and competitive advantage - are worth the effort. As you take these steps, remember that every small action contributes to a larger cultural and organizational transformation.

It's time to lead the way and inspire change, demonstrating what's possible when technology and ethics are working hand in hand.

Operationalized Progression Checklist

At the operationalized stage, leadership and culture are no longer simply enabling factors – they are embedded systems that shape how responsible AI and Power Platform initiatives evolve, scale, and sustain impact.

This checklist captures the critical behaviors, structures, and mindsets that distinguish mature organizations. Use this as a practical reference to assess your organization's progress.

External recognition and thought leadership	Leadership accountability & governance integration
☐ Motivation for external visibility is clearly defined (e.g., trust, talent, innovation).	☐ Accountability mechanisms (roles, metrics, regular reviews) are in place.
☐ Leadership speaks at events, contributes to white papers or shares case studies.	☐ RAI and Power Platform initiatives are aligned with high-priority business goals.
☐ Strategic channels are identified for outreach (conferences, social media, blogs).	☐ Governance frameworks and tools are fully established and operational.
☐ Employees are encouraged and supported to act as RAI and platform ambassadors.	☐ Ethical frameworks like W.A.I.T and C.H.A.T are used in project reviews.
☐ A strategic roadmap exists for scaling AI and Power Platform initiatives.	

Training, feedback & reinforcement

- [ ] Ongoing training programs exist, tailored by role (developer, exec, citizen dev).
- [ ] RAI adherence is part of performance reviews and/or incentive structures.
- [ ] Feedback mechanisms are in place (forums, anonymous reporting, AMAs).
- [ ] RAI champions and ethical successes are recognized and celebrated.

Sustained governance & metrics

- [ ] Transparent metrics track the effectiveness of RAI governance.
- [ ] Governance processes evolve based on feedback and AI/Power Platform advancements.
- [ ] Metrics such as RAI-reviewed solutions, ethical issue resolution time, and training completion are tracked.
- [ ] Governance is communicated as a value-enabler, not just a compliance function.

AI council or CoE integration

- [ ] A cross-functional AI council is formed with clear roles and responsibilities.
- [ ] Council meets regularly, maintains documentation, and communicates impact.
- [ ] Policies and templates are in place (e.g. AI governance policy, solution evaluations).
- [ ] Training plans and resource toolkits are developed and maintained by the council or CoE.

Cultural adaptability and continuous improvement

- [ ] Cultural signals of successes are visible (e.g. cross-team collaboration, ethical fluency).
- [ ] RAI is integrated into daily workflows, not just compliance checklists.
- [ ] Teams are empowered to make decisions within established ethical boundaries.
- [ ] Leaders model adaptability and participate in cultural transformation activities.

CHAPTER 4

Integrated Excellence: Leadership and Culture

Pioneering | Sustainable | Influential

Leadership and Culture at the Forefront of Responsible AI

What truly sets the pioneers apart from the rest? At the **integrated excellence** level of maturity, innovation is almost natural – AI initiatives not only align with organizational goals but actively redefine them.

Decisions are guided by clear ethical guidelines, and every team member from the boardroom to the frontline understands their role in undertaking responsible AI practices.

This is the pinnacle of leadership and cultural maturity in AI and Power Platform adoption; organizations here do not just participate in change – they are often leading it.

S. Jeffery, *The Low-Code AI Maturity Model*, https://doi.org/10.1007/979-8-8688-1730-4_4

To better understand what defines organizations at this level, the illustration below captures the key characteristics of integrated excellence in leadership and culture. These qualities demonstrate how responsible AI principles and innovation converge to create real value and lasting impact.

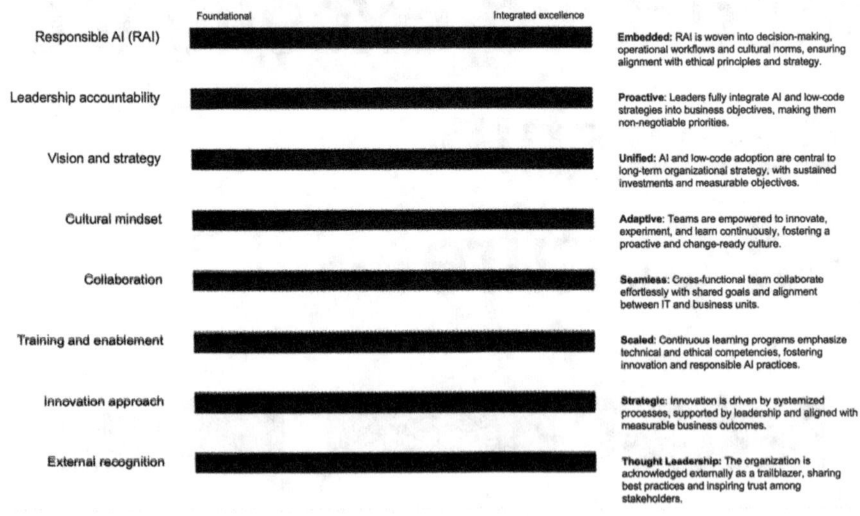

Figure 4-1. *Run state of an organization at the integrated excellence stage*

Power Platform, previously approached cautiously or experimentally, becomes the very foundation of value creation, problem-solving, and sustainable growth. At this level, I've encountered customers where I often find myself learning from their innovative practices, rather than simply advising them.

Leadership at this stage is multidimensional. It's no longer confined to internal operations or immediate business goals – it extends across organizational boundaries, influencing industries, setting standards, and driving ethical practices at a global scale. The illustration below captures this layered approach to leadership:

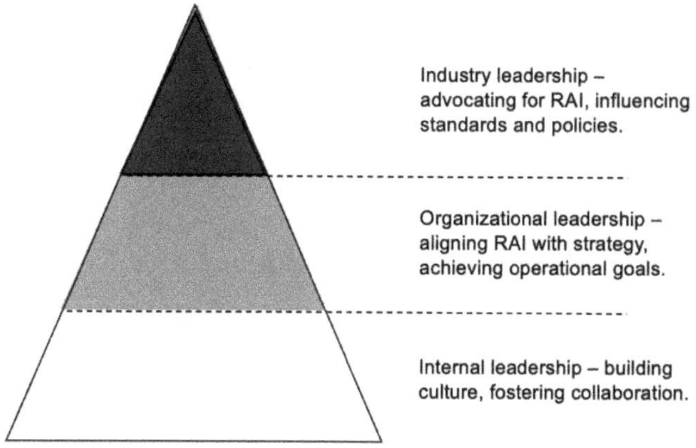

Figure 4-2. *Illustrating multidimensional leadership progression*

This illustration underscores how leadership evolves to create a lasting impact, both within and beyond the organization. Let's delve deeper into the practices and behaviors that enable this transformation.

What this chapter will explore

In this chapter, we'll explore what makes integrated excellence possible. From building a **culture of continuous responsibility** to **sustaining long-term alignment with strategic goals**, we'll uncover the key behaviors, systems, and mindsets that define leadership and culture at the highest level of maturity. Let's dive into what it means to lead with integrity and purpose and leave a legacy of responsible AI.

Building a Culture of Continuous Responsibility

What does this mean? At the heart of integrated excellence lies a culture of continuous responsibility - an environment where ethical practices are an ongoing commitment that all employees follow as part of everyday operations.[1]

[1] Schein, E.H. (2010). *Organizational culture and leadership.*

Achieving this level of maturity is the result of deliberate leadership, aligned priorities, and consistent reinforcement of RAI principles.

What defines a culture of continuous responsibility?

This culture goes beyond enforcing rules and governance – it is defined by

- **Shared ownership**: Every individual shares the feeling of responsibility for ensuring ethical AI practices, from developers building solutions to executives making strategic decisions.
- **Proactive reflection**: Teams anticipate the impact of AI-powered initiatives and solutions before implementation and continuously monitor them for unintended consequences, refining approaches as needed.
- **Embedded principles**: RAI principles: transparency, fairness, accountability, and privacy are not only documented but actively practiced and referenced in decision-making at all levels.

There is a balanced approach in this mindset that responsible innovation is as much about *creating value* as it is about *protecting trust and reputation*. This culture can withstand disruption, from rapid changes in AI capability through to organizational changes. RAI and governance frameworks are continuously adapted to evolving challenges, such as regulatory changes or technological changes.

Here are some key steps that organizations get right to build and sustain this culture:

RAI is integrated into core values

Organizations ensure RAI is a guiding principle within their mission and values. Leaders communicate this integration clearly and consistently, ensuring it is visible in both internal policies and external actions.

Example: A technology firm includes RAI as a core company value, highlighting it in recruitment, training, and all-hands meetings to ensure new hires and veterans alike understand its importance.

Active participation is encouraged

Employees are empowered to take ownership of RAI through mechanisms such as RAI committees, cross-functional AI councils, and open forums for RAI challenges. Platforms are established for employees to share concerns, raise questions, and contribute to refining RAI practices.

Example: A manufacturing company holds annual "ethics in action" workshops, inviting employees from all levels to participate in discussions about potential risks and opportunities in their projects and solutions.

They provide accessible training and resources

There is heavy focus on equipping employees with the tools they need to make informed, responsible decisions. Continuous training ensures that teams stay updated on evolving AI risks, compliance standards, and best practices.

Example: An organization rolls out an online learning platform offering courses on specific RAI principles, such as bias detection, data ethics, and RAI frameworks tailored to different roles.

Celebrating RAI successes is the norm

Recognition reinforces the behaviors that align with continuous responsibility. Organizations celebrate teams and individuals who exemplify ethical practices, tying these successes to organizational impact.

Example: A retail company awards "annual AI ethics champion" recognition to teams that successfully implement innovative, yet ethically sound, AI-powered solutions.

How leaders set the tone

Leaders play a crucial role in creating and sustaining this culture. They act as role models, consistently demonstrating ethical decision-making and ensuring RAI isn't relegated to the periphery. Humility is embraced where acknowledging mistakes and sharing lessons create an environment where continuous improvement thrives.

Leadership actions include

- Owning the successes and failure of AI initiatives
- Regularly communicating the importance of RAI and linking it to broader business goals
- Setting clear expectations for accountability across all levels of the organization

Building a culture of continuous responsibility isn't achieved overnight nor is it common. It requires sustained effort over a long period of time, a deep commitment to ethical principles, and a willingness to adapt to a rapidly changing landscape. For the few organizations operating at this level, the rewards are clear – they are trusted by stakeholders, employees, and admired by peers – they can tackle the most complex AI challenges with confidence.

This culture does not operate in isolation. It is supported with long-term visions that are aligned with AI. Let's look at how organizations align long-term planning with strategic goals.

Sustaining Long-Term Alignment with Strategic Goals

Achieving integrated excellence is the beginning of sustained alignment and maximized value. Organizations are focused on maintaining alignment between their AI initiatives and their overarching strategic goals. This alignment ensures that AI adoption continues to deliver measurable value.

AI is no longer treated as a separate initiative; it is a core enabler of business strategy that's fully integrated into the organization's planning and execution. Leaders are kept abreast of new features and capabilities – existing solutions are improved, and new innovations to maximize the potential of AI-powered processes are encouraged.

Strategic planning sessions *explicitly integrate technology as a critical factor*. For example, organizations might allocate resources for Power Platform innovations directly tied to goals like improving operational efficiency or enhancing customer experiences. This "task force" approach focuses on getting to value fast by iterating quickly to release high-impact solutions.

This shift is to ensure that the technological capabilities remain relevant to shifting market demands. It requires a forward-thinking approach, balancing adaptability with strategic intent. One key method is scenario planning, where leaders anticipate various market and operational scenarios to better prepare for potential shifts. We'll dive a little deeper into some of the approaches used later.

For example, they might explore how AI enhancements within Power Platform could enable predictive analytics in areas like inventory or logistics. By proactively identifying these opportunities, organizations can position themselves to respond effectively when the market evolves, rather than scrambling to catch up.

Another component is the use of pilot programs to validate strategic decisions. Leaders understand the importance of testing new capabilities before implementing them on a larger scale. For instance, companies might launch a Power App to streamline logistics for real-time fleet management in a specific region, allowing them to gather insights on its effectiveness, identify challenges, and refine the solution. By validating outcomes in a controlled environment, they ensure the broader implementation is aligned with goals while minimizing risk.

Lastly, external trend monitoring plays a pivotal role in sustaining long-term alignment. Leaders keep a close eye on industry developments, competitor actions, and emerging technologies to stay ahead of the curve.

By integrating scenario planning, pilot programs, and external trend monitoring into strategic efforts, leaders can remain agile and adaptable – always focusing on maximizing return on investment.

Proactively exploring potential futures is where scenario planning becomes invaluable. By imagining and preparing for different potential outcomes, organizations can align their strategic goals with their technological investments. Let's explore how you could use scenario planning as a practical tool to sustain alignment, empower your teams to innovate responsibly, and maximize value from Power Platform.

Running a Scenario Planning Exercise

Scenario planning[2] is a structured process that helps you anticipate future challenges and opportunities, allowing them to align their strategy with potential realities.

In this section, we'll walk through a guide to running an effective scenario planning exercise:

1. **Defining the scope and objectives**

   Start by clarifying what you want to achieve through scenario planning. Are you exploring new market opportunities, anticipating technological disruption, or preparing for regulatory changes? Be clear about the time horizon - e.g., 1, 2, 4 years - and the questions to address. Let's use an example:

   Objective: Explore how to enhance customer service capabilities using AI-powered tools like Microsoft Copilot Studio over the next 12–18 months.

[2] Schoemaker, P.J.H. (1995). *Scenario Planning: A tool for strategic thinking.*

Key questions:

- What innovations can improve customer interactions and operational efficiency?
- How might AI-powered tools help reduce response times or automate repetitive tasks?
- What risks or constraints should be considered when rolling out new solutions?

2. **Identify key drivers of change**

   Gather insights about the factors that could shape your organization's future. These could include technological advances, market trends, regulations, or other internal dynamics. Do this by engaging with cross-functional teams to brainstorm key drivers. Use tools like PESTLE (Political, Economic, Social, Technological, Legal, Environmental) to ensure broad coverage. Using our example of enhancing customer service:

   Drivers:

   - **Technological**: Advances in AI agents, automation features, and predictive analytics using Power Platform will enable more tailored, scalable, and proactive use of low-code solutions to meet evolving business demands.
   - **Customer expectations (social)**: Increasing demand for instant, personalized support.

- **Legal considerations**: Emerging privacy laws and AI regulations introduce both constraints and opportunities – prompting the need for transparent, auditable customer service solutions.
- **Internal dynamics**: Current limitations in staff training and technical readiness.

3. **Develop plausible scenarios**

   These future scenarios are designed to help leaders anticipate strategic responses – not just react to change but guide it responsibly. Each scenario explores different outcomes based on how these drivers may evolve.

   Using the key drivers, outline three to five plausible future scenarios. These should include

   - **Best-case scenario**: What happens if everything aligns perfectly?
   - **Worse-case scenario**: What challenges arise if risks are realized?
   - **Baseline scenario**: What is the most likely outcome based on current trends?

   Example:

   - **Best-case**: Full integration of AI-powered solutions. AI Agents handle 70% of customer inquiries, freeing up staff for high-value tasks. Compliance and training initiatives ensure smooth adoption. Customer satisfaction increases by 30%.
   - **Worse-case**: Deployment of AI-powered solution is rushed without sufficient oversight, resulting

in a serious responsible AI failure - such as a privacy breach, biased decision, or hallucinated customer response. The incident triggers regulatory scrutiny, damages the organization's reputation, and erodes internal trust in AI. Adoption stalls, and significant time and resources are diverted toward remediation and reputational repair rather than innovation.

- **Baseline**: Partial adoption of AI-powered solutions, with moderate improvement in efficiency and customer satisfaction.

4. **Analyze implications and opportunities**

   For each scenario, consider factors like

   - What are the risks, opportunities, and required actions?
   - How will this impact your organization's cultural and strategic goals?
   - Which technologies, like AI Builder, Power Apps, and Copilot Studio, play a role?

   Example:

   - *Best-case*
     - **Opportunities**: Improved customer satisfaction, reduced operational costs, increased team capacity for complex issues.
     - **Actions**: Accelerate training for customer service teams, ensure robust compliance measures, and allocate budget for scaling Power Platform.

- *Worse-case*
    - **Risks**: A misconfigured AI-powered customer service agent generates incorrect or biased responses, leading to a formal complaint and reputational damage. Internal investigation reveals a lack of responsible AI review or mitigation process. The incident results in compliance risk, stakeholder distrust, and a temporary halt in AI solution deployment.
    - **Actions**: Establish mandatory responsible AI assessments for AI-powered assets before deployment. Provide targeted training on responsible prompt design and model behavior monitoring. Engage legal and compliance early in the life cycle to validate high-risk use cases.
- *Baseline*
    - **Outcomes**: Moderate improvements, but opportunities for transformational change remain untapped.
    - **Actions**: Use baseline successes to build momentum and make the case for further investment

5. **Formulate actionable strategies**

   Turn insights into concrete strategies. Develop contingency plans that address each scenario, ensuring your organization can pivot effectively.

Example:

- **Baseline strategy**: Roll out agents to a single customer segment to prove value and refine processes before scaling.
- **Best-case strategy**: Invest in training and create a dedicated AI council to oversee implementation and ensure alignment with goals.
- **Worst-case strategy**: Develop contingency plans for delayed adoption, including alternate manual processes to handle increased customer demands.

6. **Review, refine, and communicate**

   - **Review**: Regularly revisit scenarios and adjust based on new insights or changes in the business environment.
   - **Refine**: Seek feedback from teams to improve strategies.
   - **Communicate**: Share outcomes with stakeholders at all levels to ensure alignment and engagement.

Scenario planning is useful for equipping leaders like you with the ability to proactively navigate uncertainty and keep their organization aligned with strategic goals. It's a powerful tool for anticipating change, empowering teams, and maintaining relevance in an evolving landscape.

Conclusion: Leadership and Culture

As we conclude Part 1 of this journey, you've explored the path from foundational efforts to integrated excellence in adopting Power Platform and RAI. Along the way, you've gained practical tools, actionable frameworks, and guiding principles designed to help you build momentum, overcome challenges, and realize the full potential of AI within your organization.

No matter where you find yourself on this maturity curve, remember that progress is a process, not a destination. Each step forward, whether it's refining your initial use cases or embedding RAI into the fabric of your business, brings you closer to a future where Power Platform drives meaningful impact, innovation, and sustained growth.

In Part 2, we'll turn our focus to establishing governance and trust – critical pillars for responsible and scalable Power Platform adoption. We'll explore how to build robust governance frameworks, address ethical considerations, and create a culture where trust in AI fuels both innovation and accountability. As you move into the next part, you'll find the guidance you need to ensure that your Power Platform journey remains transformative and principled.

Integrated Excellence Progress Checklist

At the Integrated Excellence stage, organizations no longer see responsible AI and Power Platform adoption as initiatives to manage – they are intrinsic to how decisions are made, how teams operate, and how value is delivered.

This final stage of leadership and cultural maturity is marked by proactive strategy, continuous responsibility, and multidimensional leadership that influences both internal operations and industry norms.

Culture of continuous responsibility

- ☐ Teams demonstrate proactive reflection – considering ethical risks before and after deployment of AI solutions.
- ☐ Every employee understands their role in upholding RAI practices.
- ☐ Open forums, AI councils, or feedback channels are in place to surface ethical concerns and improve RAI approaches.
- ☐ Continuous training on bias, transparency, privacy and AI ethics is offered and role-tailored.
- ☐ Responsible AI principles are embedded into the organization's mission, values, and internal communications.

Leadership behaviors and influence

- ☐ Leaders consistently model ethical AI decision-making and communicate its business relevance.
- ☐ Executives actively participate in RAI-focused initiatives, councils, or community efforts.
- ☐ Leadership takes responsibility for both the successes and challenges of AI adoption.
- ☐ RAI practices are reflected in long-term strategic decisions, not just project-level goals.

Strategic alignment and long-term planning

- ☐ AI and Power Platform capabilities are tightly aligned with strategic objectives (e.g., customer experience, operational efficiency).
- ☐ Scenario planning exercises are conducted regularly to anticipate future trends and disruptions.
- ☐ Pilot programs are used to test alignment between innovation and long-term goals.
- ☐ External trends and regulatory shifts are actively monitored and factored into planning.
- ☐ Responsible AI principles are embedded into the organization's mission, values, and internal communications.

External leadership and industry engagement

- ☐ The organization publicly shares its RAI practices through conferences, case studies, or thought leadership.
- ☐ Internal RAI champions and teams are supported to engage in external communities or events.
- ☐ Responsible AI is positioned as a competitive advantage and cultural differentiator.

PART II

Establishing Trust and Governance

In your organization's race to harness the full potential of Power Platform, governance all too often is seen as a brake pedal – a necessary restraint to keep the organization in control.

But what if governance could be an *accelerator* instead? What if the right frameworks weren't there just to offer protection but instead were designed to propel it forward and unlock innovation at scale while ensuring all apps, flows, agents, and AI-driven decisions are built on a foundation of trust?

Governance is often seen as the guardrail that keep innovation in check, but with the dynamic ecosystem of Power Platform, it must be more than that. Governance must be a catalyst, enabling organizations to innovate confidently while ensuring alignment with ethical, legal, organizational, and regulatory standards. This is a balancing act that is difficult but not impossible to achieve. I've seen numerous organizations achieve this with the right investment, support, planning, and time.

As Power Platform becomes a central driver of low-code and AI adoption, governance frameworks must evolve to reflect its unique role - not only in supporting business users but also in integrating with broader enterprise AI initiatives.

Traditional vs. Adaptive Governance

In my experience, traditional governance approaches - heavily reliant on centralized control and rigid structures - often struggle to meet the demands of Power Platform, where agility and user empowerment are paramount. Moreover, AI capabilities, whether embedded within Power Apps, Power Automate, Microsoft Copilot Studio, or through integrations with custom models through AI Builder or Azure AI Foundry, expand the governance scope beyond the boundaries of IT.

Today, effective governance requires the inclusion of new voices: *ethics committees* to assess impacts to employees and customers, *legal teams* to navigate compliance, *HR* to address workforce concerns, and beyond.

This new approach to governance must deliver more than oversight; it must equate to momentum. Power Platform thrives on enabling citizen developers and operational teams to solve problems quickly and at scale. A governance framework that hinders this pace will inevitably clash with the platform's core purpose. Instead, governance must provide clarity and confidence, allowing users to innovate within well-defined boundaries without unnecessary roadblocks.

In this part of the book, we'll explore how governance matures across the stages of Power Platform and AI adoption:

- **Foundational**: Establishing basic policies and processes that ensure security, compliance, and transparency
- **Emerging**: Creating adaptive frameworks that respond to the platform's growth, empowering users while maintaining oversight

- **Operationalized**: Scaling governance through cross-functional collaboration performance metrics and trust-building measures
- **Integrated excellence**: Embedding governance into the broader enterprise AI strategy, influencing ethical adoption and global best practices

Throughout, we'll examine practical strategies and real-world examples to help you build governance frameworks that don't just safeguard innovation but propel it forward. Establishing trust, creating adaptability, and aligning governance with organizational momentum, you'll unlock the full potential of Power Platform and set the stage for long-term success.

CHAPTER 5

Foundational: Establishing Trust and Governance

Reactive | Fragmented | Experimental

Building trust and governance is the first step in creating a strong foundation for Power Platform and AI adoption. At this stage, the focus is on setting up essential guardrails – policies, processes, and structures that ensure security, compliance, and transparency.[1] Without these foundational elements, innovation risks being inconsistent, unsafe, or misaligned with organizational goals.

The foundational stage of governance is about achieving or setting a clear direction. It's about ensuring that teams know where to start, what's expected, and how decisions will be made. Establishing trust and governance early on signals to the organization that low-code and AI initiatives are not experimental – they are strategic and deserve thoughtful oversight.

[1] Microsoft (n.d.), 2021. Power Platform Adoption Framework.

S. Jeffery, *The Low-Code AI Maturity Model*, https://doi.org/10.1007/979-8-8688-1730-4_5

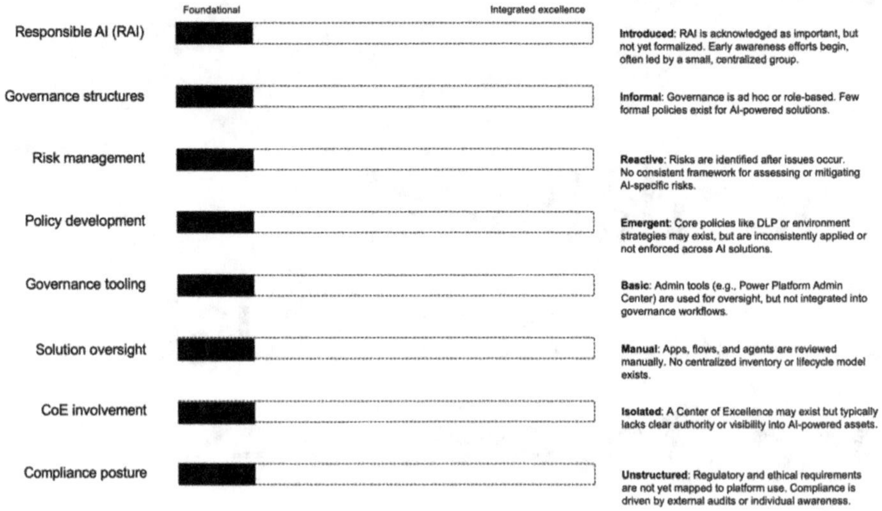

***Figure 5-1.** Run state of an organization at the foundational stage*

To help frame this foundational stage, the following mind map illustrates the key areas of focus for governance. These areas are interconnected, forming the critical building blocks that support responsible innovation and prepare your organization for the next stages of maturity:

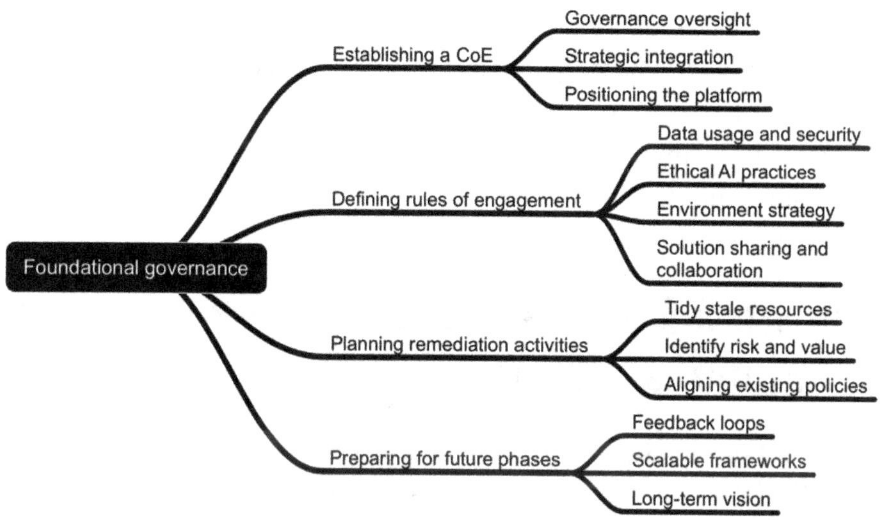

***Figure 5-2.** Foundational governance mind map*

This mind map captures the chapter's structure and the primary components of foundational governance:

- **Establishing a Center of Excellence (CoE)**: Defining roles and responsibilities, creating a hub for governance, and positioning Power Platform and AI as a strategic enabler
- **Defining rules of engagement**: Setting clear policies around data usage, ethical AI practices, environment strategy and solution sharing
- **Planning remediation activities**: Cleaning up existing resources, identifying both risk and value, and aligning with existing policies and governance goals
- **Preparing for future phases**: Introducing feedback loops, scalable frameworks, and long-term vision to ensure governance evolves with organizational needs

This chapter is intended to be a guide for organizations navigating the foundational stage, ensuring their governance efforts are both practical and acceptable. We'll explore these areas in detail, providing actionable strategies and examples to help you build a strong foundation for responsible innovation.

Why Trust Matters in the Foundational Stage

Trust is the cornerstone of adoption. Employees must trust that Power Platform is secure, that their data is protected, and that the organization supports them taking risks within defined boundaries. Equally, stakeholders must trust that innovation aligns with ethical principles and organizational value. However, at this foundational stage, many organizations may not yet have a detailed framework for what ethical principles mean in the context of AI. This is entirely natural – AI governance is an evolving discipline, and ethical considerations often become clearer as organizations gain more experience with AI-powered solutions.

This uncertainty highlights the importance of starting with broad, universally accepted principles, such as fairness, transparency, and accountability.[2] These serve as guides while the organization develops a more tailored ethical framework through collaboration, reflection, and iteration.

Reflection: *Does your organization have a shared understanding of what ethical AI looks like? If not, what steps can you take to begin defining it in a way that aligns with your values and goals?*

[2] Microsoft (n.d.), 2022. Microsoft Responsible AI Standard, v2.

Leadership plays a key role in this, setting the tone by articulating the value of governance and demonstrating a commitment to fairness, transparency, and responsibility. These efforts start the trust-building process. At this stage, organizations begin asking foundational questions like

- **Who is responsible for governance?** Identifying key roles and establishing ownership across IT, compliance, and business units.
- **What are the rules of engagement?** Defining clear policies on data usage, app sharing, and platform permissions (as a start).
- **How do we communicate trust?** Creating transparency by openly sharing governance policies, adoption goals, and how risks will be managed.

Addressing these questions sets you on a path where employees feel confident in using Power Platform and leaders are assured that innovation is happening responsibly.

Building Trust and Governance: Addressing the Key Questions

Establishing governance and building trust are foundational steps for successfully adopting the Power Platform and responsible AI. To guide you through this critical stage, here are several key questions you'll need to address as you move forward. Let's explore them together.

Who Is Responsible for Governance?

At the foundational stage, governance begins with a simple - but critical - question:

Who owns this?

Without clear ownership and structure, even the best intentions around responsible AI, security, or compliance remain fragmented. For Power Platform, establishing this clarity early ensures that the platform is not only well-managed but **recognized as a strategic enabler of AI-powered innovation.**

AI and automation are already spreading across the organization. From finance teams building forecasting models to HR departments experimenting with AI-driven interview bots, the use of AI is rarely confined to a single team or tool. This decentralization introduces complexity - **different teams, frameworks, and priorities** - which makes the governance challenge much broader than it may initially appear.

This is where Power Platform presents a unique opportunity. It democratizes AI development, giving business users the tools to automate, analyze, and solve problems independently. But with that empowerment comes responsibility. If governance structures don't evolve alongside adoption, organizations risk shadow AI, inconsistent data use, unmonitored automation, and ethical blind spots.

To meet this challenge, governance at the foundational stage must focus on **alignment** - connecting Power Platform governance with broader AI and digital initiatives. It's not just about rules - it's about creating coherence between strategy, standards, and execution across the organization.

Three key priorities emerge at this stage:

- **Defining ownership**: Appoint individuals or teams who are accountable for overseeing Power Platform growth and responsible use.

- **Establishing structure**: Create mechanisms that bring together business uses, IT, data teams, and compliance functions.
- **Foster alignment**: Ensure Power Platform governance doesn't exist in isolation but integrates with enterprise-wide AI, data, and digital strategy.

Laying the foundation for governance starts with assigning ownership and aligning stakeholders - but to sustain momentum, you need a structure that can turn strategy into action. That structure is the **Center of Excellence**. While the name suggests maturity, a CoE is just as critical at the foundational stage. It becomes the operational core of governance: shaping policies, supporting makers, and bridging the gap between innovation and oversight. Let's explore how a foundational CoE can be set up to guide Power Platform adoption with clarity, accountability, and purpose.

Establishing a Center of Excellence (CoE) for Alignment

The CoE plays a pivotal role in ensuring that Power Platform initiatives are governed effectively, align with broader AI strategies, and deliver maximum value. While its responsibilities - governance oversight, strategic integration, and platform positioning - provide a high-level framework, the process of setting up a CoE involves critical decisions that depend on the organization's structure, culture, and existing technology landscape.

One of the most important of these decisions is now the CoE itself will be structured. There is no one-size-fits-all model. Some organizations consolidate governance and delivery functions within a single central team; others distribute responsibilities across business units or regions or adopt a hybrid model that blends both. Each approach has trade-offs in speed, control, scalability, and alignment. Understanding these models – and selecting the one that fits your operating context – is a foundational step in ensuring your CoE can scale with the needs of your platform and your people.

Determining the Structure: Centralized, Decentralized, or Hybrid

The structure of the CoE should align with your organization's culture, size, and technology landscape.[3] Each approach has its advantages and challenges:

Centralized CoE

What is it? A single, unified team that oversees all Power Platform governance, strategy, and support across the organization.

Pros: Consistency in policies, streamlined decision-making, and better oversight of platform usage.

Cons: May feel disconnected from business units, slowing innovation for teams with unique needs.

Commonly best for: Smaller organizations or those with a strong preference for centralized decision-making.

[3] Forbes, (n.d.), 2024: To centralize or decentralize – a look at the Center of Excellence.

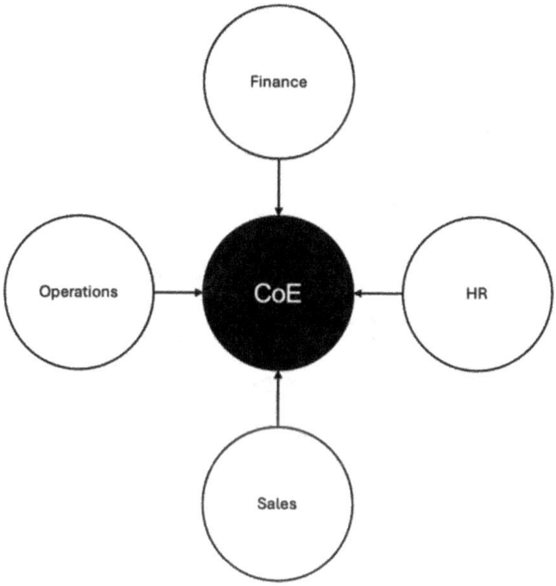

Figure 5-3. *Centralized Center of Excellence*

Decentralized CoE

What is it? Governance and support are distributed among individual business units, with each unit managing its own Power Platform initiatives.

Pros: Greater flexibility and alignment with specific departmental needs.

Cons: Higher risk of inconsistent policies, redundant efforts, and siloed practices.

Commonly best for: Organizations with highly autonomous business units or varied technology requirements.

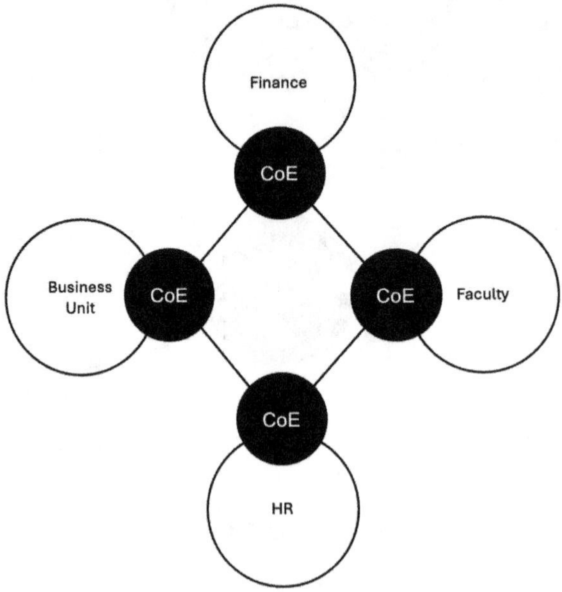

***Figure 5-4.** Decentralized Center of Excellence*

Hybrid CoE

What is it? Combines elements of centralized and decentralized models, where a core team provides overarching governance and support while business units retain some autonomy.

Pros: Balances consistency with flexibility, enabling innovation while maintaining control.

Cons: Requires clear definition of roles and responsibilities to avoid confusion.

Commonly best for: Larger organizations or those with diverse needs and a mix of central and local priorities

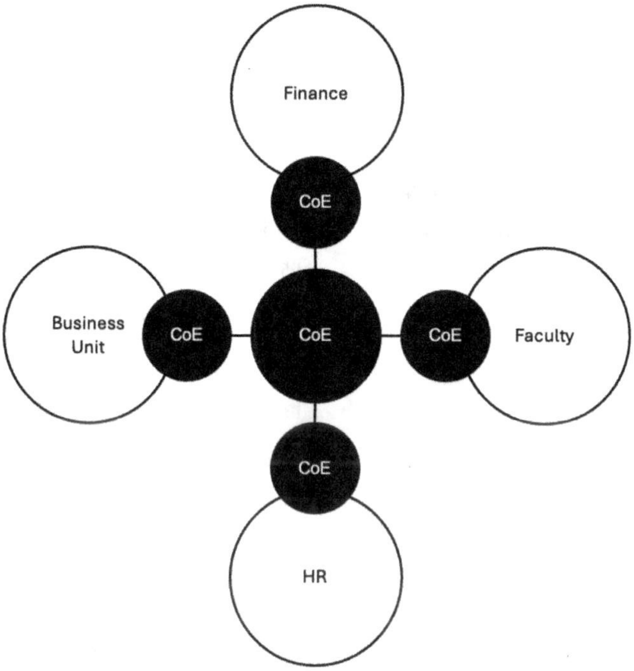

Figure 5-5. *Hybrid Center of Excellence*

Choosing the right CoE structure is not a one-size-fits-all decision. The optimal model - whether centralized, decentralized, or hybrid - depends largely on your organization's size, culture, industry, and existing governance practices. For example, some organizations prioritize central control to ensure compliance, while others need flexibility to accommodate the autonomy or diverse teams or departments.

To provide a little more clarity and context, the table below outlines examples of organization types that commonly adopt each CoE model. These examples highlight how different structures align with varying operational needs and governance priorities:

CoE model	Examples of organization types	Context/reasoning
Centralized	Government agencies Financial institutions Small to medium sized enterprises (SMEs)	Centralized control ensures strict compliance and security, which is critical for regulation heavy industries like government and finance. SMEs benefit from a single team overseeing governance, as they often lack resources to support multiple CoEs.
Decentralized	Universities Large consulting firms Multinational corporations (specific business units)	Universities often have high levels of departmental autonomy, with each faculty or department managing its own innovation efforts. Consulting firms require flexibility to tailor solutions for diverse client needs. Multinationals with distinct regional or business unit priorities prefer local decision-making.
Hybrid	Universities (larger or federated structures) Global technology companies Healthcare networks	Large universities with federated governance models may centralize strategic priorities but allow autonomy for faculties or departments. Technology companies balance innovation at the department level with alignment to global objectives. Healthcare networks often blend central policies (e.g., HIPAA compliance) with local autonomy for hospital or clinic operations.

The examples illustrate how different organization's approach their CoE structures based on their unique needs and contexts. Regardless of the model - centralized, decentralized, or hybrid - the success of the CoE ultimately depends on its ability to fulfil key responsibilities.

Key Responsibilities of the CoE

The CoE is critical to managing Power Platform governance, but it also serves as a bridge between Power Platform initiatives and the organization's broader AI strategy. By centralizing efforts, the CoE ensures Power Platform is both governed effectively and recognized as a primary strategic choice for AI-powered solutions.

Its responsibilities can be grouped into three core areas:

1. **Governance oversight**: Developing policies that align Power Platform usage with organizational standards and broader AI principles, such as fairness, transparency, and accountability
2. **Strategic integration**: Connecting Power Platform governance with existing AI initiatives to avoid duplication of efforts and foster collaboration
3. **Positioning the platform**: Highlighting the platform's capabilities in enabling rapid, accessible AI solutions through tools like Microsoft Copilot, AI Builder, Power Apps, and Power Automate

These responsibilities overlap and reinforce one another, creating a cohesive framework that supports responsible innovation and drives Power Platform's strategic value. The illustration below maps these responsibilities and their interconnections:

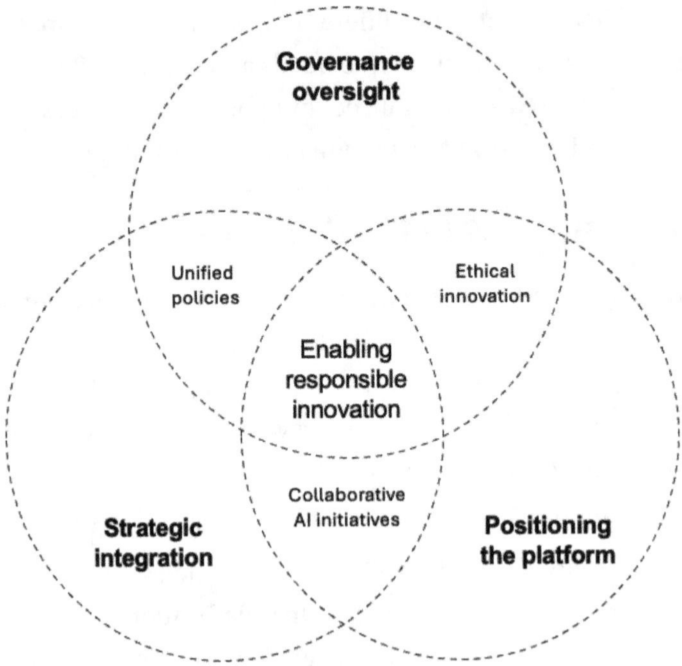

***Figure 5-6.** Venn diagram illustrating responsibilities of a CoE*

The illustration highlights how the CoE's efforts come together to enable responsible AI innovation with Power Platform. By addressing governance, integration, and positioning simultaneously, the CoE ensures that Power Platform is both governed effectively and recognized as a key enabler of AI-driven solutions.

Reflection:

- How does your current governance structure align with these responsibilities?
- Are you positioning Power Platform as a strategic option for AI solutions in your organization?
- Are there other teams you need to identify or engage in discussion with to achieve these objectives?

As you reflect on the responsibilities and opportunities outlined, it's essential to consider how Power Platform can position itself as the *go-to service* for AI-powered solutions within your organization. Its comprehensive set of AI-capabilities - ranging from AI Builder to Microsoft Copilot Studio - offers a unique proposition for enabling accessible and impactful innovation. By leveraging these tools under a strong governance framework, you can align Power Platform initiatives with broader organizational goals and set it apart as a strategic enabler.

Let's explore each responsibility in greater detail, with a focus on roles, responsibilities, and practical considerations.

Governance Oversight

Governance oversight is the backbone of the CoE, ensuring that Power Platform aligns with organizational standards and responsible AI principles. It provides the structure, accountability, and visibility needed to ensure that Power Platform use aligns with both organizational standards and responsible AI principles. At this stage, oversight doesn't require heavy bureaucracy - but it does require intentionality.

Oversight is what connects day-to-day platform usage to strategic governance goals - ensuring solutions are built securely, data is handled responsibly, and AI-powered features are reviewed with care. Without oversight, even the best policies risk being ignored or misapplied.

Key roles for oversight

While the exact structure may vary by organization, foundational oversight usually involves a small set of clearly defined roles. These roles can be part-time or distributed across existing teams at this stage:

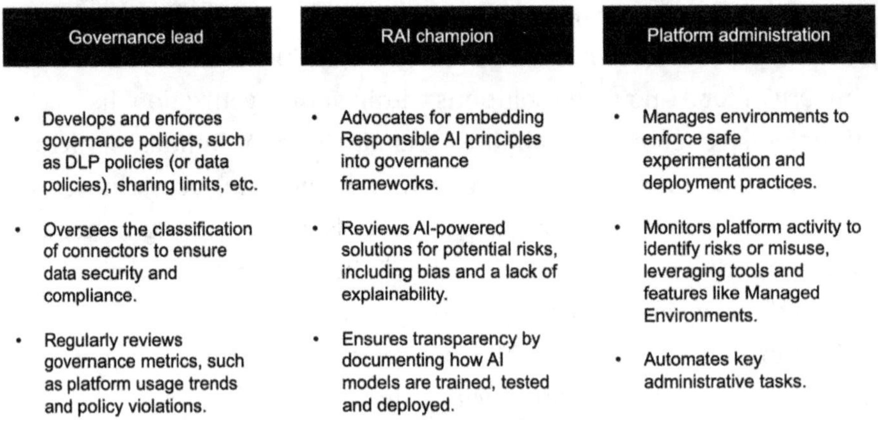

Figure 5-7. *Typical roles associated with governance oversight*

The goal here isn't to perfect oversight but to establish consistency and repeatable practices.

- **Start with practical data protection**

  Deploy baseline DLP[4] policies that protect sensitive data (e.g., SQL Server) while still allowing space for experimentation. Review usage patterns regularly (which connectors are apps, flows, and agents using most), and adjust based on risk and feedback.

- **Review AI-powered solutions early**

  Introduce a lightweight review process for new solutions that use AI Builder, custom models, grounded prompts, or Copilot Studio. Reviews can be as simple as a checklist or brief meeting focused on data use, fairness, and transparency.

[4] Microsoft - Data Loss Prevention policies.

- **Document decisions and patterns**

  Establish a shared space to document governance policies, common design patterns, exceptions, and lessons learned. This becomes a valuable resource as adoption and complexity grows.

- **Create regular check-ins**

  Even a monthly governance sync across platform admins, data owners, and CoE leads can make a huge difference. These sessions create a rhythm of oversight and encourage a feedback loop across disciplines.

As governance practices begin to take shape through oversight roles, policy reviews, and monitoring routines, it's also time to embed responsible AI (RAI) principles into how the organization approaches automation and AI development. Governance, after all, is about shaping the ethical and responsible use of AI in a way that reflects your organizational values.

At this stage, I suggest introducing a handful of practical, lightweight RAI initiatives that raise awareness, establish ownership, and encourage responsible design thinking before systems go live. These early steps do not require deep technical skill or formal RAI committees, but they do lay the groundworks for deeper maturity. We cover RAI in more depth later in this chapter.

Introducing responsible AI initiatives

Here are key RAI initiatives that fit naturally into the foundational stage of governance.[5] Each one is actionable, scalable, and designed to align with your CoE structure and Power Platform strategy:

[5] (2021), IBM – Everyday ethics for Artificial Intelligence.

1. **Nominate a responsible AI advocate/RAI champion**

   Establishing a responsible AI advocate within your CoE or governance team sends a strong signal that ethical oversight in low-code environments matters. This individual doesn't need to be deeply technical, but they should be respected and connected across departments and have a foundational understanding of responsible AI principles. Their role is to raise concerns, spot patterns, and connect data, compliance, and maker teams.

   A named advocate signals intent. It gives AI risks and ethics a seat at the table.

2. **Run a bias awareness session**

   At this stage, running a bias awareness workshop can have a profound effect. These short sessions educate business uses, makers, and IT teams on where bias can emerge - not just in data but in prompts, design choices, and workflows. Even a single 45-minute session using real examples from AI Builder or Copilot Studio can shift conversations from "is it accurate" to "is it fair."

   Bias isn't just a data issue - it's a design issue.

3. **Create an AI risk checklist**

   Establishing a lightweight risk checklist helps ensure teams ask the right questions - about data sensitivity, explainability, and potential harms - before building or releasing AI-powered solutions. It encourages consistent thinking without slowing teams down.

Adapt Microsoft's RAI Impact Assessment template,[6] and integrate it into your project intake or app documentation requirements.

4. **Define sensitive features in your data**

   Train your CoE or governance leads to identify sensitive data features (e.g., race, age, gender) in AI-powered solutions. These may be direct fields or inferred inputs. Awareness of how these fields are used allows you to set boundaries, introduce fairness checks, or exclude high-risk use entirely.

These four foundational initiatives represent a manageable but meaningful start to embedding responsible AI into your Power Platform governance model. Together, they create

- A visible advocate for ethical AI
- A shared understanding of risk and bias
- A habit of documenting decisions and asking the right questions early
- A bridge between data governance and responsible AI design

Introducing responsible AI initiatives at the foundational stage builds the right habits - raising awareness, encouraging thoughtful design, and laying the groundwork for trust. But to move from isolated improvements to sustainable impact, these practices must be strategically integrated into your broader Power Platform and organizational planning.

[6] https://msblogs.thesourcemediaassets.com/sites/5/2022/06/Microsoft-RAI-Impact-Assessment-Template.pdf

Power Platform governance and RAI can't sit on the sidelines. They need to become part of how solutions are envisioned, assessed, and deployed. This is where the governance journey begins to expand. It's no longer just about managing Power Platform responsibly; it's now about positioning the platform as a strategic contributor within the organizations broader AI ecosystem. This means aligning governance models, sharing insights across teams, and creating mechanisms for collaboration – not duplication.

Strategic Integration

Strategic integration ensures that Power Platform governance does not operate in a silo. Instead, it becomes part of a coordinated effort to deliver AI responsibly, efficiently, and consistently across the organization. In many organizations, AI efforts emerge in parallel, across departments, using a mix of tools – from enterprise data science platforms and custom machine learning models to low-code solutions built by citizen developers.

Without integration, the risk is clear: duplicated effort, fragmented risk management, disconnected policies, and uneven governance maturity. Strategic integration bridges these gaps.

Integration at this stage doesn't require a centralized AI program or formal council. It requires proactive connection points – places where Power Platform governance can plug into existing AI, data, and digital strategies, influence their decision, and benefit from their structure.

Key responsibilities in strategic integration

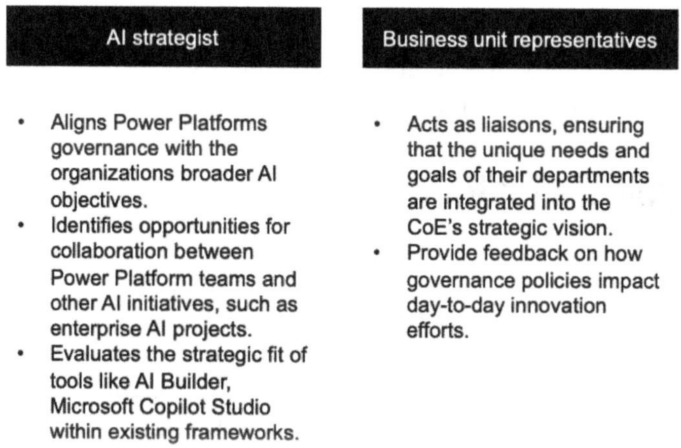

Figure 5-8. *Key roles aligned with strategic integration*

Some actionable steps to get strategic integration underway:

- Host cross-functional workshops to align Power Platform governance with broader AI strategies.
- Identify areas where Power Platform tools can complement existing AI initiatives, such as automating workflows or providing predictive insights.
- Create a centralized repository for documenting cross-team AI efforts to reduce duplication and improve collaboration.
- Celebrate or publish stories of how an integrated approach has achieved success.

Why it matters at the foundational stage

Strategic integration builds the connections – ensuring Power Platform governance aligns with broader AI programs, data policies, and compliance frameworks. But alignment alone isn't enough. For Power Platform to play a leading role in AI adoption, it must be deliberately positioned as the go-to service for building responsible, scalable, and impactful AI-powered solutions.

This is about perception, advocacy, and storytelling. The CoE must take an active role in shaping how the platform is seen, used, and prioritized. That begins with visibility.

Positioning the Platform

Positioning Power Platform as a strategic enabler of AI solutions requires more than technical excellence – it demands clear communication, stakeholder engagement, and visible success.

When business leaders, makers, and IT teams see the platform not only as safe and compliant but also as fast, flexible, and outcome-focused, adoption accelerates.[7]

At the foundational stage, this means intentionally elevating the platform's profile through early wins, compelling narratives, and a growing internal community.

Typical roles and responsibilities for platform positioning include advocacy and training leads. These roles focus on raising awareness, building confidence, and amplifying success across the organization. In some cases, the advocacy lead may also serve as the executive sponsor – particularly where senior-level support is already established.

[7] (2020) Harvard Business Review – The AI-powered organization, Wilson & Daugherty.

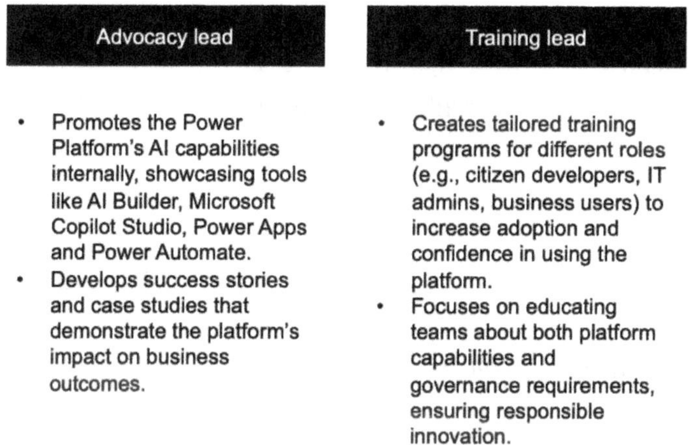

Figure 5-9. *Roles and responsibilities focused on positioning the platform*

Advocacy lead

The advocacy lead is the voice of the platform - responsible for shaping how Power Platform is perceived across the organization. This role is critical because, at the foundational stage, perception matters just as much as capability. Many business stakeholders and technical teams are still forming their understanding of what the platform can do, how it fits into their workflows, and whether it's trusted for serious AI-driven solutions. The advocacy lead helps shape this narrative by highlighting real success stories, communicating strategic value, and ensuring that key messages - like the platform's ability to support responsible, scalable innovation - are consistently reinforced.

Importantly, the advocacy lead doesn't just broadcast; they connect conversations across teams. By engaging with business units, IT, compliance, and the maker community, they create bridges that help align efforts, reduce duplicated work, and build cross-functional excitement. This role is often filled by someone with both business credibility and technical fluency, and in many organizations, it overlaps with or is championed by an executive sponsor. Their storytelling and leadership help elevate Power Platform to a strategic service, paving the way for deeper investment, broader adoption, and stronger governance buy-in.

Training lead

The **training lead** plays a quieter but equally pivotal role in positioning Power Platform as a strategic AI platform. While advocacy sparks interest, *training sustains momentum* – equipping teams with the knowledge and confidence to use the platform effectively and responsibly. At the foundational stage, this role helps remove barriers to entry, dispel fears around AI, and promote consistent understanding of both platform capabilities and governance expectations. A well-designed training program reinforces the message that Power Platform isn't just powerful – it's safe, inclusive, accessible, and aligned with organizational goals.

More than just technical instruction, the training lead *curates learning experiences that reflect business context.* They tailor sessions to different roles – makers, IT admins, business analysts – ensuring each group understands how Power Platform can meet their needs. They also play a vital role in *embedding responsible AI principles into practice,* helping teams understand concepts like bias, explainability, and transparency through hands-on scenarios. In doing so, the training lead directly supports the platform's reputation as not just easy to use but strategically sound and ethically grounded.

Here are some actions your team can undertake to get started:

- Launch pilot projects that solve high-impact business challenges using Power Platform and AI capabilities.
- Develop a communication plan to share the outcomes of these projects. Emphasizing time savings, cost reductions, or improved decision-making.
- Build an internal "Power Platform Champions" community to promote awareness and advocate for responsible AI-powered solutions.

Reflection:

- *Are your governance policies comprehensive yet flexible enough to encourage innovation?*
- *How is your CoE aligning Power Platform initiatives with broader AI strategies?*
- *What steps are you taking to position Power Platform as a leader in responsible AI-driven solutions?*
- *Are your policies well understood by new makers and conveyed within your training program?*

Defining governance roles and positioning Power Platform as a leader are foundational steps. However, it doesn't stop here. Clear, actionable rules that guide innovation while safeguarding the organization are required, which brings us to the second question: *What are the rules of engagement?*

What Are the Rules of Engagement?

At the foundational stage, defining the rules of engagement is about setting up clear, actionable guardrails that empower teams to innovate responsibly while minimizing risk. These guardrails aren't simply restrictive policies - they are thoughtfully designed frameworks that guide teams on how to use Power Platform safely, ethically, and in alignment with organizational objectives.

Integrating Responsible AI from the Start

At the heart of these rules is the commitment to responsible AI principles. Every AI-powered solution should be built with ethical considerations in mind - ensuring transparency, fairness, and accountability are baked into the design. Integrating these foundational responsible AI standards ensures that all solutions, from simple apps to complex workflows, adhere to a consistent ethical framework right from their inception.

Key Priorities in Setting the Rules of Engagement

Defining these rules involves a two-pronged approach:

Oversight and accountability: asset identification and ownership

The first step is to identify each asset across your Power Platform estate. An asset could be any digital solution or component, such as an app, an agent, a prompt, a flow, a table, or connection reference. For every asset you identify, it is critical to assign a clear owner who is responsible for its life cycle - from design and development to ongoing maintenance and eventual decommissioning. This dual step of asset identification and ownership is the foundation of accountability in your governance framework.

How can you do this?

One particularly effective approach is to leverage the native capabilities of the Power Platforms - specifically, using the Power Platform for Admins family of connectors to scan and identify assets within environments.[8] Once assets are identified, a flow can automatically send an adaptive card[9] to the asset's creator. It's important to note that the creator of an asset might not necessarily be its owner (it could be a developer); thus, this adaptive card serves as a prompt for clarification.

The adaptive card can be designed to include

- **Asset details**: Automatically populate fields, such as the asset type, name, and created date.

---

[8] The CoE Starter Kit can easily be extended for this purpose.

[9] `https://adaptivecards.io`

- **Ownership declaration**: Clear instructions prompting the recipient to declare ownership. This might include a simple check box for "I am the owner" and a text input if they wish to nominate someone else.
- **Automated update**: Once the creator responds, a flow updates or creates a new row in Dataverse, ensuring that asset ownership is captured accurately and stored centrally.

An example:

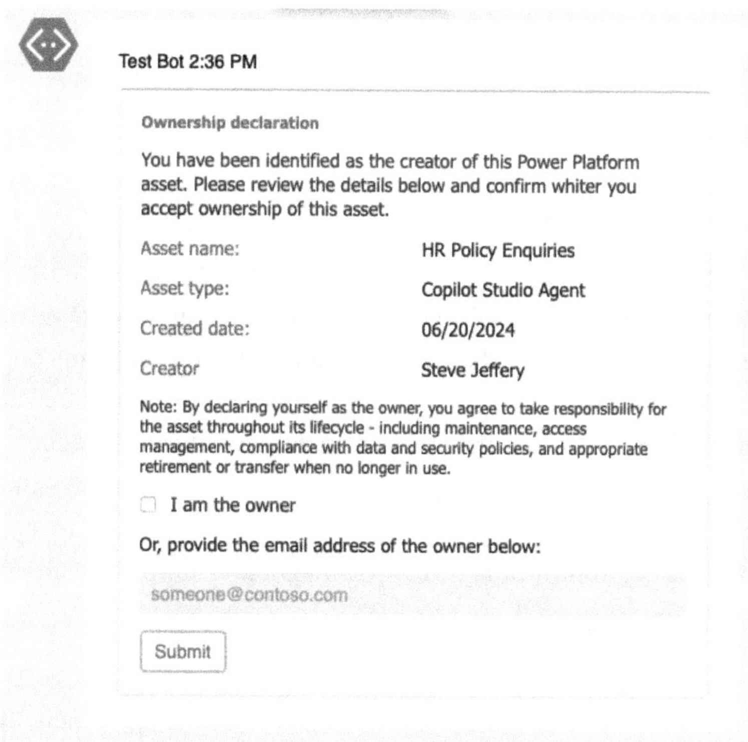

***Figure 5-10.** Asset ownership declaration adaptive card*

This method is particularly effective because adaptive cards are interactive, engaging, and seamlessly integrated into the tools your teams already use (such as Microsoft Teams). They simplify the process of data capture, making it both intuitive and efficient. The result is a real-time, centralized repository of assets and ownership details that feeds directly into your governance framework, ensuring continuous oversight and accountability.

After assets are identified and ownership is declared, the next step is to assess each asset's impact and governance complexity. This assessment helps determine the level of oversight required. For example, a personal productivity app shared with a few users might be low impact and low complexity - requiring minimal oversight - whereas an enterprise-wide AI-powered agent that makes HR decisions is likely to be high impact and high governance complexity, requiring comprehensive oversight.

This systematic approach - combining asset identification, clear ownership through adaptive cards, and an evaluation of impact and complexity - ensures that every asset is managed responsibly. It not only provides immediate clarity on where value is being generated and where risks might exist but also lays the groundwork for applying tailored governance measures.

In the following sections, we will delve deeper into governance dimensions and the impact and governance complexity matrix, which together form a structured decision-making framework to ensure each asset receives the appropriate level of oversight and support.

Governance Dimensions

Having identified your assets, their owners, and the initial guardrails, the next step is to translate those insights into a structured governance framework. This is where governance dimensions come into play. Each dimension represents a specific lens through which you can evaluate and manage risk, ensuring that every solution - whether it's a simple personal app or a complex AI-driven enterprise workflow - receives the right level of oversight.

Why dimensions matter

- **Tailored oversight**: By categorizing solutions according to dimensions such as license type, data storage, and usage complexity, you can apply **targeted governance policies** rather than one-size-fits-all rules.
- **Scalable approach**: As the Power Platform estate grows, these dimensions provide a **common language** for risk assessment, making it easier to onboard new solutions without reinventing the governance wheel.
- **Transparency and consistency**: Dimensions help ensure **consistent treatment** of similar solutions. A personal-use flow with standard connectors might fall under light governance, while an enterprise-critical, AI-powered solution demands more stringent checks.

Key governance dimensions

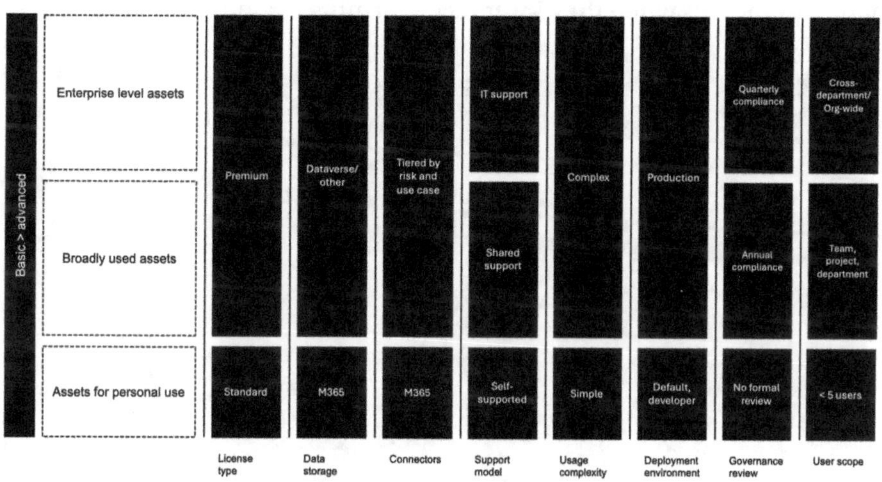

Figure 5-11. Typical governance dimensions

As you can see in the illustration above, each dimension (e.g., license type, data storage, support model) is a critical factor in determining how governance should be applied. The governance dimensions we've discussed also overlap in important ways, and their integration is critical to building a cohesive Power Platform governance strategy.

The overlap of governance dimensions

In many organizations, governance is often segmented into distinct areas:

- **Solution governance**: For example, usage complexity, connectors, and environment strategies
- **Data and security governance**: For example, DLP policies, data classification, and access controls
- **Responsible AI**: For example, fairness and transparency

While it's helpful to define these domains separately, true governance maturity emerges when they **intersect** - forming a cohesive, integrated framework.

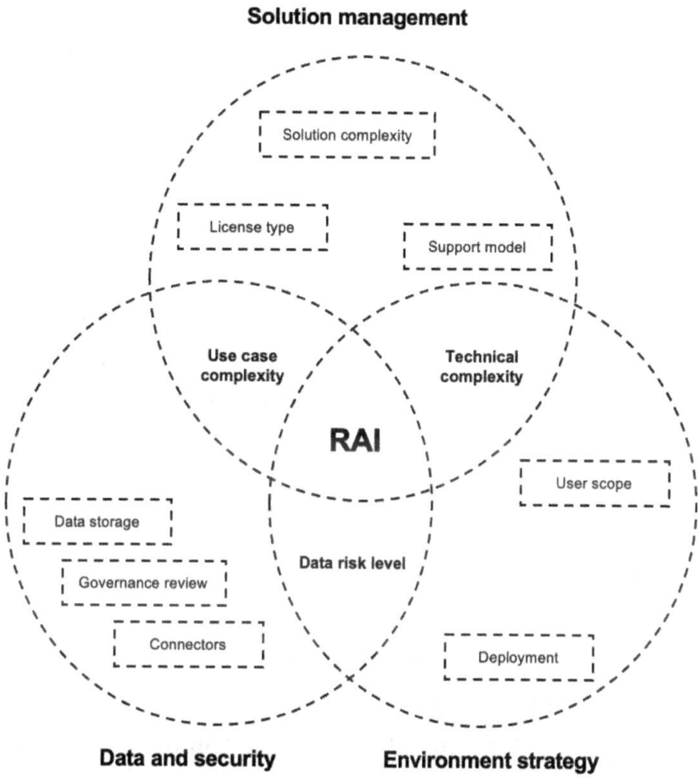

***Figure 5-12.** The overlap of governance dimensions*

Why this matters

By visualizing these domains in the diagram above, you can see immediately that no single dimension exists in isolation. A solution's risk profile might be shaped by how it accesses data, whether it employs AI, or which environment it's deployed in. Similarly, a robust data governance policy must consider how AI models or automations will use that data - and how to mitigate bias or misuse. Responsible AI, in turn, cannot function correctly without consistent guardrails for data handling and solution development.

A moment of reflection

This intersection underscores the need for collaborative governance. Each dimension brings its own policies, tools, and stakeholders (e.g., IT admins, data stewards, AI ethicists), but their work must be coordinated for governance to truly be effective.

Take a moment to consider your organization's current approach: How are these different governance domains interacting today? Are teams siloed, or is there a clear pathway for cross-departmental collaboration?

Reflect on whether your existing policies encourage dialogue between technical, ethical, and compliance perspectives. Ask yourself: Are there opportunities to better integrate these areas so that each solution is evaluated holistically? How can your teams improve their coordination to ensure that technical safeguards, data security, and responsible AI practices are mutually reinforcing? If these questions uncover barriers, involve your executive sponsor.

From governance dimensions to real-world application

With a clear understanding of governance dimensions in place, it's time to translate theory into practice. The dimensions you've explored – license type, data storage, support models, governance review, connectors, usage complexity, deployment, and user scope – intersect and overlap to form the foundation of a tailored governance framework that adjusts to the risk and value of each asset.

In the next section, we'll see how these dimensions guide real-world governance decisions. We will introduce an impact and governance complexity matrix – a tool that helps you categorize each asset based on its business impact, governance complexity, and responsible AI sensitivity. This matrix is not only a visual aid but also a decision guide, enabling your teams to apply the right "package of governance" to each solution.

Applying Governance Dimensions in Practice

Now that we've outlined the dimensions of governance, let's explore how they apply in real-world scenarios:

Dimension	Explanation	Example governance policy
License type	Standard licenses may suffice for personal or team-specific use, while premium licenses unlock capabilities like Dataverse and may require additional oversight.	• Standard licenses allow noncritical solutions with basic (M365) connectors. • Premium licenses require bespoke DLP enforcement.
Data storage	Assets utilizing external systems or sensitive data repositories like Dataverse demand stricter governance compared to internal-only solutions.	• Connecting to external systems (SAP, SQL Server) must undergo compliance reviews.
Support models	Defines who supports (handling user queries, dealing with break/fix tickets) the asset (e.g., self-supported, shared support, or IT-supported).	• Personal use assets are self-supported. • Enterprise critical apps must be supported by IT or a dedicated Power Platform team.
Governance review	Personal use assets might not require formal reviews, whereas enterprise-critical apps undergo quarterly compliance checks.	• Personal assets require no formal review. • Critical assets undergo quarterly compliance reviews for security and performance.

(continued)

Dimension	Explanation	Example governance policy
Connectors	Specifies the connectors permitted for use, categorized by risk level (low, medium, high).	• High-risk connectors (e.g., HTTP, SQL Server) must undergo security checks. • Low-risk connectors (e.g., Outlook, Planner) are unrestricted.
Usage complexity	Refers to the technical complexity of the system (e.g., simple cloud flows vs. AI-powered apps and agents).	• Simple cloud flows require minimal governance. • Complex apps require full documentation and dependency tracking.
Deployment	Indicates where and how the solution is deployed (e.g., development, default, dev/test/prod).	• Development environments allow for experimentation with relaxed policies. • Production assets require full governance and ALM.
User scope	Specifies the breadth of the audience (e.g., individual, team, cross-department, enterprise).	• Personal apps shared with <5 users need no approval. • Enterprise apps must follow strict governance protocols.

As we dive deeper into how governance dimensions apply to specific use cases, it's important to recognize that the flexibility of this model allows organizations to tailor governance based on the complexity and impact of each asset.

The illustration below categorizes use cases across two critical axes: **impact** and **governance complexity**.

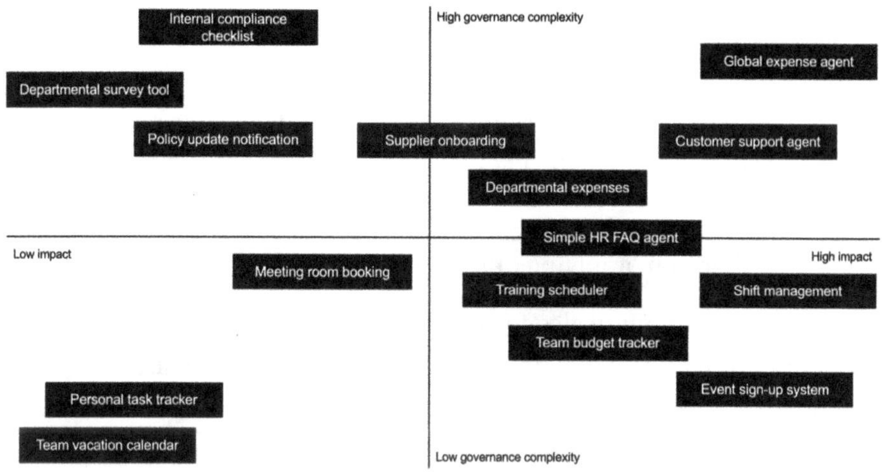

Figure 5-13. *Use cases overlaying governance complexity*

This visual representation demonstrates how solutions range from low-impact, personal productivity tools to enterprise-critical assets that require higher levels of governance oversight. Each quadrant reflects a unique governance approach designed to align the use case with organizational standards while enabling innovation.

Why you should build your own impact and complexity matrix

Understanding the real-world complexity and impact of AI-powered use cases is essential for building scalable, responsible governance. A visual matrix – like the one shown here – helps organizations categorize their solutions not just by technical difficulty or business value but by the level of oversight, risk management, and stakeholder engagement they require.

At this stage, creating your own impact and governance complexity matrix can provide enormous clarity. It allows your team to

- **Quickly assess which solutions need tighter control** (e.g., AI-driven financial decisions, customer-facing agents).

- **Identify low-risk opportunities** to pilot AI responsibly with minimal overhead.
- **Prioritize use case reviews** by combining business value with regulatory sensitivity.
- **Tailor governance policies** to different categories rather than applying one-size-fits-all rules.

This kind of mapping also fosters productive conversations between CoE leads, IT, legal, and business units - because it turns abstract governance principles into concrete, visual guidance. It becomes a shared frame of reference for deciding what guardrails to apply, how much documentation is needed, or when to involve an RAI advocate.

How to use the matrix: A decision guide

Not all Power Platform solutions are created equal. Some deliver quiet productivity gains; others shape high-stakes decisions or automate sensitive processes. And as AI capabilities become more embedded - through Copilot Studio, AI Builder, or custom integrations - the risk profile changes too.

To assess a use case and decide how it should be governed, reviewed, and classified, consider three key lenses:

- **Value**: What business or user impact will this solution deliver? How critical is the solution to the business?
- **Governance complexity**: How difficult is it to control, support, and oversee this solution?
- **RAI sensitivity**: Does it use AI, and if so, could it affect fairness, transparency, privacy, or human oversight?

These three views form the foundation of a practical decision tree:

Step 1: Assess Value/Business Impact

Ask:

- Does the solution affect decisions related to people, finance, or compliance?
- Will failure significantly disrupt business operations?
- Is this used cross-department or enterprise-wide?

Yes, to any of these: High-impact

No: Likely a low-impact or team-level use case

Step 2: Evaluate Governance Complexity

Now consider operational and architectural factors:

Governance dimension	Indicator of high complexity
License type	Uses premium capabilities (e.g., Dataverse, custom connector)
Data sources	Connects to sensitive systems (e.g., HR, finance) or external APIs
Deployment	Deployed in production environments and subject to change control or ALM
Support model	IT-supported or cross-functional team required
Connectors	Uses high-risk connectors
User scope	Shared with large groups or external users

More than two indicators: High governance complexity

Fewer or none: Low to moderate complexity

Step 3: Consider RAI Sensitivity

Now ask AI-specific risk questions:

- Does this solution generate predictions, classifications, or decisions using AI Builder, Copilot Studio, or Azure AI services?
- Could AI decisions directly affect individuals (e.g., shift assignment, approvals, routing)?
- Does it use sensitive features like age, race, gender, or health status?
- Can users see how decisions are made, or is the logic opaque?
- Would users know they're interacting with an AI system that is generating answers, ideas, or classifications?

If you answer "yes" to any, this use case has elevated RAI sensitivity and should include

- A bias and fairness review (e.g., use of Microsoft's RAI evaluation template)
- Clear documentation for explainability
- A designated owner for oversight

Putting it all together

Use the matrix and these questions to triage use cases:

RAI Sensitivity	Governance Complexity	Business Impact	Governance Approach
High	High	High	Full governance lifecycle: review gate, RAI documentation, monitoring, support
High	Low	High	Targeted RAI review, simplified governance, clear audit trail
Low	High	Low	Apply connector/DLP controls, encourage simplification
Low	Low	Low	Light-touch governance, good candidate for experimentation

***Figure 5-14.** Risk and impact management matrix*

Example: a Copilot Studio agent that answers internal HR queries might be low complexity, but high RAI sensitivity. It deserves a bias review and clear explanation strategy, even it doesn't need full architectural oversight.

Final thought

RAI isn't a separate category – it's a layer that should inform how all AI-enabled solutions are governed. Integrating it into your complexity and impact assessment ensures you're not just asking "Can we build it?" but also "Should we? And how should we do it responsibly?".

Having explored how strategic integration can align your Power Platform governance with broader AI initiatives, it's time to zoom in on the core of responsible AI. The next step is embedding ethical and risk-aware practices right from the start – before your organization scales its AI initiatives.

At the foundational stage, responsible AI is about **setting the guardrails that guide every decision**, ensuring compliance and fostering trust. By proactively incorporating relevant legislation, regulations, and frameworks – such as the EU AI Act, GDPR,[10] and the NIST AI RMF[11] – into your governance planning, you lay a robust foundation for both

[10] `https://gdpr.eu`

[11] `https://www.nist.gov/itl/ai-risk-management-framework`

ethical integrity and operational effectiveness. This is the moment to turn awareness into action, establishing clear privacy and ethical standards that not only protect sensitive data but also reflect your organizations values.

Responsible AI and Risk at the Foundational Stage

At the foundational stage, responsible AI should start becoming embedded into the governance process, even if the organization is still defining its broader ethical AI framework.

It's crucial to incorporate relevant legislation, regulations, and frameworks into your AI governance planning for compliance, ethical integrity, and operational effectiveness.

Early integration of these elements helps organizations navigate the complex landscape of AI deployment, mitigating risks and fostering trust among stakeholders.

Key legal and regulatory frameworks influencing AI governance:

Legal or regulatory framework	Overview	Governance implication
European Union Artificial Intelligence Act (EU AI Act)	Adopted in June 2024 and enforced from August 1, 2024, the EU AI Act classifies AI systems into risk categories – unacceptable, high, limited, and minimal risk – each with specific obligations.	Organizations must assess their AI systems against these categories to determine compliance requirements. High-risk systems, for instance, necessitate rigorous risk assessments, transparency documentation, and human oversight, directly influencing governance policies and team structures.

(continued)

Legal or regulatory framework	Overview	Governance implication
General Data Protection Regulation (GDPR)	Enforced since May 2018, GDPR governs data protection and privacy within the EU.	AI systems processing personal data must adhere to GDPR principles, impacting data governance policies, consent mechanisms, and data subject rights management.
NIST AI Risk Management Framework (AI RMF)	Developed by the US National Institute of Standards and Technology, this voluntary framework offers guidelines for managing AI risks throughout the AI life cycle.	Implementing the AI RMF assists organizations in identifying, assessing, and mitigating AI-related risks, shaping governance processes and accountability structures.
US Executive Order on Safe, Secure, and Trustworthy AI	Issued to promote the development and deployment of AI technologies that are safe, secure, and trustworthy.	Organizations are encouraged to implement rigorous testing and validation procedures for AI systems, influencing governance policies related to system reliability and safety.
Algorithmic Accountability Act (US)	A proposed legislation aiming to require companies to conduct impact assessments on automated decision systems to evaluate their performance and effectiveness.	If enacted, organizations would need to integrate regular impact assessments into their AI governance processes, ensuring transparency and accountability in automated decisions.

One of the most critical steps you can take now is to become familiar with the evolving landscape of legal and regulatory frameworks. Whether it's the EU AI Act, GDPR, NIST AI RMF, ISO standards, or upcoming legislation, these are not abstract references - they shape the way you must design, deploy and govern AI systems.

Reviewing these frameworks now ensures your governance efforts are proactive rather than reactive. Waiting until a regulation comes into force increases both risk and rework. Equally important is assigning responsibility.

Someone in your organization must own the task of monitoring regulatory developments, updating internal policies, and ensuring your governance practices remain aligned. Without a clear owner, this work will drift - and the consequences of misalignment may only become clear when it's too late to fix easily.

Even at this early stage, weaving legal awareness into your governance strategy sets the tone for responsible innovation and builds the habit of treating governance not as an afterthought but an essential enabler of sustainable AI adoption.

From Awareness to Action: Embedding Privacy and Ethics

Understanding the regulatory landscape and assigning ownership is just the beginning. What follows is translating that awareness into specific, actionable standards that shape how AI and automation are designed, built, and governed - starting with privacy and ethics.

In this initial phase, these guardrails will not need to be complex - but they do need to be clear. Privacy standards ensure you meet regulatory expectations like GDPR or CCPA, while ethical guidelines make sure your solutions reflect your organizations values.

Privacy standards and ethical guidelines serve as the baseline guardrails for ensuring that Power Platform solutions align with organizational values and regulatory requirements. Policies will vary by industry and geography,[12] but here are the key principles that organizations can adopt:

Privacy standards

These guidelines focus on protecting sensitive data and ensuring compliance with legal and regulatory requirements.

- **Data minimization**: Only collect and store the data that is strictly necessary for the solution to function.
- **Purpose limitation**: Data must only be used for the purpose it was collected.
- **Access control**: Define and enforce role-based access to data.
- **Retention policies**: Data should not be stored indefinitely. Set clear retention schedules and automate purging where possible.
- **Regulatory compliance**: Solutions must comply with relevant laws like GDPR, HIPAA, or CCPA depending on the geography and use case.

[12] Privacy standards, found in data protection regulations like GDPR (General Data Protection Regulation, EU), CCPA (California Consumer Privacy Act, US), and HIPAA (Health Insurance Portability and Accountability Act, US).

Ethical guidelines

These guidelines[13] ensure that AI-powered or automated solutions operate in a way that reflects fairness, transparency, and accountability.

- **Transparency**: Make solution logic clear to users and stakeholders.
  - **Example**: An HR FAQ agent should disclose when answers are generated by AI vs. sourced from preapproved policies.
- **Bias mitigation**: Regularly review AI-driven solutions to identify and mitigate potential biases.
  - **Example**: A shift management tool should ensure shifts are assigned equitably, without favoring specific groups.
- **Explainability**: Ensure that decisions or outputs from AI systems can be explained in understandable terms.
  - **Example**: A supplier onboarding app should document why a specific supplier was flagged for approval.
- **Accountability**: Assigning ownership for AI-powered solutions, making it clear who is accountable for ethical compliance.
  - **Inclusive design**: Solutions should be designed to be accessible and usable by all, including people with disabilities.

[13] Microsoft Responsible AI Standard, OECD AI Principles, *Emphasized in global AI ethics frameworks.*

These standards can be planned under the **Data readiness and Security** governance dimension and map them to specific use cases.

With foundations in responsible AI and risk management - anchored by clear privacy standards, ethical guidelines, and the integration of key regulatory frameworks - the next step is to ensure that these principles are not only implemented but also clearly communicated. All the careful work in establishing governance rules, defining oversight roles, and embedding legal and ethical considerations is only as effective as the way it's conveyed to the rest of the organization.

When governance is hidden in dense documents and complex policies, it risks being misunderstood or even bypassed by those it's meant to protect and guide. To truly unlock the potential of your responsible AI framework, you need to build trust through transparent, engaging, and tailored communication. In the following section, we'll explore how to communicate trust effectively - turning governance into an enabler of innovation and a catalyst for collaboration across diverse stakeholder groups.

How Do We Communicate Trust?

Building trust is primarily about clearly articulating the purpose and value of the governance rules you've created. When governance works effectively, it serves as a foundation for innovation rather than a barrier. Yet, without transparent communication, governance often risks appearing restrictive or overly bureaucratic.

Here's the core challenge: If employees don't understand why governance exists, how it benefits them personally, or how it aligns with organizational objectives, they're unlikely to engage - or worse, they might actively circumvent it.

Trust grows from clarity, collaboration, and consistent messaging that positions governance as an enabler of creativity and accountability.

This section explores strategies to communicate trust effectively, turning governance into a proactive tool that builds confidence, fosters innovation, and maintains accountability.

Introducing the Approach to Communicating Trust

Too often, I've encountered governance reduced to a lengthy document buried deep within a SharePoint site. These hefty policy documents outlining acceptable usage of Power Platform rarely attract readers - not even those who authored them. As a result, CoE's find themselves perpetually firefighting, reactively chasing policy breaches rather than proactively building trust and understanding.

Effective governance communication requires a thoughtful, tailored strategy that resonates across diverse organizational audiences - from casual app users and citizen developers to compliance teams and senior executives. At the foundational stage, your communication approach should clearly address each stakeholder groups' unique priorities, concerns, and expectations.

Audience analysis

Each audience interacts with governance uniquely. Recognizing what matters most to each group enables you to craft targeted messages that resonate. For instance, *developers* seek clarity on acceptable connectors, whereas *business leaders* prioritize alignment with strategic goals.

Step 1: Identifying Your Key Stakeholder Groups

Governance touches numerous stakeholders, but at a foundational stage, prioritize these primary groups:

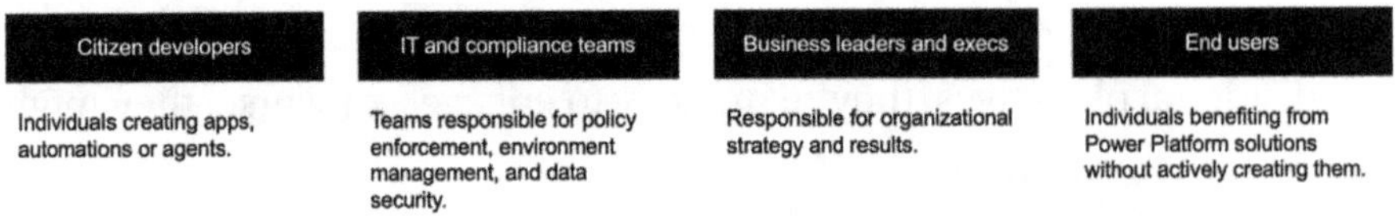

Figure 5-15. *Typical audiences for governance communications*

Step 2: Understanding What They Care About

For each group, pinpoint what matters most to them to craft targeted messaging.

Group	What they care about
Citizen developers	Innovation opportunities, ease of compliance, and clarity on what's allowed
IT/compliance teams	Risk mitigation, visibility into data flows, and alignment with organizational policies
Business leaders	ROI, strategic alignment, and ensuring governance enables innovation
End users	Data privacy, transparency in decisions, and trust in AI-driven processes

Step 3: Choosing the Right Channels

Each audience interacts with information differently, so selecting the right communication channels ensures your messages reach them effectively:

Group	Preferred channels
Citizen developers	Governance portal, training videos, templates, and Q&A forums on platforms like Teams
IT/compliance teams	Detailed dashboards, policy documentation, and incident reporting tools
Business leaders	Executive reports, Power BI dashboards, and leadership briefings
End users	FAQs, user-friendly notifications, and walkthroughs embedded in the apps they use

Step 4: Mapping Needs to Messaging

Now you've identified your audience, ensure your messages address their specific concerns and interests; here are some examples illustrating the key points, concerns, and tone of voice:

- **Citizen developers**: "Here's how you can innovate freely while staying compliant. We've simplified the rules and created templates to get you started."
- **IT/compliance teams**: "These policies provide the visibility and control needed to safeguard our data and ensure compliance."
- **Business leaders**: "Governance supports faster app development and increased ROI by reducing delays caused by compliance issues."
- **End users**: "Our governance ensures that your data is protected and all decisions made by AI solutions are transparent and fair."

Step 5: Iterating Through Feedback

Governance communication isn't static - nor is governance. Create mechanisms to gather input from each audience. Consider feedback forms for citizen developers to share their challenges with compliance, quarterly governance roundtables for IT, and compliance teams to discuss policy effectiveness and one-on-one briefings for executives to review governance outcomes and strategic alignment.

Conclusion: Reflections on the Foundational Stage

This chapter has laid out the critical importance of building a robust foundation for trust and governance in your Power Platform and AI initiatives. At this foundational stage, every decision - from defining clear roles and responsibilities to integrating complex legal and ethical frameworks - plays a pivotal role in ensuring that innovation is not only rapid and agile but also secure, compliant, and sustainable over time.

By embedding governance early on, you transform what might seem like a burdensome administrative process into an enabler that safeguards your organization while empowering creativity.

One of the central themes of this chapter is the necessity of clear ownership and structured governance. Without defined responsibilities, even the best policies can become fragmented, leading to shadow IT and inconsistent practices. Establishing dedicated roles and aligning them with a centralized or hybrid Center of Excellence (CoE) ensures that there is always accountability at every level – from day-to-day operations to strategic decision-making. This clarity not only minimizes risk but also builds internal trust, as teams recognize that there are robust checks and balances in place to support their efforts.

Integrating responsible AI principles into your governance framework is another crucial element. Even if your organization is still shaping its broader ethical AI framework, it is imperative to start with universally accepted principles such as fairness, transparency, and accountability.

By proactively incorporating frameworks like the EU AI Act, GDPR, and the NIST AI RMF, you set the guardrails that will prevent future compliance issues and foster an environment of ethical innovation. Practical initiatives – such as bias awareness sessions, AI risk checklists, and defining sensitive data features – ensure that responsible AI practices become part of your everyday operations, instilling confidence among stakeholders.

Moreover, we emphasized the importance of strategic integration – ensuring that Power Platform governance aligns with broader organizational AI initiatives, digital strategies, and compliance frameworks. This strategic alignment bridges the gap between siloed efforts and enterprise-wide collaboration. Tools like the impact and governance complexity matrix not only help you evaluate the risk and value of individual use cases but also serve as a visual road map for how governance should evolve as the platform scales.

In essence, while the foundational stage of governance demands considerable upfront effort – from meticulous policy design and legal review to role assignment and cultural change – it is precisely this rigorous approach that makes subsequent stages more manageable. By investing the time and thought required at this early phase, you create an adaptable, transparent, and resilient governance framework. This framework will not only mitigate risks and ensure regulatory compliance but also support innovation by making it clear that governance is a tool for enabling creativity and ensuring sustainable, responsible growth.

As you reflect on the concepts presented in this chapter, consider the long-term benefits: establishing trust and governance now means fewer reactive fixes later and a smoother pathway for scaling your low-code and AI initiatives. Ultimately, this early and significant investment transforms governance from a perceived hurdle to a powerful asset, one that underpins all future successes in your digital transformation journey.

Foundational stage progression checklist

The foundational stage is where governance begins – not as a barrier but as an enabler of safe, strategic innovation. If you've made it this far, you've already started defining ownership, aligning stakeholders, and establishing your first policies and roles. But how do you know if you're progressing in a sustainable way?

The checklist below serves as a reflective tool – designed to help you assess your current progress and identify any gaps as you prepare to scale. It's not a pass/fail assessment. Instead, it provides clear markers of maturity across the core elements we've explored: CoE setup, policy creation, responsible AI practices, integration with broader AI initiatives, and communication strategies.

Use this checklist as a guidepost. Revisit it periodically as your governance model matures. If there are areas not complete, that's normal – what matters is that you have a plan to address them as your platform grows.

Center of Excellence structure

- ☐ A Center of Excellence (CoE) structure is established (centralized, decentralized, hybrid) and aligned with organizational needs.
- ☐ Roles for governance, training, advocacy and oversight are defined and assigned.
- ☐ A shared repository exists to document policies, decisions, patterns and guidance.

Governance oversight and asset ownership

- ☐ Baseline governance policies are in place (e.g., app sharing, data use, connector restrictions).
- ☐ Power Platform assets are being identified using admin connectors or inventory tools.
- ☐ Asset ownership is declared and recorded (e.g., via adaptive card + Dataverse workflow).
- ☐ Governance dimensions are defined and used to guide oversight and support decisions.

Responsible AI foundations

- ☐ A Responsible AI advocate or champion has been appointed within the CoE or platform team.
- ☐ A lightweight risk checklist or review process is used for AI-enabled solutions.
- ☐ Bias awareness or ethical AI workshops have been delivered to raise understanding.

Strategic integration

- ☐ Power Platform governance is linked to broader AI, data, or digital governance efforts.
- ☐ Early AI-powered use cases are documented and shared to build visibility and credibility.
- ☐ An internal community of champions or advocates is forming or actively supported.

Privacy, ethics and regulatory awareness

- ☐ Key frameworks (e.g., EU AI Act) are reviewed and aligned with governance policies.
- ☐ Privacy and ethics standards (e.g., data minimization, fairness, explainability) are defined.
- ☐ Someone is assigned responsibility for monitoring legal and regulatory developments.

Communication and trust

- ☐ Governance communications are tailored for each audience (citizen developers, IT, business leaders, end-users).
- ☐ Governance is positioned as an enabler of innovation – not a blocker.
- ☐ Feedback mechanisms are in place (e.g., roundtables, forms, support tickets) to adapt governance based on real-world experience.

CHAPTER 6

Emerging: Establishing Trust and Governance

Exploratory | Structured | Evolving

The emerging stage represents a pivotal point in your governance journey. Building on the foundational guardrails, this stage focuses on scaling, formalizing processes, and aligning governance with organizational strategy. Now, you will focus on evolving the CoE into a more structured entity that operationalizes governance frameworks while driving alignment across departments and teams.

Something you'll notice as you enter this stage is that the complexity and scope of responsibilities grow. The focus shifts from establishing foundational principles to scaling, operationalizing responsible AI (RAI), and fostering collaboration across teams. To guide this progression, three key questions come into focus:

1. **How do we scale governance across the organization?**

   Scaling governance means expanding policies, automating processes, and ensuring they are robust enough to handle enterprise-wide use cases while maintaining flexibility for innovation.

S. Jeffery, *The Low-Code AI Maturity Model*, https://doi.org/10.1007/979-8-8688-1730-4_6

2. **How do we operationalize responsible AI?**

   RAI must move from habit forming and theory to action. The challenge lies in embedding RAI principles, such as fairness and transparency into governance frameworks in ways that are measurable and enforceable.

3. **How do we foster collaboration and alignment?**

   Governance is not a siloed activity. It requires cross-functional alignment and collaboration between IT, business units, compliance teams, and leadership to ensure consistency and widespread adoption.

We build on top of the foundational work already completed, driving your organization forwards toward a model of governance that scales with growth, empowers innovation, and aligns with ethical principles. Let's look deeper.

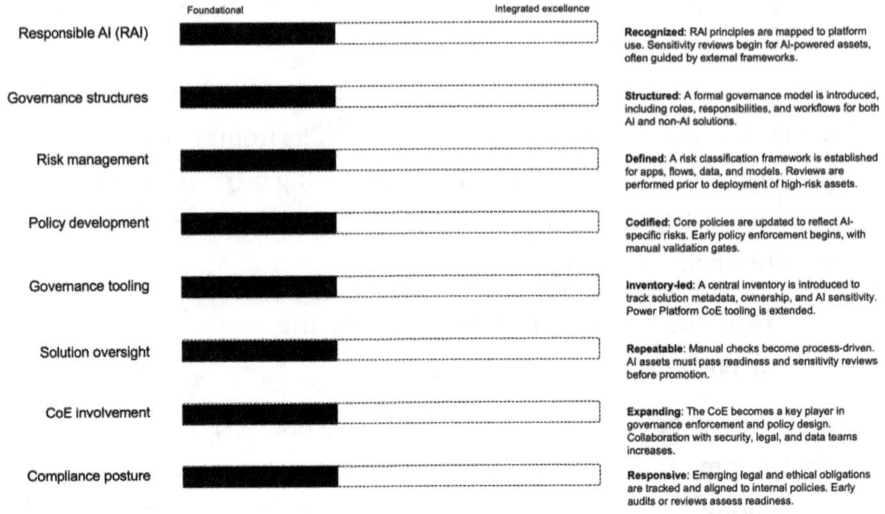

***Figure 6-1.** Run state of an organization at the emerging stage*

Introduction to Scaling Governance

Scaling governance is the most critical challenge your organization will face in the emerging stage. Governance needs to evolve and scale to accommodate:

- **An increase in complexity**: The number and diversity of Power Platform assets grow as adoption spreads across teams and departments.
- **Broader use cases**: Governance must address a wider range of scenarios, from personal productivity tools to enterprise-critical AI-powered applications.
- **Higher stakes**: As more teams rely on the platform for their daily operations, the risks of misaligned governance - including compliance violations or mismanagement of AI capabilities - increases significantly.

Scaling governance requires that a delicate balance between control and creativity is maintained. To achieve this, several key elements must remain top of mind:

1. **Efficiency**: *Streamline governance processes to avoid bottlenecks.*
   As governance processes are rolled out and used more widely, bottlenecks can emerge, especially in areas requiring multiteam collaboration. For instance, requesting approval for a new connector to be added or for use of a custom connector might involve other departments, such as procurement for licensing or security for data processing reviews. These dependencies can slow down workflows if they aren't accounted for during the design phase.

Tip Involve related teams, such as procurement, legal, and security, during the design of governance workflows. Map out the end-to-end process to identify potential bottlenecks early, and establish clear escalation paths for expedited approvals.

Reflection: *Are there governance processes in your organization that could benefit from simplification or automation?*

2. **Coverage:** *Ensure that policies are applied across all use cases, from low-risk to high-impact solutions.* Governance must account for the entire spectrum of Power Platform use cases. From personal productivity apps to enterprise-critical AI solutions, each scenario presents unique risks and opportunities. Policies should be scalable and adaptable, providing proportional oversight based on the complexity and impact of the solution. **Example**: A low-risk personal app might require minimal governance (e.g., basic compliance with data privacy rules), while an enterprise-critical solution could demand extensive reviews, documentation, and ongoing monitoring.

Tip *Create a tiered governance model that categorizes asset by use case type, aligning oversight levels with their associated risks.*

---

3. **Flexibility:** *Allow teams to innovate while staying aligned with organizational standards.*

   Governance should act as an enabler of innovation, not a barrier. Rigid policies that don't accommodate diverse use cases or experimental scenarios risk stifling creativity. Conversely, overly lax governance can lead to mismanagement and risk exposure. The key is finding a middle ground. I will repeat this several times throughout this section; it's crucial.

Tip Introduce "*safe-to-fail*" environments (personal developer) or sandbox environments where teams can experiment with new ideas while adhering to essential safeguards, such as restricted data access and preapproved connectors. We'll cover this in more detail when we cover environment strategies.

Reflection: How does your current governance framework encourage innovation while ensuring compliance?

Balancing efficiency, coverage, and flexibility is critical to scaling governance successfully. By streamlining processes, tailoring policies to use cases, and empowering teams with adaptive frameworks, you can create a governance strategy that supports responsible innovation and drives organizational growth.

Approaches to Scaling Governance

Establishing a tiered governance framework

Tiered frameworks are by far the most common approach to scaling governance. The reason they work is that they offer varying levels of oversight based on the complexity, criticality, and scope of an asset. In the illustration below, we'll look at how this can be applied to a variety of use cases.

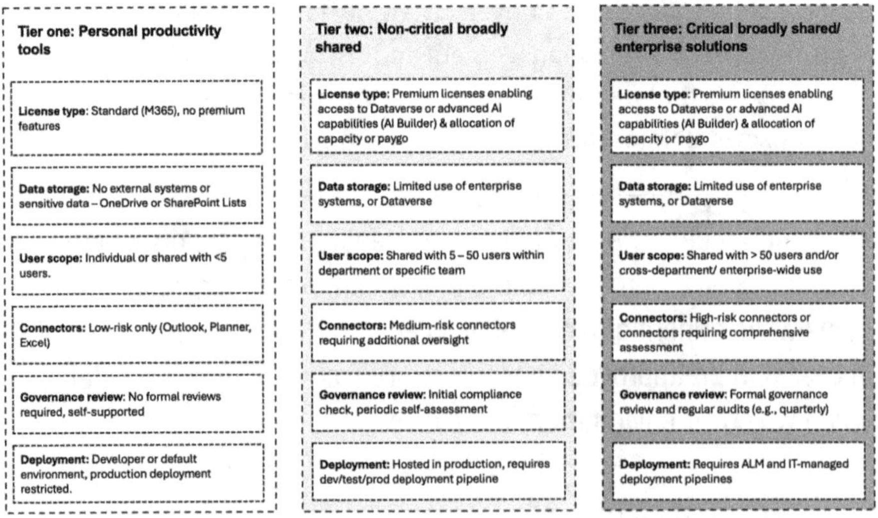

Figure 6-2. *Example tiered governance framework illustrating dimensions as entry points*

This framework approach ensures

- **Appropriate oversight**: High-risk solutions get more scrutiny, while low-risk solutions avoid unnecessary complexity.
- **Scalability**: Policies adapt to the complexity and criticality of use cases, enabling governance to grow with the organization.
- **Flexibility for innovation**: Lower tiers empower experimentation, while upper tiers provide robust safeguards for critical solutions.

With a tiered governance framework in place, the next logical step is to define how your Power Platform environments align with those tiers. Environments act as the foundation for governance, providing the physical spaces where solutions are built, tested, and deployed. A well-structured environment strategy ensures that governance policies are applied consistently across tiers while empowering teams to innovate responsibly.

Traditional Environment Strategies: Balancing Structure and Flexibility

Traditionally, environment strategies provide a structured approach to managing Power Platform assets and aligning them with governance goals. Organizations at the emerging stage often adopt different strategies depending on their priorities, operational complexity, and risk tolerance.

Here are the most common approaches to traditional environment strategies:

- **Business unit or function-based environments**

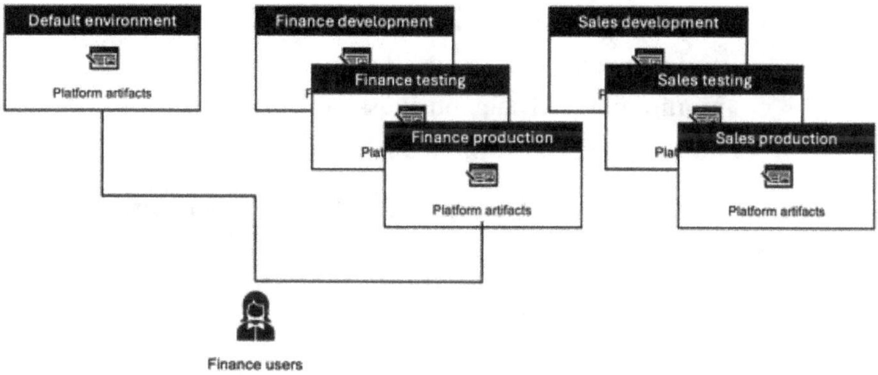

Figure 6-3. *A typical "flat" business unit/function-based environment strategy*

What it is: Environments are segmented by organizational departments, such as Finance, HR, or Marketing. Each business unit is provided with dedicated environments to house their Power Platform assets, such as apps, flows, agents, and AI models.

Why it works: This approach aligns governance and resources directly with the unique needs of each department, ensuring a tailored and effective governance framework.

- *Key benefits*
    - **Easier administration**: Each department or function is assigned a clearly defined environment, simplifying administration and ensuring better control over resources.
    - **Improved usability:** Teams gain autonomy within their environments, allowing them to innovate freely without navigating unnecessary restrictions means for other departments.
    - **Tailored governance:** Policies, DLP rules, and connectors can be customized to fit the requirements of each business unit, ensuring a more effective governance strategy.
- **Challenges**: While this approach offers clear benefits, it requires
    - Coordination across departments to align governance policies
    - Consistent monitoring to avoid siloed efforts or duplicated resources
    - Inconsistent environment configurations at the tenant level

- **Security/isolation strategy**

  In this approach, environments are segmented based on the risk and sensitivity of the solutions they host.

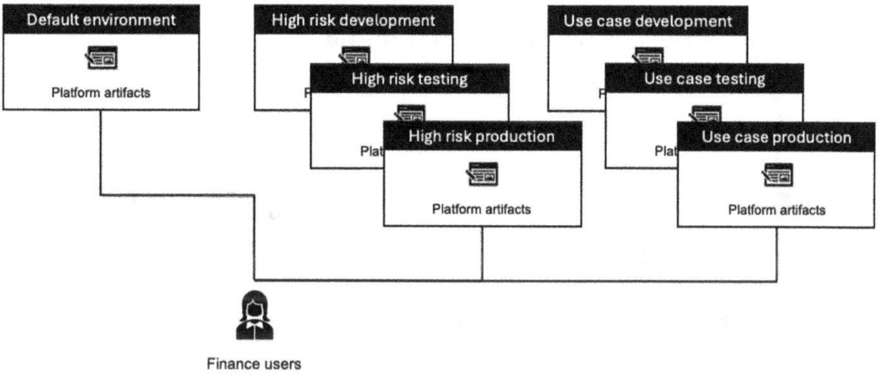

Figure 6-4. *Isolation strategy based on risk of assets*

Key benefits

- **Enhanced security**: Isolating high-risk assets minimizes exposure and provides stricter controls.
- **Risk mitigation**: Low-risk environments allow for innovation without jeopardizing critical systems.
- **Compliance readiness**: Simplifies audits by isolating sensitive data into environments with strict policies.

Challenges

- More complex environment management, especially as the number of isolated environments grows
- Increased risk of inconsistent policy enforcement if security roles or classifications are not standardized
- Potential friction for solution builders who need cross-environment access or data sharing

Example

- Secure environment: For processing sensitive customer data or using a combination of connectors that should not be broadly used outside of the use case
- Shared environment: General-purpose space for departmental tools that don't require heavy restrictions

Geographically distributed strategy

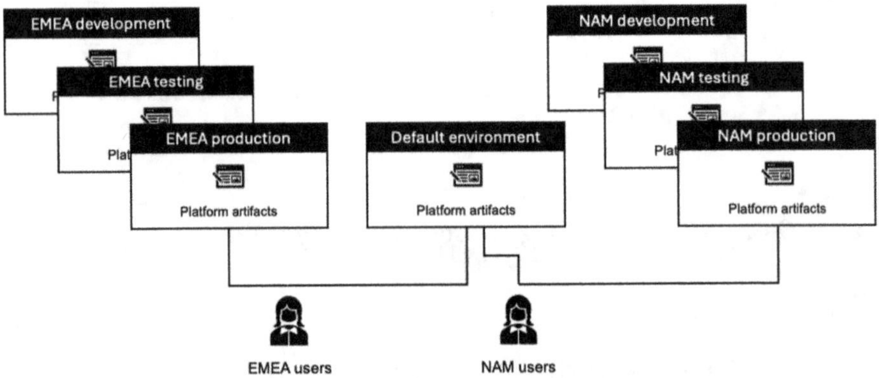

Figure 6-5. *Geographically distributed strategy*

This strategy groups environments by geographic region, ensuring compliance with local laws and data residency requirements.

Key benefits

- **Regional compliance**: Environments are tailored to meet local regulations like GDPR (EU) or CCPA (California).
- **Data residency**: Ensures data is stored within the designated regions to avoid breaches of international law.
- **Reduced latency**: Users accessing assets in regions that are closer to them reduces any possible latency.

Challenges

- Ongoing effort to align regional environments with evolving local data residency laws
- Duplication of governance processes across geographies, which can lead to inefficiencies
- Additional coordination between regional and central IT teams to maintain consistency in oversight

Environment Strategies: Combining Approaches for Optimal Governance

While individual environment strategies - such as organizing by business unit, security needs, or geography - each offer unique advantages, many organizations adopt a **hybrid approach** that combines elements from multiple strategies.

This flexibility allows organizations to address complex governance requirements, geographical challenges, and unique security requirements while maintaining the adaptability needed for innovation.

Personal Developer Environments and the Default Environment

The default environment often becomes a catch-all space for development and production apps, leading to ungoverned, siloed usage and a mix of critical and experimental work.

In contrast, **personal developer environments**[1] offer a more structured and flexible solution. By providing individual makers with their own individual spaces, personal developer environments enable controlled experimentation and rapid prototyping without risking the integrity of production systems.

---

[1] https://learn.microsoft.com/en-us/power-platform/developer/plan

This separation not only enhances accountability - since each environment is tied to a specific developer and purpose - but also streamlines the transition from experimental to production work. As we move forward, we'll explore how adopting personal developer environments can lead to better governance outcomes, improved security, and ultimately, a more agile innovation process.

Key points for personal developer environments:

- **Innovation without risk**: These environments allow for rapid experimentation while segregating developmental work from enterprise-critical systems.
- **Clear ownership**: Although a maker creates and uses the environment, governance processes (such as adaptive cards for asset identification) ensure proper tracking and later integration with broader governance frameworks.
- **Defined boundaries**: Personal environments are typically earmarked for testing and learning and are not intended for production use. They should eventually be migrated to formal environments once they meet production criteria and governance standards.

Applying Use Case Governance Across Environments

Once environments are structured, governance must be applied in a way that reflects the use cases the environments support. This means aligning governance dimensions - such as license type, data storage, connector access, and user scope - with the intended purpose of each environment. A well-structured approach ensures consistency while maintaining the flexibility needed for innovation.

In practice, organizations often manage **dozens, if not hundreds of environments**, depending on factors like company size, Power Platform adoption, and licensing models. Without an efficient system in place, managing environment requests and enforcing governance at scale can become a bottleneck. The key to maintaining efficiency is **automation** - ensuring that environment provisioning, governance policies, and security measures are applied dynamically based on clear rules.

One of the most effective ways to streamline this process is by using automated environment request workflows. The Center of Excellence (CoE) Starter Kit[2] provides a prebuilt solution for managing environment requests,[3] allowing organizations to establish a standardized intake and approval process. Even if the CoE Starter Kit doesn't fully align with your organization's requirements, exploring it offers valuable insight into best practices already adopted by thousands worldwide.

For organizations looking to customize their approach, Copilot Studio provides an opportunity to develop a conversational agent capable of

- Capturing environment request details through natural language interactions
- Applying governance logic to determine the appropriate governance dimensions
- Triggering automated provisioning to create environments, configure security access groups, and apply DLP policies

By integrating automation into environment management, organizations can reduce friction, improve compliance, and enable governance at scale - ensuring that new environments align with both operational needs and responsible AI (RAI) principles from the outset.

---

[2] Microsoft (N.D), The Center of Excellence Starter Kit

[3] `https://learn.microsoft.com/en-us/power-platform/guidance/coe/setup-environment-components`

Modernizing Environment Strategies with Managed Platform

Managed Platform introduce a premium governance layer that transforms how organizations balance innovation with oversight.

This supports the journey toward responsible AI by embedding proactive controls, visibility, and scalability into the fabric of Power Platform governance. Here's how:

- **Enhanced oversight with built-in monitoring**: Managed Platform provide administrators with detailed insights into usage patterns, app performance, and flow activities. This visibility helps ensure that AI-powered solutions, such as those built with AI Builder capabilities or integrated via Copilot Studio, operate within defined boundaries, reducing the risk of unintended consequences.
- **Proactive governance controls**: Features like data loss prevention (DLP)[4] policy enforcement, sharing limits, and usage insights[5] allow organizations to set guardrails that align with both operational goals and responsible AI principles. For example, restricting connectors to high-risk AI services in low-security environments ensures compliance without stifling creativity elsewhere.

[4] DLP policies are not limited to Managed Platform.

[5] Sharing limits and usage insights are a feature of Managed Platform.

- **Sustainability at scale**: As adoption grows, Managed Platform reduce administrative overhead by automating routine tasks - like usage reporting - freeing teams to focus on strategic priorities like RAI compliance and risk management.
- **Scalable segmentation using environment groups**: Environment Groups allow administrators to apply consistent policies, roles, and governance settings across multiple environments that share a common purpose - such as experimentation, production, or AI-sensitive development.

The shift to Managed Platform doesn't replace traditional strategies; it can be used to enhance them. A hybrid model might pair business unit-based environments with Managed Platform for high-risk AI use cases or layer geographic strategies with premium governance for compliance-heavy regions. This evolution ensures organizations can scale responsibly, embedding security and oversight into every layer of their Power Platform ecosystem.

At this level, however, governance needs to be focused on more than tools; you also need to focus on principles.

Defining and Applying Responsible AI Principles in Power Platform Governance

Microsoft's six responsible AI principles - fairness, reliability and safety, privacy and security, inclusiveness, transparency, and accountability - provide a foundational lens for governing AI within Power Platform. These principles represent actionable mandates that ensure AI-driven solutions - whether AI Builder models or actions, Copilot Studio agents, or Power Automate flows - deliver value responsibly.

However, Power Platform rarely operates in a vacuum. It's part of a larger organizational ecosystem, interacting with enterprise AI tools like Azure Foundry, third-party SaaS platforms, and custom-built solutions.

This interconnectedness means Power Platform governance must not only adopt these RAI principles but also integrate with and influence the organization's wider AI governance framework. Global frameworks like the EU AI Act elevate the stakes by mandating strict compliance for high-risk AI systems. As these legal requirements intersect with your governance strategy, they underscore the importance of aligning internal practices with external standards.

Let's now turn our attention to a primer on the EU AI Act, where we will unpack its risk classification framework and key elements. We can then review how this might impact governance more broadly

Primer: The EU AI Act – Risk Classification and Key Elements

The EU AI Act[6] is a comprehensive law designed to regulate artificial intelligence based on its potential risks to society, safety, and fundamental rights. Adopted in June 2024 and enforced from August 1, 2024, it classifies AI systems into four risk tiers, each with tailored obligations:

- **Unacceptable risk:** AI systems posing severe threats, like social scoring or real-time facial recognition for mass surveillance, are banned outright. These prohibitions took effect in February 2025, giving organizations a tight window to eliminate noncompliant use cases.

[6] https://www.europarl.europa.eu/topics/en/article/20230601STO93804/eu-ai-act-first-regulation-on-artificial-intelligence

- **High risk:** Systems affecting safety or rights – e.g., AI in hiring, credit scoring, or critical infrastructure – are subject to strict requirements. These include risk assessments, transparency documentation, and human oversight, with compliance deadlines starting on August 2026. Power Platform assets processing sensitive data or integrating with enterprise systems might fall here – demanding robust governance.
- **Limited risk:** AI with transparency risks, like chatbots, requires users to be informed of AI interaction. This applies to Copilot Studio agents where disclosure is key (important for user trust), with obligations effective in February 2025.
- **Minimal risk:** Low-impact AI, like spam filters, basic recommendation engines, or simple Power Automate flows using AI Builder, would be considered minimal risk under the Act. No specific requirements apply, leaving these systems free to operate.

Who Does It Apply To?

The EU AI Act casts a wide net! Targeting anyone involved in the AI life cycle within the EU market:

Providers: Those developing or placing AI systems on the market bear the heaviest load, from startups to tech giants. Here is a real-world example:

- Microsoft is the provider of Copilot Studio.
- If your organization builds a conversational agent using Copilot Studio and offers it across business units (as a simple example), your organization also becomes a provider for that AI system.

Users: Businesses, public bodies, or individuals deploying AI must comply with usage rules, especially for high-risk systems. This includes Power Platform adopters leveraging Copilot or custom AI models (this includes your organization).

Importers and distributors: They also face obligations, ensuring compliance flows through supply chains. The Act's extraterritorial reach is also notable - non-EU providers targeting EU users or affecting EU citizens (e.g., via data processing) are in scope. Exemptions are slim! Military AI and purely research-based systems are excepted - but for most, even if you're not in the EU, the rules may apply to you. Here is an example:

- If a Microsoft partner outside the EU develops an AI-powered solution and sells it to customers within the EU, the partner is the provider, and a reseller in the EU becomes the importer.

Leadership must grasp this scope to align culture; governance must enforce it to build trust.

How to Check the Assessment?

Determining your AI-systems risk tier is a structured process. Start with the Act's Annexes (I–III), which list high-risk use cases like biometric identification or education scoring - cross-check your solutions here. Next, self-assess:[7] Does this system influence rights, safety, or transparency? Tools like conversational agents are typically straightforward (limited risk), but a Power App screening job candidates (for example) is likely high risk and demands deeper scrutiny.

[7] https://artificialintelligenceact.eu/assessment/eu-ai-act-compliance-checker/

One key point here is the difference in classification. High-risk AI-systems have a special set of rules, including

- **Risk management system**: Implement a continuous, iterative risk management process throughout the AI system's life cycle. This involves identifying, assessing, and mitigating risks associated with the AI system's operation.
- **Data governance**: Ensure the quality and relevance of datasets used to train AI systems. This includes implementing measures to minimize bias and errors, thereby enhancing the system's performance and fairness.
- **Technical documentation**: Maintain comprehensive documentation detailing the AI system's design, development, and operational processes. This documentation should enable authorities to assess the systems compliance with the AI Act.
- **Record-keeping**: Establish mechanisms to automatically record events during the AI-system's operation. These logs are crucial for traceability and facilitating post-market monitoring and audits.
- **Transparency and provision of information**: Design AI systems to operate transparently, providing users with clear information about the system's capabilities and limitations. This ensures that users can make informed decisions when interacting with the AI system.

- **Human oversight**: Implement measures that allow human operators to oversee the AI system effectively. This includes the ability to intervene or deactivate the system if necessary, ensuring that human judgement remains central in critical decisions.
- **Accuracy, robustness, and cybersecurity**: Ensure that AI systems achieve high levels of accuracy, are resilient against errors, and are protected against cybersecurity threats. This involves regular testing and validation to maintain the system's integrity and reliability.

These obligations point toward a broader vision of **governance excellence** - one in which oversight, transparency, and trust are embedded across the life cycle of AI solutions. This is an opportunity to build durable, scalable governance structures that serve both innovation and accountability.

The Impact of the EU AI Act

The EU AI Act is perhaps the most transformative external influence at the time of writing. With strict risk classification and tailored obligations, the Act requires that high-risk AI systems undergo rigorous risk assessments, maintain detailed technical documentation, and implement robust human oversight.

In practical terms, this means that what started as a basic risk assessment in the foundational stage must now evolve into a continuous, iterative, and proactive process. This is a shift compelling organizations to adopt a new approach, ensuring that their governance framework is not static but evolving with regulatory expectations. The obligations imposed by the EU AI Act, while rigorous, are ultimately a positive force that pushes organizations to elevate their oversight and governance practices to new heights. With this framework in mind, we now turn to a deeper dive into Microsoft's six responsible AI principles. These principles not only align with the Act's mandates but also offer actionable guidance for implementing ethical AI solutions that deliver value responsibly.

A Deeper Dive: Microsoft's Six Principles of Responsible AI

Microsoft's responsible AI principles are commitments to building solutions that genuinely enhance human experiences, protect individuals, and uphold societal values. Let's explore what each principle truly means, why it matters to real people, and how it reshapes your governance strategy:

1. **Fairness**

   Ensures AI systems treat everyone equitably, without bias or unjustified discrimination. AI must respect the diversity of the users and communities it serves, reflecting fairness in outcomes, not just intentions.

   Bias or unfairness in AI doesn't just result in flawed technology – it impacts real lives. When hiring systems inadvertently exclude qualified candidates based on gender, race, or age or when credit-scoring algorithms systematically disadvantage certain communities, trust is damaged and ethical standards compromised.

   How it reshapes governance

   Bias assessments: You need to establish clear processes for verifying training datasets to identify and address potential biases (we'll address this topic deeply in Part 3 of this book).

Mitigation tracking: Maintain documentation proving that bias mitigation methods (e.g., dataset balancing, fairness audits, fairness metrics) have been systematically applied.

Ongoing fairness reviews: Regularly revisit fairness assessments throughout an AI model's life cycle, especially when models evolve or data changes.

2. **Reliability and safety**

   AI systems must perform consistently, safely, and accurately, even under unusual or unforeseen conditions. This means predicting, identifying, and managing potential failures or vulnerabilities proactively.

   AI systems increasingly handle sensitive or high-stakes operations – from medical diagnostics to autonomous vehicles. An unreliable AI model or unhandled edge can lead not only to operational disruption but also to serious harm or loss of trust among users.

   How it reshapes governance

   Robust testing procedures: Implementing comprehensive edge-case analysis and stress testing during model development and deployment.

   Incident response planning: Clearly define how to handle unexpected failures or disruptions, ensuring rapid responses and recovery.

   Model monitoring and retraining: Establish mechanisms to continuously monitor model performance and rapidly adapt or retrain as needed.

3. **Privacy and security**

   AI must respect and protect user data, ensuring confidentiality, data integrity, and protection against unauthorized access or misuse. Privacy and security form the bedrock of trusted AI interactions.

   AI's reliance on vast amounts of data exposes sensitive information to potential risks. Data breaches, misuse, or accidental disclosures can irreparably damage trust, cause legal ramifications, and severely affect an organization's credibility.

   How it reshapes governance

   Data handling protocols: Implement strict standards governing data collection, storage, and use, adhering closely to privacy laws like GDPR or CCPA.

   Access and identity management: Apply rigorous role-based access controls, ensuring only authorized personnel can access sensitive data or models.

   Audit trails and accountability: Ensure comprehensive logging and auditing mechanism to detect, trace, and respond rapidly to unauthorized activities.

4. **Inclusiveness**

   AI should empower everyone equally, regardless of background, ability, or language. Inclusive AI prioritizes accessibility, usability, and representation, making technology beneficial to a diverse spectrum of users.

When AI excludes groups – whether through inaccessible design or cultural bias – it limits technology's potential, disenfranchises communities, and risks reputational harm. Inclusive design maximizes the positive impact of AI by ensuring it reaches and empowers everyone.

How it reshapes governance

Inclusive design standards: Implement organizational standards ensuring projects meet accessibility and cultural inclusion guidelines from the outset.

Diverse user testing and validation: Incorporate diverse stakeholder and user group testing early and regularly throughout AI development.

Inclusive feedback channels: Establish clear channels for diverse communities to provide ongoing feedback, ensuring continuous inclusivity improvements.

5. **Transparency**

   Ensures users understand how AI systems operate, make decisions, and affect their lives. Explainability and openness in AI operations help build trust and accountability.

   Opaque AI systems lead to uncertainty, confusion, and mistrust among users and stakeholders. If people cannot understand how or why AI decisions are made – particularly critical decisions – they may reject or mistrust the technology, hindering adoption, or even worse, work around the system entirely.

How it reshapes governance

Documentation and explainability: Develop clear documentation on how AI decisions are made, including training data, decision criteria, and any limitations.

User-centric explainability: Implement mechanisms (such as explainable AI tools or model cards) to present AI decisions clearly and understandably to end users and stakeholders.

Disclosure practices: Establish guidelines to transparently disclose when users interact with AI systems.

6. **Accountability**

   Humans remain accountable for AI systems, overseeing their development, deployment, and operations. Accountability ensures that clear responsibilities are defined and that oversight and governance continuously enforce adherence to ethical standards.

   Without clear accountability, responsibility becomes diffuse, and ethical violations or failures can occur unnoticed or unchecked. Defined accountability maintains human oversight, ensuring responsible use and immediate corrective actions when issues arise.

   How this reshapes governance

   Clear roles and responsibilities: Define who is accountable for each stage of the AI life cycle – from development and deployment to monitoring and incident management.

AI governance committees: Establish cross-functional committees or working groups to review high-impact AI systems regularly and respond swiftly to emerging concerns.

Incident accountability framework: Create clear escalation paths and procedures for addressing ethical concerns or AI-related incidents promptly and transparently.

Implementing Microsoft's responsible AI (RAI) principles - fairness, reliability and safety, privacy and security, inclusiveness, transparency, and accountability - is not a one-time effort. It's a continuous journey that evolves alongside your organization's AI maturity and regulatory landscapes.

At the foundational stage, your organization sets the groundwork by becoming familiar with these principles and understanding their implications. However, as you progress, then expectations become more rigorous, and the principles require deeper integration into your governance frameworks.

It now becomes critical to understand the concept of **RAI sensitivity**. Not every solution presents the same risks or requires identical governance attention. RAI sensitivity acknowledges that certain AI-powered assets carry greater ethical, legal, and operational risks, thus requiring enhanced oversight and tailored governance practices. Moreover, we now have a two-stream governance model

Introducing a Two-Stream Governance Model

Now that your governance strategy is maturing from foundational to emerging, it's important to acknowledge that there is no longer a single approach or a one-size-fits-all approach to solutions.

Instead, your governance structure naturally branches into two distinct themes, based on a critical initial question:

"Is this asset AI-powered?"

This single question ensures that every new asset receives appropriate governance oversight from the outset - balancing innovation, risk management, and compliance efficiency.

Stream 1: Non-AI-Powered Solutions

For non-AI-powered solutions, governance remains straightforward and leverages a consistent model.

Overview

- **Initial assessment**: Perform an impact and governance complexity assessment.
- **Governance classification**: Based on the initial assessment, assign the asset into predefined governance packages (e.g., personal productivity, broadly used departmental apps, or enterprise-critical solutions).
- **Implementation**: Apply governance practices - such as DLP policies, security reviews, and environment assignments - aligned to the complexity and impact identified.

This stream continues to focus on traditional operational concerns - security, data handling, support models, and risk management - without needing deep ethical or fairness reviews.

Stream 2: AI-Powered Solutions

For AI-powered solutions, governance requires additional scrutiny. You now introduce a critical new dimension, one that we briefly touched upon in the foundational stage – **responsible AI sensitivity**.

Responsible AI sensitivity evaluates how deeply an AI solution could impact users, stakeholders, or compliance obligations.

The earlier you evaluate AI solutions for responsible AI sensitivity, the better. Waiting until assets are deployed into production increases both complexity and risk. Problems identified late often require costly rework, risk compliance violations, or lead to unintended ethical consequences.

A proactive governance strategy identifies and assesses AI-powered solutions **as early as possible** – ideally during initial planning, prototyping, or development phases. At this early stage, you have the greatest leverage for effective intervention, allowing teams to build responsible design, transparency, fairness, and safety directly into their solutions.

While it's important to be aware of the EU AI Act and its implications, integrating its concepts into your internal responsible AI governance is best done with legal oversight. One of the key Act's requirements is the completion of a Fundamental Rights Impact Assessment (FRIA) for high-risk AI systems – a process that examines how AI may affect users' rights, freedoms, and safety.

There is a publicly available EU AI Act compliance checker[8] that helps classify AI systems according to the Act's risk categories. However, interpreting whether your solutions – or your organization – fall under these definitions is not always straightforward.

My recommendation is simple: do not attempt to make this determination alone. Engage your organization's legal or compliance teams to assess applicability and obligations under the Act.

[8] https://artificialintelligenceact.eu/assessment/eu-ai-act-compliance-checker/

Internal evaluation is precisely where responsible AI sensitivity becomes crucial. Unlike external regulatory assessments, which often require specialist legal guidance, the RAI sensitivity evaluation offers a structured yet straightforward internal mechanism. It empowers organizations to systematically identify, understand, and migrate ethical and operational risks associated with their AI solutions at every stage of development.

To facilitate this, a clear, practical sensitivity assessment based explicitly on Microsoft's six responsible AI principles is recommended. Let's explore how to carry out this assessment within your governance framework.

The responsible AI sensitivity evaluation

This evaluation is designed to be completed at the earliest stages of development (ideally during planning or prototyping). It does not require legal expertise but offers a guided framework that complements your governance strategy.

Each question is scored either

- **Yes**: Indicating a higher sensitivity
- **No**: Indicating a lower sensitivity
- **Unclear**: Flag for further investigation or expert review

Some questions are critical flags: a single "yes" may elevate the sensitivity tier automatically.

RAI sensitivity questions

1. Does the system influence decisions related to individuals' rights, access, or status?

   Yes: Critical sensitivity

   Unclear: Escalate to ethics/legal review

2. Could the output of the system result in significant harm if it fails or is misused? (This includes both physical and psychological harm.)

   Yes: Critical sensitivity

   No: Low sensitivity

   Unclear: Escalate for review

3. Is the system interacting with the public or external users (e.g., customers, citizens)?

   Yes: High sensitivity

4. Does the system generate content or make decisions that are not easily explainable to users? (Opaque logic increases need for explainability mechanisms in place.)

   Yes: Medium sensitivity

   No: Low sensitivity

5. Is this system trained on user data, personal data, or potentially sensitive datasets? (This includes data sourced from customers, employees, or unverified third parties.)

   Yes: High sensitivity

   No: Lower sensitivity

   (Critical flag is PII is involved without safeguards.)

6. Will the system operate in an unpredictable environment? (e.g., social platforms, open data input, rapid contextual shifts)

   Yes: High sensitivity

   No: Lower sensitivity

7. Does the system make autonomous decisions without human review? (Even if oversight is available, automation increases oversight requirements.)

   Yes: Medium to high sensitivity

   No: Lower sensitivity

8. Are marginalized or vulnerable groups likely to be affected by this solution? (This includes systems influencing education, welfare, healthcare, legal status, etc.)

   Yes: High sensitivity

   No: Low sensitivity

Scoring and tiering guidance

Score or pattern	Tier	Guidance
Any critical flag = Yes	High sensitivity	Requires full RAI documentation, risk reviews, and bias mitigation plan. Escalate to RAI lead or AI committee.
4+ "Yes" answers (non-critical)	Medium sensitivity	Requires targeted mitigation (e.g., fairness audit, explainability testing, privacy assessment). Document review outcomes.
0–3 "Yes" answers	Low sensitivity	May proceed with lightweight governance (e.g., owner declaration, simple review process). Document decisions.
Any "Unclear"	Review required	Consultation with SME, legal, or ethics group before deployment.

The RAI sensitivity evaluation helps you answer one core question:

> "What level of scrutiny, governance, and oversight does this AI solution require?"

By evaluating a set of focused questions about data use, human impact, explainability, and deployment context, you categorize the solution into a sensitivity tier:

> **Low sensitivity**: Light-touch governance, asset should still be documented and owned, but minimal additional oversight is needed.
>
> **Medium sensitivity**: Requires targeted mitigation steps - e.g., bias checks, explainability, or stakeholder communication.
>
> **High sensitivity**: Triggers formal governance - including responsible AI documentation, stakeholder reviews, testing, and ethical oversight.

Importantly, this is not a **one-time activity**. Sensitivity may evolve over time - what begins as a low-risk internal tool may become high-impact if integrated with new data sources or deployed externally. This underscores the need for **ongoing governance and regular reassessment**.

Once an AI-powered asset has undergone a responsible AI sensitivity evaluation, the next step is developing mitigation strategies (outlined in the "guidance" column of the table above). These strategies help reduce the likelihood of harm and ensure the solution aligns with both regulatory requirements and your organizational values.

Rather than attempting to create one-size-fits-all rules for AI risks, it's more effective to anchor your mitigation planning around the six responsible AI principles. Each principle reflects a category of risk that must be addressed through governance, oversight, and design safeguards.

Let's explore each principle with a focus on the kinds of mitigations, controls, or design strategies that should be in place to reduce risk and demonstrate responsible AI implementation:

Mitigation Strategies by Responsible AI Principle

Fairness

Risk addressed: Bias in data, logic, or outcomes that may result in discrimination or unequal treatment.

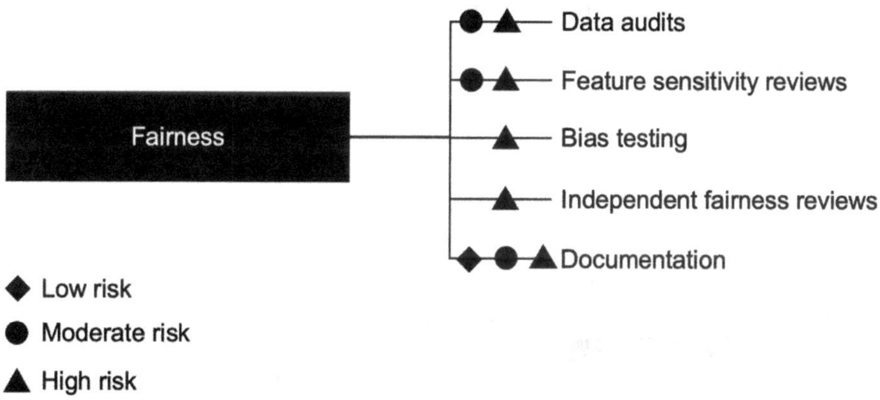

Figure 6-6. *Fairness mitigation strategies*

Key mitigations

- **Data audits**: Review and document datasets for representativeness. Are certain populations over- or underrepresented?
- **Bias testing**: Apply tools like Microsoft Fairlearn[9] (more on this in Part 3) or AIF360[10] to test for disparate impact across sensitive features (e.g., age, gender, race).

[9] https://fairlearn.org

[10] https://github.com/Trusted-AI/AIF360

- **Feature sensitivity reviews**: Evaluate whether sensitive attributes are used directly or indirectly (proxies).
- **Independent fairness reviews**: Involve stakeholders not involved in the design phase - such as data stewards, DEI advocates, or ethics boards - to evaluate potential bias.
- **Documentation**: Record fairness checks and decisions as part of solution design artifacts. This becomes vital for later audits or remediation.

Transparency

Risks addressed: Users or stakeholders don't understand how the system works or that AI is involved at all.

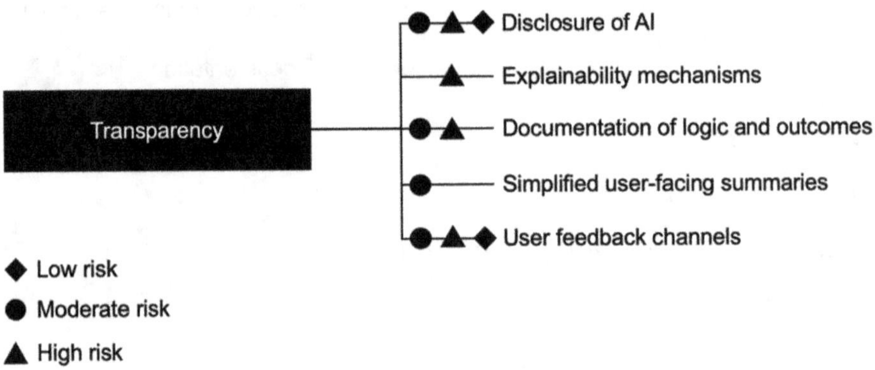

***Figure 6-7.** Transparency mitigation strategies*

Key mitigations

Disclosure of AI: Make it clear when users are interacting with an AI system. This can be a banner, notification pop-up, or disclaimer in AI generated content.

Explainability mechanisms: Provide meaningful insight into how outputs are generated.

Documentation of logic and outcomes: Describe the decision model in understandable terms.

Simplified user-facing summaries: Make system purpose and data usage clear.

User feedback channels: Allow for complaints, questions, or appeals.

Accountability

Risks addressed: Lack of ownership or traceability if something goes wrong.

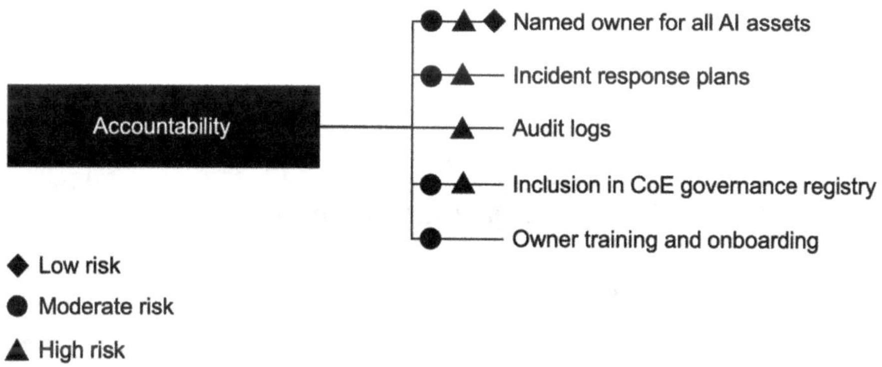

Figure 6-8. *Accountability mitigation strategies*

Key mitigations

Named owners for all AI assets: Technical and ethical accountability.

Incident response plans: Define how failures or risks are escalated.

Audit logs: Store traceable logs for inputs, decisions, and outcomes.

Inclusion in CoE governance registry: Ensure every solution is catalogued and reviewed.

Owner training and onboarding: Ensure owners understand their governance responsibilities.

Reliability and safety

Risks addressed: The system behaves inconsistently, crashes, or causes harm through incorrect outputs.

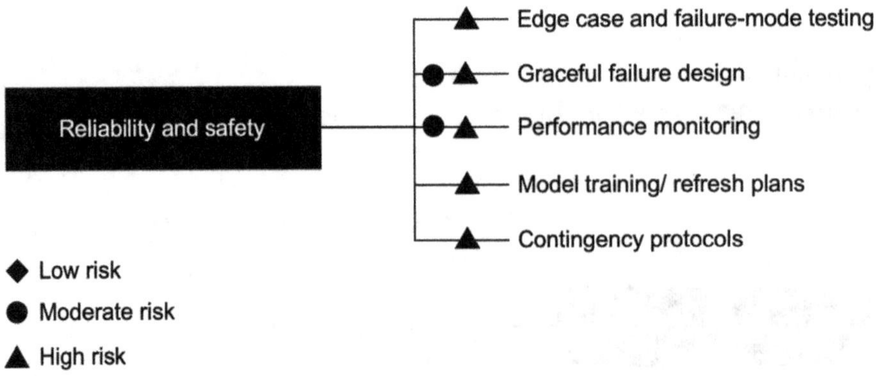

***Figure 6-9.** Reliability and safety mitigation strategies*

Edge-case and failure mode testing: Test unexpected or unusual scenarios.

Graceful failure design: Build fallback logic or human-in-the-loop processes.

Performance metrics: Set thresholds and alerts for drift of degradation.

Model retraining/refresh plans: Regularly evaluate and improve the system.

Contingency protocols: Document what to do if the model misfires.

Privacy and security

Risks addressed: Exposure of personal data, unauthorized access, or use beyond intended purposes.

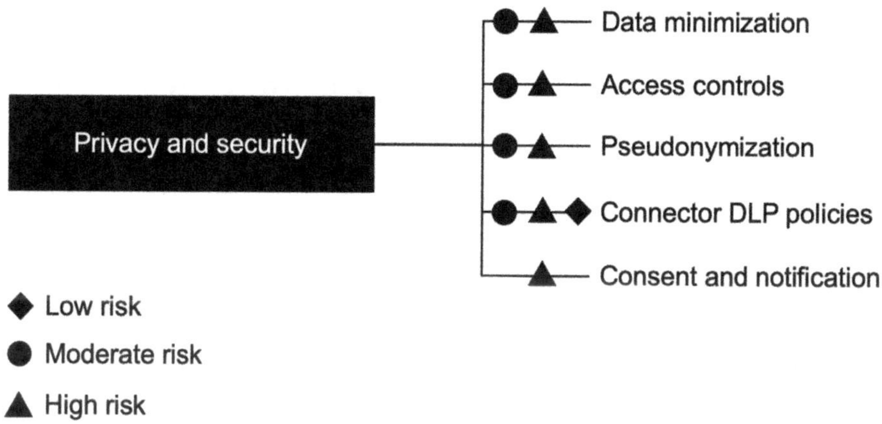

***Figure 6-10.** Privacy and security mitigation strategies*

Data minimization: Remove unnecessary fields from use and storage.

Access controls: Use role-based access control and restrict access to training and production datasets.

Pseudonymization: Replace identifying data where possible.

Connector DLP policies: Prevent movement of sensitive data through risky connectors.

Consent and notification: Where needed, let users know how their data is used.

Inclusiveness

Risks addressed: The system excludes, alienates, or disadvantages certain groups.

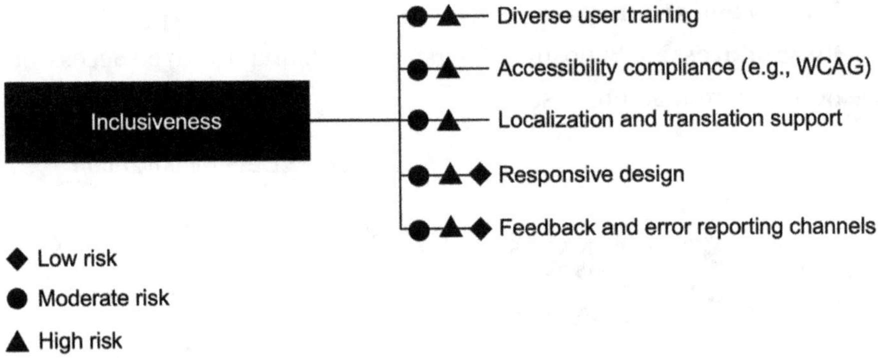

***Figure 6-11.** Inclusiveness mitigation strategies*

Diverse user testing: Engage real users across backgrounds, languages, and abilities.

Accessibility compliance: Build for screen readers, contrast, etc.

Localization and translation support: Build multilingual input/output paths.

Responsive design: Test across devices, tech literacy levels, and user roles.

Feedback and error reporting tools: Enable inclusive user feedback channels.

The mitigation strategies mapped to each responsible AI principle are more than checklists - they represent a mindset shift in how your organization prepares for and responds to the challenges of AI. These strategies serve a dual purpose in your governance model.

First, they act as proactive design tools. When applied early - during planning, prototyping, or development - they help teams think ahead, anticipate potential risks, and embed responsible practices directly into the solution life cycle. Whether it's running a fairness check on training data or ensuring explainability is built into user-facing outputs, these interventions help prevent harm before it ever occurs. This proactive stance is critical not only for ethical alignment but for compliance with frameworks like the EU AI Act and expectations set by internal governance bodies.

Second, they lay the foundations for effective response plans. When issues do arise - whether due to bias, unexpected outputs, or ethical concerns - these same strategies can be leveraged as part of a rapid, structured response. Audit logs, explainability documentation, and named ownership already allow your teams to quickly investigate what happened, who is responsible, and how mitigation was (or wasn't) applied. In this way, mitigation strategies create infrastructure for accountability and continuous improvement, reducing cost and impact of errors.

Conclusion of Emerging Stage: From Foundations to Frameworks

At the emerging stage, your organization has moved beyond the basics of Power Platform and AI governance. You've gone from establishing initial guardrails to building a more adaptive and connected governance framework - one that integrates cross-functional ownership, regulatory awareness, responsible AI principles, and structured processes for assessing and managing risk.

This stage is where foundational ideas take form as repeatable practices.

You've built a growing understanding of your asset landscape. You've begun to identify what's in production, who owns it, and how it aligns (or doesn't) with enterprise policies. For AI-powered solutions, you've introduced a second stream of governance - one that accounts for RAI sensitivity, asks deeper questions about risk, and begins to shape processes around fairness, transparency, accountability, and safety.

You've also introduced the idea of tiered assessments - that different assets deserve different levels of scrutiny based on value, complexity, and RAI sensitivity. This is a critical concept that will underpin everything that follows as governance matures.

Mitigation strategies linked to RAI principles now serve a dual role: guiding design and strengthening your readiness to respond. With these mechanisms in place, you're better positioned to manage the AI life cycle responsibly - from ideation to retirement.

However, the emerging stage also comes with warning signs: manual processes start to show limitations, oversight can become inconsistent, and the pace of innovation can outstrip governance capacity. You've built solid structures, but scaling those structures is your next challenge.

Emerging Stage Progression Checklist

Use this checklist to confirm whether your organization has successfully moved from foundational setup to a scalable, structured approach to governance and responsible AI. Each item represents a capability or practice that should be in place before moving into the operationalized stage.

Policies and rules of engagement

- [ ] Governance policies are mapped to dimensions such as license type, data sources, connectors, support model, and deployment types.
- [ ] These governance dimensions are documented and used to automatically apply or recommend policies.
- [ ] You've introduced a use case matrix that maps governance expectations based on complexity and risk.
- [ ] Guidance exists for how to remediate non-compliance or unknown assets.

Responsible AI integration

- [ ] RAI principles have been formally introduced and discussed across teams.
- [ ] You've developed and tested a RAI sensitivity questionnaire, with documented guidance for interpreting results.
- [ ] Mitigation strategies are defined for each Microsoft RAI principle (fairness, privacy, Transparency, etc.).
- [ ] Mitigation planning is proactive, not reactive – applied during development and not just post-deployment.
- [ ] Teams are aware of key regulatory frameworks such as the EU AI Act, GDPR, etc.

Governance structure and oversight

- [ ] A formal Center of Excellence has been established or enhanced to include roles related to AI governance.
- [ ] Power Platform governance is no longer siloed; it is now connected to broader AI, data, and digital strategy teams.
- [ ] Governance responsibilities are defined and distributed across relevant functions (IT< business units, compliance, etc.).
- [ ] You've developed or adopted an environment strategy that aligns with organizational size, geography and risk profile.
- [ ] A method exists to identify, track and assign ownership for all Power Platform assets.

AI and asset classification

- [ ] Assets are being classified using impact and governance complexity dimensions.
- [ ] A clear distinction is made between AI-powered and non-AI-powered solutions.
- [ ] You have implemented a RAI sensitivity evaluation process for AI-powered solutions, preferably triggered in early development.
- [ ] Solutions are triaged into governance tiers based on complexity, impact and RAI sensitivity.
- [ ] Asset records include RAI assessment outcomes, owners, and governance requirements.

Transparency and communication	Readiness for scaling
☐ Governance rules are documented in a central, discoverable location (not hidden in a SharePoint folder).	☐ You've identified which governance processes are manual and fragile – and are planning for automation.
☐ You've identified stakeholder groups and created a communication strategy that meets their needs (e.g., makers, business leads).	☐ The team understands which practices need to scale as adoption increases (e.g., RAI reviews, documentation, monitoring).
☐ There are visible internal champions for Responsible AI and platform governance.	☐ You've started exploring tools and services to support automation (e.g., Managed Environments, CoE Starter Kit, custom tooling).
☐ Training materials reflect RAI considerations and governance expectations.	☐ Governance performance is being tracked with basic metrics or scorecards.

If most of these boxes are ticked, you're ready to operationalize – to take the best parts of your governance strategy and scale them across teams, environments, and use cases.

CHAPTER 7

Operationalized: Establishing Trust and Governance

Strategic | Scalable | Integrated

Governance doesn't get easier – it gets smarter.

By the time you reach the operationalized stage, the chaos of experimentation has been replaced with something more deliberate: repeatable practices, shared language, embedded ethics.

Organizations haven't stumbled into responsible AI (RAI); it has been earned through intentional effort, transforming the emerging stage's scattered experiments into a disciplined, proactive system.

This is not a light lift: moving from ad hoc trials to streamlined operation demands sustained focus, structure, buy-in, and strategy.

Microsoft's six RAI principles – fairness, reliability and safety, privacy and security, inclusiveness, transparency, and accountability – shift from ambitions and taglines to operational bedrock, woven into processes and daily workflows.

S. Jeffery, *The Low-Code AI Maturity Model*, https://doi.org/10.1007/979-8-8688-1730-4_7

For Power Platform adopters, this stage marks a tipping point:

- **Enterprise-wide adoption accelerates**: AI and automation scale beyond departmental use cases and into mission-critical workflows.
- **Governance and compliance frameworks move from reactive to proactive**: Tools like Power Platform Admin Center, Managed Platform, the CoE Starter Kit, and custom tooling provide insights, automated risk assessments, and enforced policies at scale.
- **RAI** is now operationalized.
- The **Center of Excellence** is now a cross-functional hub with dedicated roles.

In this chapter, we'll dive into the core processes that make operationalized governance tick. Here's what to expect:

- **Formalizing governance structures**: Turning your CoE into a cross-functional powerhouse with clear roles and responsibilities
- **Automating governance workflows**: Using tools like Power Platform to govern itself, streamlining approvals, compliance checks, and more
- **Defining and tracking governance metrics**: Measuring success with KPIs that show governance isn't all red tape
- **Embedding Responsible AI (RAI) into solution development**: Making ethical AI a core part of development workflow, not an afterthought

- **Training and awareness at scale**: Ensuring everyone - from developers to end users - knows their role in responsible innovation
- **Automating RAI compliance**: Leveraging automation to enforce ethical standards without hitting the brakes on progress

As you work your way through this content, you'll see governance evolving into (hopefully) a catalyst for growth. It's not the end of the road. Once you've nailed the operationalized stage, you'll be ready to take things to the next level: integrated excellence. This is where governance becomes more than processes - it really becomes a core part of the organization's DNA. We'll explore that in the next chapter, so keep reading!

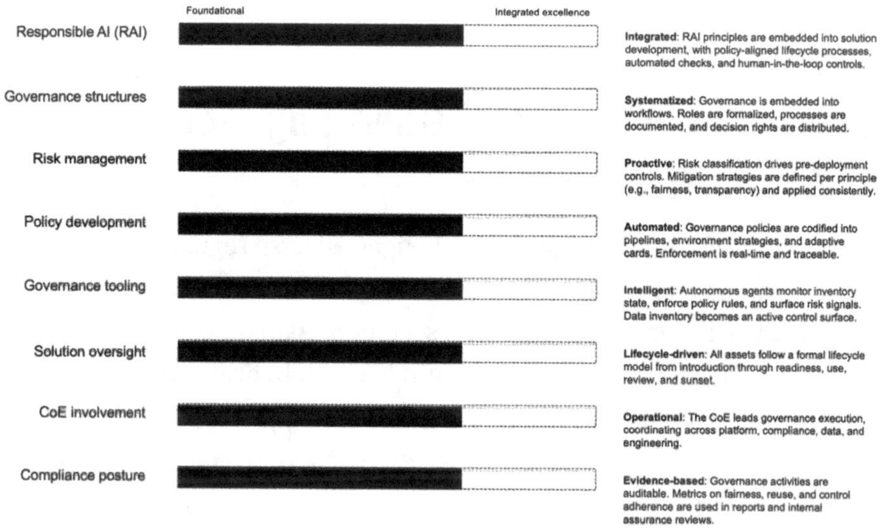

***Figure 7-1.** Run state of organization at the operationalized stage*

The CoE's Broader Influence: Shaping AI Governance

At the operationalized stage, governance structures need to evolve beyond the initial frameworks established earlier in your journey. We're not revisiting foundational concepts, like setting up a Center of Excellence, or running basic training programs – those are already in place.

Instead, we need to talk about refining and formalizing these structures to handle enterprise-wide adoption, particularly for organizations with global operations, and ensuring that you are exerting influence over broader AI governance strategies.

Based on my experience working with diverse organizations, I've seen how these structures evolve depending on the operating model and how the CoE can extend its reach beyond Power Platform. Let's dive in.

Tailoring Governance to Operating Models in Global Organizations

Global organizations face unique challenges: diverse regulatory landscapes, varying departmental priorities, and the need to balance consistency with local adaptability. It's a tough ask! The governance structure – whether centralized, decentralized, or hybrid – must be formalized to address these complexities while supporting scalable, accessible innovation. We'll delve into how each model adapts at this stage, with specific implementation strategies to ensure that governance is aligned with operational realities.

Centralized Governance: Unified Oversight with Regional Adaptation

In a centralized model, the CoE serves as the single authority for Power Platform governance, enforcing policies across all regions and business units. At the operationalized stage, this structure shifts from establishing basic rules to implementing sophisticated, scalable processes that account for global diversity.

Implementation ideas

- **Policy customization framework**: Develop a tiered policy system where core standards (e.g., data security, RAI compliance) are globally mandated, but regional variations are permitted within defined parameters. For instance, a global data loss prevention (DLP) policy might restrict high-risk connectors universally while allowing EU teams to add GDPR-specific controls.
- **Regional governance representation**: Assign dedicated CoE members to oversee regional compliance, equipped with tools like Power Platform Admin Center to monitor usage and enforce policies. These liaisons act as bridges, ensuring global standards are applied while addressing local needs.
- **Global consideration**
    - Time zones demand structure reviews - like quarterly virtual summits - to keep everyone aligned.
    - Metrics matter - tracking compliance rates centrally becomes a key indicator of success, giving leadership confidence without micromanaging local teams.

Decentralized Governance: Localized Control with Strategic Alignment

In a decentralized setup, each business unit or region runs its own CoE or governance team, with autonomy to tailor policies. Based on my experience, at this stage, the focus shifts to formalizing this autonomy while tying it back to enterprise goals.

Implementation ideas

- **Standardized governance templates**: I've seen organizations provide local CoEs with templates for policies (including RAI sensitivity evaluations) to provide some consistency.
- **Cross-regional governance council**: A council of local leaders can keep things cohesive. I once worked with a university where faculty leaders met bimonthly to collaborate and track decisions to avoid silos and share best practices.
- **Global considerations**
    - Preventing silos is key. I've seen success with a minimum set of KPIs (like the % of apps and flows audited) to keep everyone on track. Meeting regularly also helps reduce siloed efforts.
    - A centralized repository works well, sharing policies, governance practices, best practices, etc.

Hybrid Governance: Balancing Central Direction and Local Flexibility

The hybrid model blends centralized oversight with decentralized execution, and at this stage, the focus is on formalizing both layers for scale. Based on what I've seen, this approach works well when clarity and adaptability are equally critical.

Implementation ideas

- **Governance blueprint**: The Central CoE defines a blueprint outlining mandatory policies (e.g., RAI principles, security baselines) and optional guidelines (e.g., approval workflows, licensing chargeback models). Local teams implement the blueprint, customizing optional elements - such as setting shorter review cycles for high-risk solutions, especially in regulated markets (e.g., healthcare).
- **Environment-specific governance**: Using Managed Platform to enforce central policies (e.g., sharing limits, DLP enforcement) while allowing local teams to configure sandbox environments for experimentation. For instance, a European team might create a sandbox allowing additional connectors to test that is subject to a central review.
- **Integrated reporting**: Dashboards that aggregate data from all regions, offering both a global view (e.g., total compliance incidents) and local breakdowns (e.g., app usage in EMEA). This ensures central visibility without stifling local innovation.

- **Global considerations**
    - Define clear escalation paths for conflicts - like routing a regional policy deviation to the central CoE for approval within 48 hours.
    - Leverage Copilot Studio to build a governance conversational agent that answers local queries (e.g., "What's the DLP policy for SQL Server connectors?") while reinforcing central standards.

As AI becomes foundational across an organization's services, the risk of governance fragmentation grows. Without a unified approach, different teams may apply inconsistent RAI standards, compliance strategies, or ethical interpretations. The CoE, with its hard-won expertise in implementing scalable, practical governance for Power Platform, is now perfectly positioned to help the organization create a shared AI governance framework.

In many organizations, the Power Platform CoE becomes the proving ground for responsible innovation. By stepping into a leadership role, your CoE not only secures ethical AI within its own domain - it becomes the model for how the rest of the enterprise governs AI responsibly.

Broader AI Governance

Influence in governance can be a shark tank. I've seen it a few times in organizations where there are multiple technologies involved in similar realms. Each fights to position itself as the default option and exert influence. It's important to ensure that Power Platform is strategically positioned to ensure that broader objectives can be served.

At this level, the CoE should be a strategic influencer in your organization's broader AI governance ecosystem. This is crucial: as AI adoption spreads beyond Power Platform - into Azure Machine Learning, third-party tools, or custom solutions - the CoE's expertise positions it as a leader in shaping enterprise-wide AI strategies. Let's look closely at how this influence manifests and how to implement it effectively.

Establishing a unified AI governance framework

The CoE leverages its Power Platform governance experience to create a unified framework that applies to all AI initiatives, ensuring consistency and reducing redundancy. Here are some ideas to achieve this:

- **AI governance policy template**: Develop a comprehensive policy that integrates Power Platform practices - like RAI audits - with broader AI needs. For example, mandate bias assessments, whether built as custom models in AI Builder, or Azure ML, or any other stack, using a standard checklist adapted from Microsoft's Fairlearn or the Responsible AI Institute's governance template.
- **Cross-platform integrations mapping**: Create a central hub to track AI assets across platforms.

Facilitating cross-functional collaboration

The CoE should drive collaboration between IT, legal, compliance, and business units, ensuring that AI governance reflects the diverse perspectives and organizational priorities. Some ways that this can be implemented:

- **AI governance committee:** Formalizing a standing committee led by the CoE, meeting monthly to align AI policies. Include reps from legal (for regulatory compliance), HR (for workforce impact), and IT (for technical feasibility and security).
- **Scenario-based workshops:** Host frequent workshops to address real-world AI challenges - like ensuring a customer-facing Power App with privacy laws. Record

outcomes as case studies, shared via a governance portal, to guide future efforts. Use the reverse brainstorming technique that we covered in Part One of the book!

- **Feedback mechanisms:** Implement a feedback loop that allows teams to report governance gaps or overlaps.

The expertise that you have in the CoE can establish enterprise-wide ethical AI practices, further reinforcing it as a strategic enabler and influencing how all AI solutions are designed and deployed. Consider creating

- **An RAI certification process:** Certifications work well. One organization that I worked with took this approach with general compliance of Power Platform assets. The certification was seen as a mark of quality internally while keeping the organization compliant. Using an approach for RAI compliance would work equally well and could be extended to any AI-powered system.
- **Regulatory alignment:** Mapping RAI to regulations and laws like the EU AI Act is critical. Taking the lead on alignment will save other areas of your organization time and positions the CoE as a leader internally (and even possibly externally).

Formalizing governance structures at the operationalized stage is where you create a robust, adaptable framework that scales with your organization's needs - whether centralized, decentralized, or hybrid. For global organizations, this means implementing tailored policies, automation, and reporting to balance unity and flexibility.

Meanwhile, the CoE's influence extends now beyond Power Platform, driving a unified AI governance strategy that enhances trust, compliance, and innovation across the enterprise.

Formalizing Risk Management in Operationalized Governance

Governance needs to shift from identifying risks to managing them systematically. You can no longer focus on spotting isolated issues - your challenge is keeping pace with risks at scale, across hundreds or thousands of AI-powered assets. This requires a structured, repeatable life cycle that embeds risk management directly into development, governance operations, and post-deployment activities.

The diagram below outlines this life cycle model. It's not linear - it's circular and iterative, because governance must evolve alongside the systems it manages.

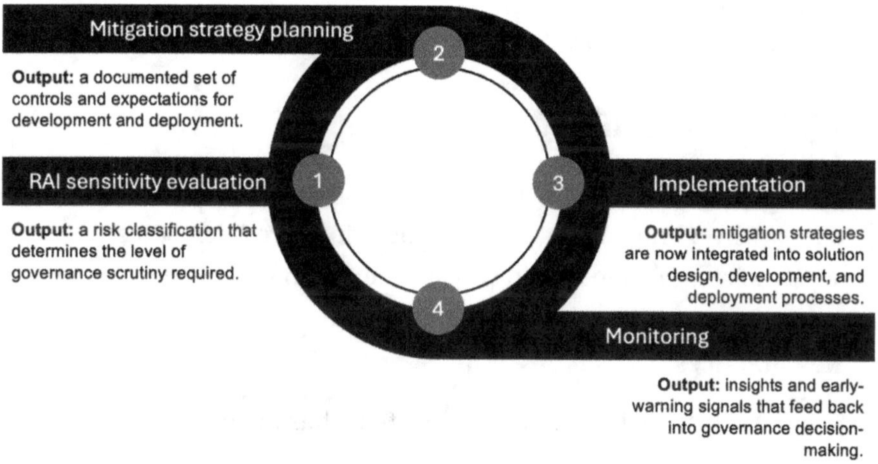

Figure 7-2. Circular model for operationalized risk management

Let's break it down.

RAI Sensitivity Evaluation

At scale, AI systems vary dramatically in complexity and potential harm. Some are simple automations; others influence hiring, finance, or public-facing content. Understanding the sensitivity of each system is the foundation for applying the right level of oversight.

Key practices

- Trigger evaluations early - ideally at asset creation of prototype stage. We discussed this earlier in the book - having a cloud flow that runs daily to detect any new Power Platform AI-powered assets created in the past 24 hours.
- Use your established RAI sensitivity evaluation to assess factors like human impact, opacity of decision-making, and reliance on sensitive data.
- Define risk tiers and log results alongside the asset in your inventory system.

Mitigation Planning

Once sensitivity is known, the next step is to reduce risk before the system goes live. Mitigation strategies operationalize Microsoft's six principles of responsible AI and convert them into specific design and documentation steps.

Key practices

- Apply principle-specific mitigations based on risk tier. For example:
- **Fairness**: High-risk systems require evidence of bias mitigation and diverse testing data.

- **Transparency**: Even mid-tier systems should clearly label AI-generated content.
- **Accountability**: Assign named owners for oversight at launch.

Implementation

Mitigation plans are only valuable when they're implemented - and this is where RAI becomes part of the solution life cycle, not just a review check box.

Key practices

- Use automation where possible. For example, C.H.A.T. assessments during build stages, to ensure solutions stay aligned with original intent.
- Enforce mitigation actions through tooling (e.g., requiring explainability banners in outputs).
- Track completion status for all mitigation steps - this becomes your audit trail.

Monitoring

Systems drift, AI behaviors evolve as data changes, user patterns shift, or new integrations are introduced. Monitoring ensures mitigations remain effective.

Key practices

- Use telemetry to monitor run failures, unusual usage patterns, or escalated support tickets.
- Configure autonomous agents to identify exceptions - for instance, flagging a cloud flow with high run volumes.

- Define alert thresholds based on risk tier - higher risk solutions should be checked more frequently or automatically escalated.

Walkthrough: A Conversational Agent for Product Advice

Let's take this out of the abstract. Imagine your organization is building a Copilot Studio agent for a customer-facing website that provides product advice based on user questions. It pulls data from internal product databases and CRM systems and can help customers compare features or recommended items.

How does this agent move through the four stages of operationalized risk management?

1. **RAI sensitivity evaluation**

   The agent is assessed early in development. A flow detected this agent within 24 hours of creation automatically. The RAI evaluation identifies:

   - AI is used in recommending products (nondeterministic logic).
   - Users are interacting with it directly (customer facing).
   - It impacts customer perception and decision-making.

   Based on the evaluation, the agent is classified as medium sensitivity. It's not making credit decisions, or HR actions, but it does shape experiences, could reinforce bias in product suggestions, and needs transparency.

2. **Mitigation planning**

   Given the classification, the following mitigation strategies are selected:

   - **Transparency**: Configure the agent to introduce itself as an AI agent and include a "Why did you suggest this?" link using embedded explanation prompts.
   - **Privacy**: Strip CRM fields of personal identifiers before ingestion; store logs masking personal data.
   - **Fairness**: Review data to ensure certain categories or brands aren't overrepresented in default answers.
   - **Accountability**: Identify a named business owner from the product team and a technical owner from IT.

3. **Implementation**

   The CoE is notified automatically, which logs the mitigation strategy and asset ID in Dataverse.
   Developers

   - Apply required transparency language.
   - Use the C.H.A.T framework at 30% and 90% development completion to review intent and check assumptions.
   - Configure product usage tracking using custom events in Azure Application Insights (e.g., which product pages are linked to from the chat).

4. **Monitoring**

   Once deployed:

   - Alerting in Azure Application Insights tracks the agent's error logs and triggers an escalation if failure rates spike.

- Usage data is reviewed monthly to ensure product recommendations stay balanced.
- If the agent is expanded to support product purchase transactions or offer financing options, its sensitivity level is re-evaluated.

All of this should be tracked in an RAI governance dashboard - owners, mitigations, telemetry status, and next review dates. Governance is embedded, not burdensome.

Defining and Tracking Governance Metrics

Automation streamlines governance workflows - whether through deterministic rules or adaptive agents - it's the metrics derived from these processes that reveal their true impact. At the operationalized stage, governance must move beyond execution to evaluation, quantifying compliance, efficiency, and risk mitigation to ensure Power Platform delivers on its enterprise promise.

My experience with organizations at this maturity level shows that well-defined and tracked metrics transform governance from an operational necessity into a strategic asset, providing clarity on what works and where adjustments are needed.

The role of metrics in governance

Metrics serve as the **backbone of effective governance** at scale. Without them, efforts like automated approvals or compliance monitoring lack tangible proof of success, leaving everyone around the governance domain guessing about their value. At this stage, where Power Platform adoption spans departments, regions, or even global operations, metrics provide visibility into whether governance policies are enforced, resources are optimized, and responsible AI principles upheld.

Why metrics matter

At this stage, governance is likely overseeing a complex ecosystem - hundreds or thousands of apps, flows, and AI-driven solutions - each with potential risks and benefits. Metrics matter because they

- **Demonstrate compliance**: They confirm that policies, such as DLP rules, or RAI standards are consistently applied, reducing exposure to regulatory or ethical violations.
- **Measure efficiency**: They quantify time saved through automation - e.g., faster approvals or provisioning - freeing resources for strategic, proactive priorities.
- **Mitigate risks**: They highlight issues like noncompliant apps or over-provisioned licenses, enabling proactive resolution before escalation.
- **Build trust**: They provide leadership with evidence of governance's value, fostering confidence in the platform's enterprise role.

For example, tracking the percentage of AI-powered apps, flows, and agents that are adhering to RAI principles ensures trust scales with adoption. Without metrics, governance remains anecdotal, unable to justify its role, effectiveness, or growing demands.

Establishing governance objectives

Before defining metrics, it's worth pausing to consider: Do you have clear governance objectives? What are they? At this stage, metrics must align with specific goals to be meaningful - tracking for the sake of tracking wastes effort and dilutes focus. My experience shows that organizations often leap to measurement without first clarifying what they aim to achieve, leading to misaligned or irrelevant data.

To establish effective governance objectives, start by aligning with your organization's broader priorities – whether that's compliance, cost efficiency, user trust, or innovation enablement. Here's how to approach it:

- **Engage stakeholders**: Convene your CoE, IT, legal, and business unit leaders to identify priorities. For example, legal might emphasize regulatory compliance (e.g., GDPR), while business units focus on rapid solution deployment. A half-day workshop can surface these goals.
- **Prioritize focus areas**: Narrow to three to five objectives based on scale and impact. Common ones include "Ensure 95% compliance on app registration." This focuses effort on what matters most.
- **Set measurable targets**: Translate objectives into quantifiable outcomes. For instance, "Minimize governance incidents to fewer than five per month" provides a clear benchmark. This step requires minimal effort but ensures metrics tie directly to goals.
- **Document and communicate**: Record objectives in a shared Dataverse table or governance portal, ensuring all teams understand them. This fosters alignment.

Establishing governance objectives

Once objectives are set, metrics can be tailored to measure progress against them. Effective metrics are specific, measurable, and tied to these goals. Based on my experience, organizations at this stage benefit from a balanced set of metrics covering compliance, efficiency, and risk management. Here are key metrics to consider within Power Platform context, assuming objectives like those above:

- **Compliance rate by environment**: The percentage of apps and flows meeting RAI policies (e.g., transparency disclosures) per environment. If your objective is "Ensure 95% compliance," this tracks success, with production environments often targeting that threshold or higher.
- **Approval turnaround time**: The average time from app submission to approval, ideally reduced to hours via automation. For an objective like "Reduce approval times to under 24 hours," this quantifies efficiency gains as request volumes rise, using audit data for accuracy.
- **Incident frequency**: The number of governance violations (e.g., AI-powered solutions not registered with RAI compliance) per month, assessing risk assessment and automation effectiveness. Tied to "minimize incidents to fewer than five," it highlights areas needing refinement, sourced from telemetry.
- **RAI feedback score**: User-reported satisfaction with RAI principles (e.g., transparency, fairness), collected via surveys or forms. If your objective is "Maintain RAI transparency," this ensures ethical alignment scales with adoption, using direct user input.

These are a few examples of metrics that align with Power Platform data – usage logs, policy outcomes, user feedback – ensuring they're actionable and relevant at this stage. Metrics also reveal where gaps persist – such as inconsistent RAI adherence flagged by user feedback or incident trends. Addressing these gaps requires embedding responsible AI principles directly into the development process, ensuring ethical considerations are proactive rather than reactive.

Automating Governance Workflows

Governance needs to keep pace with the enterprise-wide adoption of Power Platform. This pace quickly outgrows manual processes. Early-stage governance often relies on emails, spreadsheets, and ad hoc reviews, but as the number of apps, flows, agents, and AI-powered solutions climbs into the thousands, this approach becomes untenable.

My experience with organizations at this maturity level underscores a critical shift: automating governance workflows is a necessity for maintaining control, ensuring compliance, and enabling scalable innovation. By leveraging the full suite of Power Platform tools, including Power Automate, Managed Platform, and Copilot Studio's conversational and autonomous agents, organizations can (and do) streamline oversight while addressing common governance challenges effectively.

The imperative for automation

Manual governance workflows - such as reviewing app permissions, approving solutions, or tracking compliance - consume a disproportionate amount of time and resources as adoption scaled. A single app approval might involve multiple stakeholders exchanging emails, cross-referencing policies in a document, and manually updating a tracking sheet, taking hours or even days. Multiply this across dozens of requests weekly and the inefficiency is compounded. This increase delays, errors, and oversight gaps.

Automation transforms this by embedding governance into repeatable, technology-driven processes, reducing human effort while improving consistency and auditability. The value lies in freeing governance teams to focus on high-value priorities rather than repetitive clerical tasks.

Identifying candidates for automation

Effective automation targets tasks that are repetitive, high-volume, and governed by clear criteria, where technology can deliver consistency and speed. Several processes stand out as prime targets within Power Platform ecosystem:

- **App, flow, and agent approvals**: Reviewing new solutions for compliance policies, RAI standards, or licensing requirements is a frequent task as adoption grows. These approvals often follow predictable criteria, such as connector usage or sharing scope - this makes them ideal for automation. The high volume of requests at this stage amplifies the time savings.
- **Environment provisioning**: Creating new environments for development, testing, or production involves applying standard configurations (e.g., DLP policies, security settings). Manually setting up each environment risks inconsistency, but automation ensures uniformity with minimal effort, especially as team multiply across regions.
- **Compliance monitoring**: Tracking adherence to governance policies - like ensuring apps use approved connectors or flows avoid sensitive data - requires ongoing vigilance. This process scales poorly manually but thrives under automation, given the repetitive nature and clear rule set.

Deterministic tasks, with their predictable rules, align with structured automation, while nondeterministic tasks, needing judgement or adaptation, benefit from intelligent agents. Identifying this fit ensures automation maximizes governance impact.

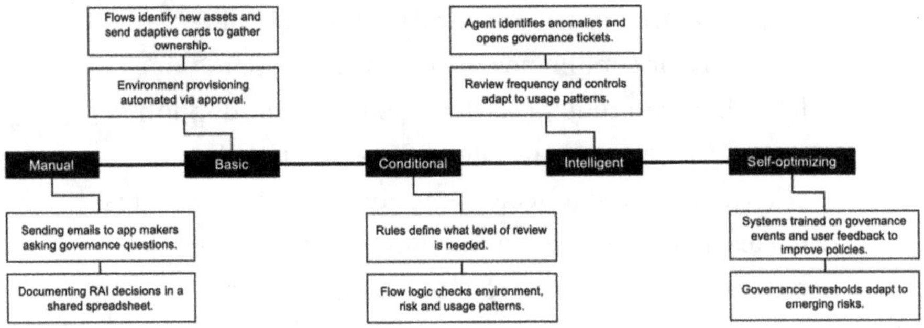

Figure 7-3. *Scale of automation*

The Power Platform "stack" of capabilities excels at automating deterministic workflows – those with fixed, repeatable steps and outcomes.

For app and flow approvals, a flow can extract metadata relating to apps, flows, agents, and more and escalate exceptions to your governance policies to the CoE. This eliminates manual reviews, ensuring every app meets the same standard. Implementation involves creating the flow logic – a moderate effort – but the reward scales with volume: hundreds of reviews processed consistently in minutes, not days.

I recall working with one organization that implemented processes to identify apps that had been using a particular connector. Within hours of creating the flow logic, they had a list containing every instance, enabling them to contact app owners for more information. This process, had it been approached manually, would likely have taken days.

Environment provisioning

Environment provisioning offers another deterministic fit. A flow triggered by a request (e.g., via Microsoft Forms) applies predefined settings – DLP, security groups – and creates the environment using the admin connector.

This ensures uniformity across dozens of environments, a task that grows unwieldy manually as teams expand. Collecting the required information does not need to be Forms, however.

I worked with one organization that used a conversational agent (known as a chatbot when this was implemented) to collect information in a conversational way as they required more information, including how long the environment was needed (e.g., was it temporary (for a project or experiment)), who the backup owner would be, etc. Consider this when you design your processes – trying to make the requests friction free makes your users less likely to try and find workarounds.

Copilot Studio: Adaptability for nondeterministic processes

At the time of writing this book, Copilot Studio's autonomous agents have recently been released. These tackle nondeterministic tasks, where outcomes vary based on context or require adaptive decision-making.

This new capability also opens numerous opportunities for new governance policies. Let's walk through one example of how autonomous agents can change the way you think about governance.

Imagine you were able to provide clear definitions of use cases. Let's use a cloud flow as an example. You create a word document containing the characteristics of use cases; this document states that cloud flows for personal use should not run more than 50 times per day.

Now imagine, you've invested some time in creating an agent that is instructed to look for exceptions – it runs daily and identifies a flow in a personal productivity environment that is running more than 50 times. This is where the agent can excel. It can contact the flow owner in a conversational way to gather more information – there may be valid reasons for the high number of flow runs, or it could be poorly implemented logic or unawareness of policy – an easy fix where the agent can provide some best practices using knowledge that you either provide or it finds. This simple intervention makes governance enforcement less binary, more approachable, and (in my opinion) more likely to be adopted.

This simple example demonstrates how using a nondeterministic approach to a typically deterministic process opens several advantages and possibilities that were not until now possible. Consider approaching these types of process with a view that sometimes compliance isn't binary and that an approach such as this balances efficiency with adaptability. This results in a framework that supports and encourages innovation while upholding standards - this is a hallmark of operationalized maturity.

Considerations for implementation

- **Iterate on nondeterministic**: Design agents with modular logic - e.g., a compliance agent checks connectors first, then usage patterns - tested in a sandbox. Include guardrails, like escalation to human review for ambiguous cases (e.g., "unsure if 'the high number of runs' is intentional"), to prevent false positives. Initial effort is higher (logic crafting, instruction writing, testing), but iterative refinement ensures adaptability scales with governance needs.
- **Use platform data**: Feed agents with structured data, from the admin connectors or the CoE Starter Kit (e.g., flow run logs, app metadata). This minimizes custom data collection - low effort - while maximizing accuracy, as agents operate on near real-time, verifiable inputs.
- **Define clear objectives and scope**: Start by specifying the agent's purpose - e.g., monitoring compliance or managing environment requests - and limit its scope to avoid overreach. For example, focus on specific metrics (e.g., connector usage, RAI principles) rather than broad analysis. This reduces complexity - moderate effort in planning - and ensures focused actionable outcomes that enhance reliability.
- **Monitor and tune performance**: Use Copilot Studio analytics to track agent outcomes and adjust logic during maintenance windows (e.g., every four weeks).

- **Ensure auditability**: Log agent decisions in Dataverse (e.g., "Flow X flagged for review: reason high run rate") with timestamps and rationales. This adds minimal effort – configuring logging – but ensures transparency, a key governance requirement.

These practices ensure autonomous agents deliver adaptive governance[1] – handling nuanced tasks like anomaly detection – while remaining reliable and scalable. This effort is higher than deterministic automation, but the reward is proactive risk management that evolves with the platform.

However, automation alone isn't enough to validate governance success; it generates the data that must be measured and analyzed to demonstrate value and guide continuous improvement.

Embedding RAI into Solution Development

As Power Platform adoption scales, solutions – whether Power Apps, flows, agents, or AI Builder models – touch critical processes (employee scheduling, customer interactions, financial approvals).

At this point, the risks of AI misuse increase substantially: biased AI outputs can skew decisions, opaque models can erode user trust, and weak data handling practices can lead to serious privacy violations.

Risk management frameworks define what *should* happen. Development practices decide what *does* happen. To embed responsible AI in an operationalized Power Platform environment, governance structures must meet developers where they build – inside the life cycle of every app, flow, agent, and AI-enhanced solution.

[1] Gartner Research (2022). Innovation insight for adaptive governance.

The goal here is to embed and align RAI principles into the design, build, test, and deployment phases. This helps developers avoid downstream compliance issues and supports a culture of trust and ethical design.

Integrating RAI into development involves embedding checks, tools, and processes at every stage - design, build, test, and deploy. Here's how an organization might achieve this at scale.

Design Phase

The design process is your first - and best - opportunity to embed responsible AI into a solution. By this point, your organization should already be performing a formal RAI sensitivity evaluation. This phase is not about duplicating that process; it's about enabling designers, makers, and teams to embed ethical awareness and inclusive thinking directly into the earliest decisions they make.

Responsible AI in design isn't something you layer on later - it's a mindset and habit that starts here. To support this, CoEs can define a design guardrail checklist or integrate these ideas into solution briefs and intake templates. You're not asking for heavy documentation - you're setting expectations that good design means designing with responsibility in mind.

Here are practical best practices that can guide design teams at this stage:

Starting with a user lens: Focus on who the AI solution will impact, directly and indirectly. Encourage teams to map out primary users, edge users, and any groups that could be inadvertently excluded or harmed. Encourage empathy-driven design tools like personas and journey mapping to surface potential fairness or inclusiveness concerns early.

Clarify the problem AI is solving: Not every problem requires AI. Ask design teams to articulate why AI is needed and whether automation will improve the outcome for users. Misapplying AI introduces both ethical and technical debt and cost.

Use responsible design prompts: Create frictionless prompts during design reviews or intake forms:

- "How will the user know this content or decision came from AI?"
- "Could the model unintentionally prioritize one group over another?"
- "What happens if the AI makes the wrong call – who gets impacted and how?"

These small questions lead to more thoughtful outcomes.

Assign accountability early: Every AI-powered solution should have a named business or domain owner responsible for oversight, not just a builder. Add a field in your solution intake process to capture this. Ownership needs to begin at the idea stage, not just post-launch.

Provide ethical UI components: Equip makers with standard UI elements like

- A transparency banner ("This content was generated by AI")
- Consent prompts for automated decisions
- A feedback look ("Was this helpful?" tied to training feedback)

These can live in your design system or CoE site.

Schedule an early feedback touchpoint: Before build work begins, create a space to gather feedback from users, peers, or RAI advocates. It doesn't need to be formal.

These best practices aren't heavy governance. They're lightweight, repeatable habits that embed Microsoft's RAI principles – like fairness, transparency, and inclusiveness – into the earliest stages of solution development. As your organization matures, consider integrating these prompts into templates, solution intake processes, or provisioning agents to make responsible design second nature.

Build Phase

This is where good intentions meet technical reality. At this point, design principles start to manifest in solution logic – data is connected, AI models are selected, and decision-making processes are wired up.

The goal in the build phase is to **ensure those decisions remain aligned with responsible AI practices** as the technical implementation takes shape.

To help make this practical, the CoE should provide **templates, reusable components, and tools** that nudge responsible behaviors without slowing teams down.

Best practices for responsible AI development

Enforce component-level transparency: If AI-generated content appears anywhere in an app or chat, make transparency effortless. Provide developers with a prebuilt banner, tool tip, or modal that states: "This content was generated by AI – let us know if it doesn't feel right."

These components can be published in your design system or solution starter templates.

Use RAI-compliant solution templates: Whether using Power Apps, Power Automate, or Copilot Studio – template solutions should include

- Metadata fields for ownership and purpose
- Consent management components
- A prefilled CHAT (Change, Harm, Accountability, Transparency) assessment block
- Placeholder flows for feedback capture and RAI exception handling

Templates should embody your RAI governance framework and help make responsible development the default.

Support safe experimentation with developer environments and DLP: During the build phase, innovation and risk often spike. Developer environments and DLP help contain this by enforcing connector restrictions.

Provide RAI reference logic blocks: Many citizen developers copy from existing flows. Offer reusable logic fragments or building blocks that include

- Audit log entries (e.g., log when a prediction or AI-generated suggestion is shown)
- Escalation logic (e.g., send to human review if an AI confidence score is below threshold)
- Conditional branching based on risk

This empowers developers to build in governance without needing to invent it from scratch.

Require metadata tagging: Enforce a tagging system (stored in Dataverse) to track

- RAI sensitivity level (from the initial evaluation)
- Owner and support contact
- Intended user population and deployment scope

This metadata feeds downstream monitoring and review processes, and it enables automation to scale.

Test Phase

Testing is where governance gets real. This phase validates whether the principles of fairness, privacy, and transparency were preserved - not just in theory but in practice.

The goal isn't just to pass QA – it's to answer the question: "*Does this AI solution behave responsibly under real conditions?*"

Testing for responsible AI means expanding the scope of your usual test plan to include ethical and compliance checks. This can be supported by structured test templates, crowd-sourced test cases, and defined escalation paths when risks are flagged.

Best practices for RAI testing

Run user simulations with edge cases: Include scenarios that test

- Users with limited digital literacy
- Nonstandard data inputs (e.g., slang, typos, multilingual queries)
- Situations where the AI might make biased assumptions

This helps surface risks that would be invisible in standard happy-path tests.

Perform fairness reviews on AI models: If your solution uses custom models (AI Builder or custom ML models):

- Validate training data diversity.
- Check for skewed results based on input characteristics (e.g., gender, age, job title).
- Use tools like Microsoft Fairlearn or similar fairness dashboards to audit impact.

Document the findings in a consistent format – even if no issues are found.

Validate human-in-the-loop handling: Ensure manual override pathways function reliably. For example, if an AI system makes a recommendation of flags an item for review, test:

- Who gets notified?
- Can they intervene promptly?
- Is their decision logged and auditable?

Tag and store RAI test outcomes: Save the results of ethical and RAI testing alongside the solution record (e.g., in Dataverse or a solution governance log). This helps with traceability, audits, and post-deployment monitoring.

Deploy Phase

Deployment is when AI systems meet the real world - and all its complexity. A responsible deployment is about ensuring visibility, ownership, and continuous oversight.

The deployment phase closes the loop from intent to impact and lays the foundation for monitoring, maintenance, and remediation if issues emerge.

Best practices for responsible deployment

Establish ongoing ownership: Every AI-powered solution must have a named individual or team responsible for

- Responding to ethical concerns
- Reviewing model drift or accuracy over time
- Handling exceptions or harm reports

Store the named individual in your asset registry.

Enable monitoring and alerting: Leverage tools like

- Power Platform telemetry
- Copilot Studio analytics
- Azure Application Insights (custom events, etc.)

Set alerts for anomalies: For example, high error rates, spikes in usage, drops in use satisfaction, or other patterns.

Maintain a user feedback loop: Even post-deployment, users should be able to

- Understand that AI is involved.
- Provide feedback on outputs.
- Flag unexpected or inappropriate behavior.

Route this input to the governance team, RAI advocate, and asset owner.

Schedule ethical reviews: Set a cadence for reviewing high-sensitivity solutions. This could be

- Every three to six months for AI-powered flows with high user impact
- Triggered by a threshold of usage, error rate, or negative feedback

Reviews don't need to be heavy – a 15-minute checklist review in the CoE can catch problems early.

Ensure audit readiness: Log decisions, versioning history, approvals, and overrides. Use Dataverse or audit logs to retain these records – in line with your RAI documentation policies. This becomes vital if a solution is ever challenged, investigated, or audited externally.

Post-deployment Oversight and Continuous Improvement

The design, build, test, and deploy phases may feel like the end of the journey, but both in reality and responsible AI, they are only the beginning. Once a Power Platform solution - whether an app, flow, agent goes live, a new set of responsibilities begins. Post-deployment is where theory meets reality.

Post-deployment oversight is about resilience. Well-governed AI systems earn user trust, adapt to changing requirements, and avoid the reputational or regulatory damage that can follow AI missteps. The reward in getting this right is responsible innovation that is both durable and scalable.

Core Responsibilities in Post-deployment Oversight

You governance framework must account for the following key pillars post-deployment:

- Ongoing monitoring
- Automated feedback loops
- Continuous improvement
- Incident management and response
- Life cycle governance reviews

Let's unpack each of these, with frameworks, tools, and implementation ideas.

Comprehensive Telemetry

Monitoring is the foundation of responsible operations. It transforms static governance policies into living, evolving practices. At the operationalized stage, this monitoring must move beyond manual review and into scalable, automated observability.

Key capabilities to implement

Real-time telemetry: Use Power Platform Admin connectors alongside Azure Application Insights to stream data about key performance metrics, like flow run frequency, duration, AI Builder success/failure rates, user engagement, etc.

Custom audit events: Embed events in Power Apps and Copilot Studio logic to track decision points, feature use, and exceptions. For example, log when AI-generated advice is displayed or accepted.

RAI tracking tags: Tag assets with their RAI sensitivity score and include it in logs. This helps you correlate high-sensitivity assets with downstream incidents or feedback.

Example framework:

Component	Framework	Frequency	Purpose
Telemetry collection	Azure Monitor, App Insights	Continuous	Capture behavior, errors, anomalies, and run frequency.
Metadata analysis	Power Platform Admin Connector, CoE Starter Kit	Weekly	Review AI usage and RAI tags.
Escalation trigger	Power Automate + adaptive cards	Event-based	Notify CoE or owners of issues.

Automated Feedback Loops

User feedback is one of the richest sources of risk signals - and one of the most underused. At scale, manual collection isn't practical or enough. Build structured, automated loops to gather and respond to feedback.

Implementation ideas

- Use adaptive cards in Teams or in-app prompts, end of conversation topics asking: "Was this decision helpful?".
- Route feedback to a Dataverse table, tagged by asset ID, developer team, and sensitivity level.
- Trigger Power Automate flows for
    - **Positive feedback**: Optional reinforcement (e.g., "You're doing great" notifications for developers)
    - **Negative feedback**: Route to a governance triage board for review

Why it matters: If fairness or reliability issues surface through feedback before performance metrics detect them, you've bought yourself time - and saved yourself damage.

Continuous Improvement

Compliance, especially RAI compliance, isn't binary. It requires iterative refinement as usage grows, data evolves, and context shifts.

Key processes: Post-deployment RAI health checks

Every AI-powered asset (medium or high RAI sensitivity) should be reviewed at regular intervals - typically every 6 or 12 weeks, depending on risk level.

Checklist items

- Does the asset still meet its original intent?
- Has the input data changed (e.g., new users, new formats)?
- Is the documentation up-to-date?
- Has any feedback suggested bias, confusion, or misalignment?

Tools

- Embed these checks into a governance calendar.
- Use flows to notify asset owners and trigger guided health check forms.
- Build dashboards to track completion and overdue reviews.

These processes reduce drift - the slow, unnoticed degradation of quality and alignment - and allow teams to make proactive enhancements rather than reactive fixes.

Incident Management and Response

Even with all the right processes, things can go wrong. A customer may receive incorrect advice. A flow might leak sensitive data. You need a well-oiled incident response process that reflects responsible AI expectations.

Incident management should become a structured, principle-based discipline that prepares your organization to detect, respond to, and learn from AI-related issues in real time – especially those that affect fairness, privacy, safety, or trust.

Structure of the incident response framework

I propose a modular approach with responses tailored to

1. The violated RAI principle(s)
2. The RAI sensitivity level of the asset
3. The stakeholder impact (internal, external, or systemic)

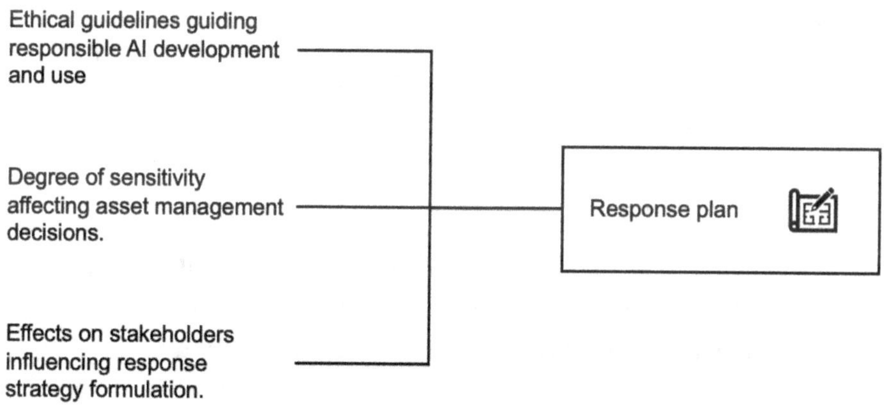

***Figure 7-4.** Converging principles, impact, and sensitivity levels drive the response*

Each response includes four stages:

1. **Detection**: Identifying when and where something has gone wrong
2. **Assessment**: Evaluating the scope and severity of the issue
3. **Response**: Executing immediate actions to contain and remediate the issue

4. **Resolution and learning**: Recording outcomes, auditing, and improving design

Detection

Detection begins with embedding visibility into AI-powered systems. If Power Platform solutions rely on AI Builder, Copilot Studio, or third-party integrations, you need to capture usage and behavior signals that flag anomalies early.

Your goal: Build detection into the asset life cycle – so issues are found by design, not by accident.

How to do it

- **Instrument AI-powered apps and agents** with telemetry hooks using Azure Application Insights, especially for high-sensitivity assets. Define custom events to drive alerting and feedback.
- **Create proactive alerting rules** in your environment. For example, set Power Automate to trigger alerts when a conversational agent exceeds its expected response volume or receives a surge of low-confidence responses.
- **Integrate feedback channels** directly into apps, agents, and interfaces – not just surveys but quick "Was this answer helpful?" toggles. For high-impact solutions, make feedback accessible from any step.
- **Ensure human oversight is active**: In scenarios flagged as medium-to-high-sensitivity (e.g., employee decisions, finance, or legal workflows), assign named reviewers who monitor performance or log escalations weekly.

Operationalizing detection reduces the time between an issue occurring and it being noticed - which is often the most dangerous delay in AI risk exposure.

Assessment

Once an issue is detected, you need a structured process to assess its scope, impact, and root cause - while mapping it back to both the Microsoft RAI principles and your RAI sensitivity score.

Your goal: Identify what went wrong, who or what was affected, and which principles may have been violated - so your response is grounded in real context.

The **RAI sensitivity evaluation × RAI principle response matrix** serves as a guide for identifying appropriate, principle-aligned responses during the assessment phase. It helps governance teams ask the right questions early and consistently across their portfolio:

- If a solution scores high on fairness sensitivity, what should we plan for?
- If transparency risks are moderate, what level of documentation or oversight is appropriate?
- If a system is low-sensitivity but an issue impacts inclusiveness, what response might we make?

The matrix bridges the gap between recognition and readiness. It offers organizations a structured way to think about what kinds of controls, documentation, or escalation paths might be needed before a system goes live - or in this case, when something goes wrong post-deployment.

This is a great proactive design and risk planning asset - a framework that helps organizations develop proportionate, thoughtful responses rooted in the values that matter most.

	Low sensitivity	Medium sensitivity	High sensitivity
Fairness	Log issue for review. No immediate action unless public-facing.	Initiate fairness review. Consider re-training or adjusting model thresholds.	Suspend system or feature. Initiate full audit. Engage legal and ethics leads.
Transparency	Add in-product explanation or disclosure during next update cycle.	Implement clearer explanations or flag for policy review. Inform CoE.	Immediate disclosure required. Provide decision explanations or roll back functionality.
Accountability	Ensure asset owner is registered and informed. No escalation required.	Trigger internal CoE review to clarify roles and responsibilities.	Initiate executive-level review. Assign interim owner and initiate root cause analysis.
Privacy and security	Check data handling settings. Update consent messages or data masking if needed.	Conduct limited impact assessment. Correct issue and notify affected users if personal data was affected.	Trigger data breach protocol. Notify regulators and affected parties as needed.
Inclusiveness	Log user feedback. Consider accessibility improvements in future updates.	Review with diversity/inclusion lens. Initiate accessibility testing or feedback loop.	Suspend deployment. Engage community or experts to address discrimination or exclusion.
Reliability and safety	Monitor over time. Add fallback logic in next revision if instability persists.	Temporarily limit use. Assign technical owner to remediate and test.	Disable affected feature. Run emergency QA and issue communications.

Figure 7-5. *RAI principle response matrix aligned to RAI principles*

Response

If the assessment phase is about understanding risk and planning for responsible action, the response phase is where the preparation pays off. Whether an agent produces a biased result, a custom AI Builder model makes a questionable classification, or a Power App inadvertently exposes private information – your organization's credibility, trustworthiness, and ethical posture will be judged not just by what happened but how you responded.

The good news is that if you invested in structured risk assessment and principled planning, your response would not need to be invented in the heat of the moment. It will already be in motion.

However, the response still needs structure.

How to build a response plan based on RAI principle and sensitivity

Start with the RAI sensitivity × RAI principle matrix from the assessment phase. This tool becomes even more valuable during response.

Here's how:

1. **Use the matrix as a triage map**

   When an incident is detected, use the matrix to

   a. Identify the primary RAI principle affected (e.g., fairness, privacy).

   b. Check the assessed RAI sensitivity level for that solution (low, medium, high).

   c. Look at the predefined response guidance from your assessment planning.

   This gives you a proportionate, pre-aligned response starting point.

2. **Trigger the response package**

   Based on the matrix output, trigger one (or many) of your predefined response pathways.

3. **Include a principle-specific action guide**

   For each principle, your response should include tailored guidance. Examples:

   Fairness: Conduct immediate bias analysis (e.g., using Fairlearn), notify affected groups, and reaudit training data.

   Transparency: Provide plain-language explanation to users, update system prompts, and log missing disclosure failures.

   Privacy: Activate breach protocols, notify DPO, audit access logs, and secure exposed data.

   Accountability: Identify responsible owner, assign corrective actions, and update accountability record in Dataverse.

Reliability and safety: Validate error path, simulate edge conditions, and rollback to safe version.

Inclusiveness: Engage affected groups, adapt UI or outputs, and document accessibility failure.

These are the heart of meaningful, value-driven governance - not extra steps.

Ideally, response should be partly automated. This means

- Detection (e.g., unexpected flow behavior, AI failure patterns, etc.) can trigger alerts.
- An autonomous agent can collect initial context (from users, developers, service desk incidents).
- Predefined response playbooks can be initiated automatically, with human intervention only where judgement is required.

This not only speeds time to action - it ensures consistency, traceability, and RAI principle alignment even when human responders are under pressure.

Resolution and learning

Once an AI-related incident has been triaged and an appropriate response has been carried out, the work isn't finished.

Resolution ensures the root cause is addressed, the system is corrected, and any harm is mitigated - but that's only half the story.

Learning turns the incident into an opportunity. It's where an organization transforms a reactive fix into a governance asset - one that improves future design, strengthens processes, and reinforces trust across the board.

Goals of the resolution and learning phase:

- Ensure the incident is fully resolved from technical, ethical, and process perspectives.
- Capture the incident and response in a way that informs future decisions.
- Update documentation, training, and processes to prevent recurrence.
- Share insights with stakeholders, fostering transparency and organizational learning.

In my opinion, resolution isn't just about correcting the immediate issue – it's about identifying and addressing the conditions that allowed it to happen.

Understanding the components of resolution

To resolve an incident responsibly, we need to address it from multiple angles – not just patch the technology but examine the governance, ethical, and communication aspect too.

Why?

Each area tells us something about where our systems may be fragile:

- Technical fixes solve the surface-level issue.
- Process corrections reveal gaps in how we build and review solutions.
- RAI principle alignment brings ethics back to the center.
- Documentation creates traceability and organizational memory.
- Stakeholder communication restores trust and supports transparency.

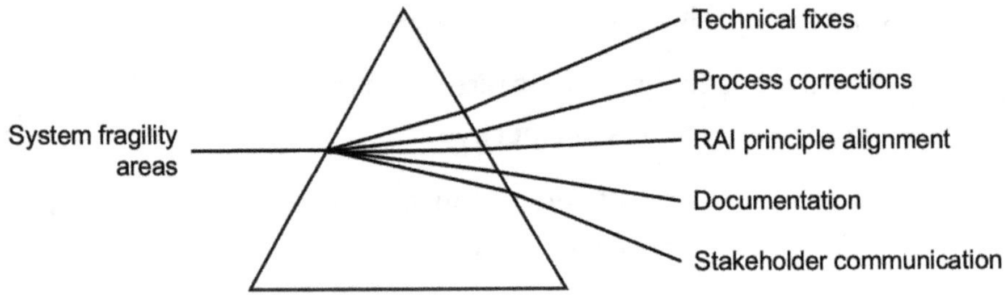

Figure 7-6. *System fragility reviews involve multiple areas*

Together, these actions ensure that the resolution is comprehensive – not a bandage but a realignment of people, processes, and systems. None of this progress can stick however, unless it's shared. A lesson undocumented is quickly forgotten; a lesson untaught is rarely applied.

This is where training becomes essential. Not just as a compliance requirement but as the key mechanism for embedding what we've learned back into our development culture, project practices, and governance expectations.

In the next section, we'll explore how to scale this knowledge across teams, roles, and regions, using tailored, practical, and responsive training approaches that reflect both the organization's maturity and its real-world experience managing RAI risks.

Training and Awareness at Scale

The success of responsible AI isn't sealed in code – it lives in the decisions made by people.

No matter how robust your technical controls are, it's your developers, makers, data handlers, and product owners who decide whether AI is used safely, ethically, and effectively. At this stage, Power Platform is in the hands of hundreds – maybe thousands – of users. And with that scale comes a critical challenge: ensuring every one of them has the right awareness, instincts, and knowledge to build responsibly.

The technical integration of RAI into solution development establishes a robust foundation, but its success depends on the people behind the platform. As Power Platform adoption spans enterprises, empowering citizen developers, professional developers, and end users alike - training and awareness must scale to ensure everyone understands their role in maintaining governance and ethical standards.

My experience with organizations at this level underscores that comprehensive, enterprise-wide education transforms any technical or governance requirement into a cultural norm, including RAI, and this sustains the trust and scalability achieved through development processes.

At this stage, organizations likely have foundational training underway - introducing RAI principles like fairness and transparency to developers or educating end users on basic platform use. However, as adoption deepens and solutions scale across departments or regions, training must evolve beyond initial knowledge to address advanced needs and persistent gaps in maker knowledge. For instance, makers introduced to fairness may still struggle to identify subtle forms of bias. Without deeper training on practical bias identification and mitigation, deployed solutions might risk perpetuating unintended inequities. Unless training expands to cover interpreting performance or rigorously testing predictions, trust might erode and impede widespread adoption. Thus, the focus is now about reinforcing accountability, refining expertise, and embedding a mindset that aligns with enterprise goals.

This matters because of the following:

- **Sustaining RAI**: Technical safeguards falter without a workforce that consistently applies them, risking erosion of trust.
- **Scaling expertise**: Citizen developers, now prolific, need advanced guidance to navigate complex RAI scenarios.

- **Assuring stakeholders**: Robust training signals to leadership, regulators, and users that the organization prioritizes responsible and trustworthy AI, bolstering credibility.

Beyond imparting knowledge, training acts as a visible commitment – an assurance that ethical AI isn't an afterthought but a core value, critical for maintaining enterprise trust as the platform's impact grows.

Identifying and Addressing Gaps

While foundational training may cover RAI basics, gaps often persist at this maturity level. I've seen organizations handle education at scale, and without pinpointing these gaps, training risks becoming a one-size-fits-all exercise, failing to adapt to the diverse needs of an enterprise workforce.

To bridge these gaps effectively, organizations must employ targeted strategies to uncover them. Here are some practical approaches for identifying training gaps:

- **Post-training surveys**: Deploy Microsoft Forms surveys after training sessions to gather immediate feedback. Questions like "How confident are you in identifying bias in AI Builder models?" or "Can you explain how transparency applies to Copilot Studio outputs?" reveal comprehension levels – e.g., 30% of citizen developers rating bias confidence below 3/5 signal a gap.
- **Focus groups**: Convene small, role-specific groups (e.g., five to ten citizen developers, professional developers, or end users) for moderated discussions. Ask open-ended questions, like "How clear are

AI-driven outputs to you?" – uncover nuanced gaps, like developers missing inclusiveness in multilingual agents. Focus groups, designing discussions, and analysis require moderate effort but yield deep, qualitative understanding that surveys might miss.

- **Performance metrics analysis**: Leverage Power BI to correlate training completion with governance metrics (e.g., "Incident frequency," "RAI feedback score"). A spike in incidents post-training, say, AI-powered solutions without accountability, might indicate a need for more precise training, while low feedback scores suggest context may need improving.
- **Interactive quizzes**: Embed quizzes in training modules. Low scores highlight specific weaknesses, like accountability confusion among professional developers.
- **Support ticket review**: Analyze support tickets for recurring issues.

These strategies ensure gaps are systematically and proactively identified, enabling training to target advanced needs – like refining bias expertise or deepening privacy awareness – rather than repeating basics. By leveraging tools like post-training surveys, focus groups, and performance metrics, organizations can pinpoint where understanding falters, ensuring that citizen developers, professional developers, and end users alike are equipped to uphold RAI principles effectively.

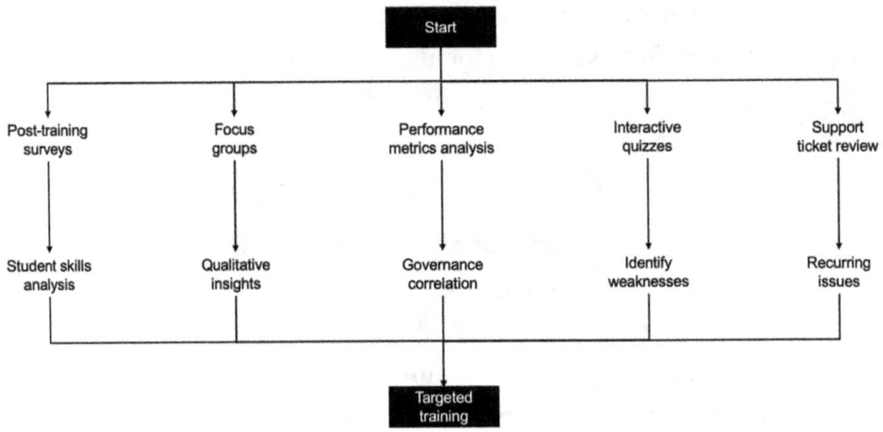

Figure 7-7. *Identifying gaps in training approaches*

Identifying skill gaps is only half the equation. To drive real behavioral change, training must be role-specific, need-specific, and adaptable to individual learning styles. Consider approaches such as

- **Targeted learning modules**: Deliver content based on user persona (e.g., citizen developers vs. pro developer) or based upon gaps identified in skill quizzes.
- **Role-based RAI certifications**: Provide different paths for builders, reviewers, and approvers, each with scenario-driven assessments.
- **Microlearning moments**: Create targeted content, short videos, and Microsoft Teams messages, e.g., "Did you consider user transparency?" when they publish an agent (or other events).

These small adjustments ensure that training becomes a continuous part of the Power Platform development experience, not just a once-a-year exercise.

However, while identifying and addressing these gaps through training builds a knowledgeable workforce, the scale of Power Platform adoption at this stage demands more than manual diligence. To enforce RAI consistently across thousands of solutions, automation becomes essential, and this brings us to the next critical focus: automating RAI compliance.

Automating RAI Compliance

At the operationalized stage, Power Platform is driving enterprise operations with thousands of flows, apps, and agents across global teams. This scale creates a relentless wave of demands: stakeholders expect ethical consistency, compliance teams require rigorous oversight, regulators enforce strict standards, and laws like the EU AI Act impose firm legal boundaries.

These pressures converge, making automation the only proactive way to manage risk and ensure responsible AI compliance across a vast ecosystem. The NIST AI RMF offers a structured approach - automation turns it into action.

The demands driving automation

A complex set of forces are fueling the need for automation; each intensifies the others:

- **Stakeholders**: Executives, employees, and customers demand transparency and fairness in AI outputs. A biased model or an unclear response risks undermining trust - that is all it takes. Automation enforces ethical standards consistently across all solutions.
- **Compliance teams**: Internal governance calls for detailed audits, policy enforcement, and risk management of thousands of Power Platform assets. Manual reviews can't keep pace - automation provides the precision and scalability to meet these expectations.

- **Regulators and legislation**: Laws like GDPR, CCPA, and the EU AI Act mandate privacy, transparency, and bias mitigation. Regulators expect hard evidence of compliance. Automation embeds these requirements into operations, ensuring legal standards are met.

These pressures are not operating in isolation. To meet the demands of this scale, organizations need not only automation but also visibility. This is where metrics come in - not simply to report on usage but to track alignment with RAI principles, identify gaps, and trigger interventions when needed. Below, we explore how organizations can leverage Power Platform metadata to monitor compliance with key RAI principles.

Metrics for Responsible AI (RAI) Compliance

As we've illustrated, measuring responsible AI (RAI) compliance is critical.

Organizations need robust, scalable metrics to ensure adherence to RAI principles without compromising privacy or security. The following metrics leverage out-of-the-box metadata from Power Platform for Admins, Power Apps for Admins, and Power Automate for Admins connectors. Each metric is measurable, practical, and designed to deliver actionable insights for governance teams.

These metrics focus on "indirect indicators" of compliance, using administrative metadata to monitor RAI principles, across Power Platform. They avoid accessing sensitive data, ensuring feasibility and alignment with privacy standards.

Transparency: Tracking AI usage

Transparency ensures users understand where AI influences decisions. These metrics highlight AI deployment to guide oversight efforts.

- **Percentage of apps and flows using AI capabilities**
    - **Description:** Measures the proportion of apps and flows using the AI Builder connector or flows using "generative actions,"[2] pinpointing where transparency checks are needed.
    - **Data source**
        - **Power Apps for Admins**: Use "Get Apps as Admin" to list apps and their connectors.
        - **Power Automate Management**: Use "List flows as Admin (V2)" to list flows and their connectors (this needs to iterate through all environments).
        - Cloud flows using the "Run a generative action" connector[3] (RunAIFlow).
    - **Calculation**: (AI Builder Assets / Total Assets) × 100
    - **Implementation**: Query apps and flows for AI Builder connectors and visualize in Power BI.
    - **Target**: Establish a baseline (e.g., 10%) and monitor growth.
    - **Why it matters**: Identifies AI-heavy areas, ensuring users are informed about automated decisions or output.

[2] https://learn.microsoft.com/en-us/power-automate/generative-actions-overview

[3] https://learn.microsoft.com/en-us/connectors/copilotflow/

- **Number of environments with AI Builder usage**
    - **Description:** Tracks how many environments contain apps or flows with AI Builder, offering a broad view of AI adoption.
    - **Data source:** "Get Apps As Admin" and "List Flows As Admin (V2)" return environment data in their response. Use these actions to cross-reference environment metadata.
    - **Implementation:** Aggregate usage by environment, and display trends in Power BI.
    - **Target:** Flag high-use environments for governance focus.
    - **Why it matters:** Helps prioritize transparency resources where AI is most active.

Fairness: Monitoring AI deployment scope

Fairness ensures equitable AI performance across groups. These metrics track AI distribution to support fairness reviews and highlight areas in the organization where AI usage is high.

- **Number of assets per department using AI capabilities**
    - **Description**: Counts AI Builder apps and flows by department, highlighting areas needing fairness oversight.
    - **Data source**
        - **Power Apps for Admins and Power Automate Management**: Retrieve app/flow metadata, including owner details.

    - Map owners to departments via Microsoft Entra ID.
    - Cloud flows using the "Run a generative action" connector[4] (RunAIFlow).
  - **Implementation**: Group assets by department and visualize in Power BI.
  - **Target**: Monitor for balanced distribution – Is it used widely across the organization?
  - **Why it matters**: Uneven AI usage could indicate training gaps or access inequities, prompting targeted action.

Reliability and safety: Assessing performance

Reliability ensures consistent AI performance. These metrics use run history metadata to monitor stability. My suggestion with this measure is instead of broadly measuring the flow error rate, consider narrowing the metric for failure rates specifically of flows identified as business-critical or high-impact.

- **Error rate of AI Builder flows**
  - **Description**: Measures the failure rate of business critical flows with AI capabilities.
  - **Data source**: Power Automate Management – Use "List Flow Runs as Admin" for run history and errors.
  - **Implementation**: Calculate error rate – (Failed runs / Total runs) × 100.
  - **Target**: <2% error rate.
  - **Why it matters**: High errors undermine trust in AI-powered processes.

[4] https://learn.microsoft.com/en-us/connectors/copilotflow/

Privacy and security

Privacy and security ensure that AI-powered solutions do not expose sensitive information. Given the complexity of identifying broad sharing scopes through nested groups and recursive API calls, I would suggest that this measure targets assets in tier two environments (noncritical, broadly shared) with a high AI sensitivity:

- **Data sharing scope of AI-powered assets**
    - **Description**: Identifies AI-powered assets that are broadly shared (e.g., with large groups), increasing privacy exposure.
    - **Data source**
        - **For Power Apps**: Use the "Get App as Admin" action.
        - **For Power Automate**: Use "ListFlowOwners" and "ListRunOnlyUsers."
    - **Implementation**: Filter AI-enabled assets with high-RAI sensitivity, and categorize by sharing level. To identify how broadly an App has been shared:
        - Use the "Get App as Admin" action:
            - **SharedUsersCount**: Number of individual users explicitly granted access
            - **SharedGroupsCount**: Number of Entra security groups with access
        - If the SharedGroupsCount is greater than 0, additional steps are required:

        - Use the Get App Role Assignments as Admin action to retrieve role assignments.
        - Iterate through these role assignments to identify Entra security groups
        - For each identified group, use the Get group member count action to determine membership size.
    - Total user access count is then the sum of
        - Individual users (SharedUsersCount)
        - Members within each identified security group
- **Target**: <10% of AI-powered assets shared broadly without review.
- **Why it matters**: Broad sharing increases risk of privacy breaches or unauthorized access to AI outputs.

Accountability: Assigning ownership and oversight

Accountability ensures that every AI-powered solution has a clear owner responsible for its behavior, outputs, and life cycle.

- **Named owner for every AI-powered solution**
    - **Description**: Ensure each AI-powered asset has a named, active owner (and ideally, a backup owner) documented.
    - **Data source**: Asset inventory/compliance catalog.
    - **Implementation**: Query metadata to assets using AI; verify owner fields and activity status.

- **Target**: 100% of AI assets with an active, named owner and backup.
- **Why it matters**: Ensures someone is accountable for maintenance, bias issues, or ethical concerns - critical for audit and escalation.

- **RAI review timestamp tracking**
    - **Description**: Monitor the last time an AI-powered solution underwent an RAI sensitivity evaluation.
    - **Data source**: Custom Dataverse table or CoE Starter Kit extension with review logs.
    - **Implementation**: Tag AI assets with RAI review metadata (e.g., review date, risk tier, outcome).
    - **Target:** All high- and medium-risk assets reviewed within the last 6–12 months.
    - **Why it matters**: Keeps oversight current and ensures that accountability isn't just a one-time check box.

Conclusion of Operationalized Stage

These metrics offer a practical approach to tracking responsible AI compliance using metadata. They help ensure that governance evolves in lockstep with the rapid adoption of Power Platform - and crucially, they reveal opportunities for refinement. For example, usage data might surface a need to adjust policy thresholds, while noncompliance trends can highlight where targeted training or updated guidelines are required. This data-driven approach transforms governance from a reactive process into a measurable discipline - one that enables organizational growth while preserving trust and ethical accountability.

As we conclude our exploration of the operationalized stage, it's clear that governance has matured into a scalable, proactive system. With the help of automation, structured metrics, and embedded RAI practices, organizations have shifted governance from a collection of ad hoc rules into an integrated capability - consistent, responsive, and aligned to risk.

Technologies like Power Automate, Managed Platform, and autonomous agents can play an essential roles in enforcing standards across thousands of apps, flows, and AI-driven solutions. Meanwhile, governance metrics provide the critical visibility needed to evaluate adoption, improve compliance, and reduce risk - without slowing innovation.

Yet, for all its progress, governance at this level still operates somewhat in parallel to broader strategy. It's structured and strategic - but not yet intrinsic. To reach the pinnacle of maturity, governance must take one final leap: to become embedded into the organization's culture, its values, and its decision-making DNA.

This is one of the goals of the next and final stage: integrated excellence - where governance helps shape the business.

Operationalized Stage Progression Checklist

As governance moves from scattered experimentation intro structured, scalable operations, clarity becomes your strongest ally. The operationalized stage progression checklist serves as a mirror - a tool to help you assess how far you've come and where there's still work to do.

Governance structure and strategy

- [ ] A formalized CoE structure is in place, with defined roles across IT, legal, security, and business units.
- [ ] Governance policies are codified and version-controlled (e.g., RAI policy, asset lifecycle management, review cadence).
- [ ] Governance model (centralized, decentralized, hybrid) has been selected and adapted to regional/ global needs.
- [ ] A governance blueprint exists to define mandatory and optional practices across all environments.
- [ ] Governance guidance is shared via a centralized, accessible knowledge base or portal.

Metrics and reporting

- [ ] Governance KPIs are defined, aligned to enterprise goals (e.g., approval turnaround time, compliance rates, RAI coverage).
- [ ] Dashboards are used to track adoption, compliance, and risk across environments and departments.
- [ ] RAI metrics are monitored and reported at the environment and department level (e.g., fairness, reliability, transparency).
- [ ] Data is used to inform governance refinements (e.g., adjusting thresholds or retraining needs).

Responsible AI integration

- [ ] RAI sensitivity evaluation is embedded as a required early step for all AI-powered assets.
- [ ] Mitigation strategies are defined for each Microsoft RAI principle and mapped to sensitivity tiers.
- [ ] RAI is embedded in development workflows (design, build, test, deploy) with checklists or automated gate reviews.
- [ ] RAI sensitivity evaluations are consistently documented for all medium and high sensitivity assets.

Governance automation

- [ ] Core governance workflows are automated using Power Automate (e.g., approvals, environment provisioning, DLP enforcement).
- [ ] Autonomous agents are deployed to support governance in non-deterministic scenarios.
- [ ] Exception handling and escalation logic is built into automated processes to allow human oversight.

Incident management and continuous improvement

- [ ] A documented incident response plan exists for AI-related governance issues, categorized by RAI principle and sensitivity level.
- [ ] Resolution and learning loops are in place, with cross-functional reviews of critical incidents.
- [ ] Post-incident processes feed back into training, risk assessment processes, and mitigation strategy updates.
- [ ] Continuous improvement practices include scheduled governance retrospectives or maturity reviews.

Training and awareness

- [ ] RAI training is mandatory for all developers (citizen and pro), solution designers, and CoE staff.
- [ ] Training covers all RAI principles and common risks related to Power Platform features (e.g., AI Builder, Copilot Studio).
- [ ] RAI comprehension is assessed through quizzes, surveys, or practical exercises.
- [ ] Training gaps are identified using performance metrics, focus groups, and feedback loops.

Scaling responsibly

- [ ] Governance scales across departments, regions, and environments without compromising oversight.
- [ ] Responsible AI compliance is embedded into workflows at scale, not handled manually.
- [ ] The CoE acts as a strategic influencer in enterprise-wide AI governance (e.g., cross-platform alignment, certification, regulation mapping).

If you've checked most of these boxes, congratulations – your governance efforts are no longer reactive mechanisms; they're proactive, scalable systems that support trust and growth.

CHAPTER 8

Integrated Excellence: Establishing Trust and Governance

Pioneering | Sustainable | Influential

The integrated excellence stage represents the zenith of governance maturity. Here, governance transcends traditional oversight and becomes seamlessly woven into the fabric of the organization. It is no longer a standalone process but a core component of how the organization operates, anticipates risks, and drives responsible innovation. At this stage, governance is not purely about maintaining control – it is also about enabling the organization to innovate rapidly while upholding the highest standards of trust, compliance, and ethical responsibility.

What sets integrated excellence apart is its holistic integration:

- **Cultural integration**: Governance becomes a shared responsibility across all levels, from leadership to citizen developers, not just the Center of Excellence (CoE).

S. Jeffery, *The Low-Code AI Maturity Model*, https://doi.org/10.1007/979-8-8688-1730-4_8

- **Proactive mastery**: Risks are anticipated and addressed before they arise, using advanced tools and foresight.
- **Strategic alignment**: Governance directly supports and enhances enterprise objectives like innovation, trust, and compliance.
- **Continuous evolution**: Policies and practices adapt dynamically based on real-time insights and feedback.

This stage is not without its challenges. Achieving integrated excellence requires a cultural shift, executive buy-in, and the adoption of cutting-edge tools. However, the rewards are transformative: an organization that can innovate at speed while maintaining unwavering trust, compliance, and ethical integrity.

What to expect in this section

In the pages that follow, we will explore

- The key characteristics that define integrated excellence and set it apart from earlier stages
- Advanced governance practices, including AI-driven governance, predictive analytics, and cross-platform integration
- Implementation strategies to embed governance into organizational culture and strategy
- Challenges and solutions to help you navigate the path to integrated excellence

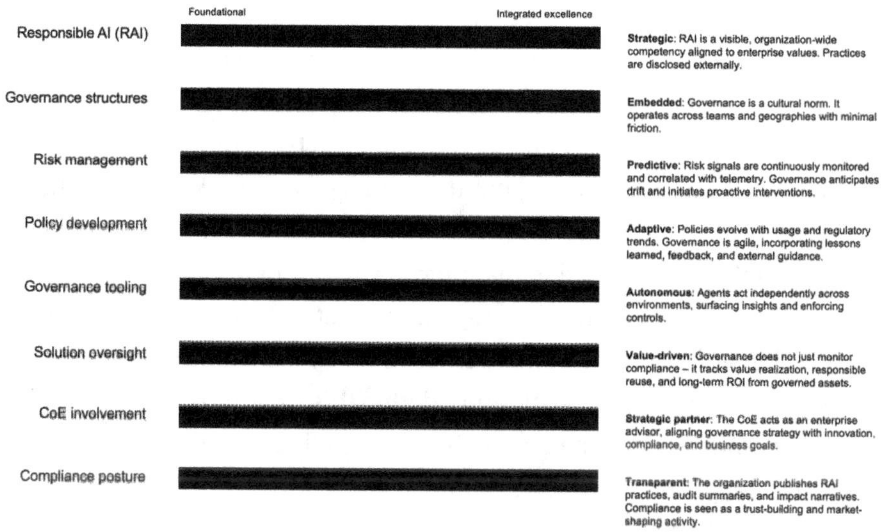

Figure 8-1. *Run state of an organization at the integrated excellence stage*

By the end of this section, you will have a clear road map to achieving governance mastery - a state where governance is a strategic asset that drives your organization's success. Let's begin the journey.

Key Characteristics of Integrated Excellence

Congratulations - you've reached integrated excellence. In my opinion, the pinnacle of governance maturity in the low-code AI maturity model. After investing significant effort to build, scale, and refine your governance practices, you're now at a stage where governance is a strategic asset. This is the ultimate in governance capabilities, where it's organized, bought-in, integrated, and efficient, empowering your organization to lead with agility, ethical compliance, and adaptability to new AI capabilities and legislation.

Where earlier stages focused on experimentation (foundational), process alignment (emerging), and scalable enforcement (operationalized), this stage marks the moment governance becomes fully embedded - no longer an external function but a deeply internalized capability that guides behavior, culture, and decision-making across the organization.

This section is designed to inspire you while providing practical, achievable, and value-driven guidance. Integrated excellence is a realistic goal you can attain with the right mindset, tools, team, and strategies. Below are the key characteristics that define this stage, grounded in real-world relevance and focused on delivering tangible benefits like trust, efficiency, and a competitive advantage.

Cultural Integration

Governance is no longer solely the domain of a Center of Excellence. It must be a shared responsibility that is embraced across your organization. When we talk about cultural integration within the context of the integrated excellence stage, we're describing a transformative state where governance isn't just a policy or a process - it becomes a living, breathing part of an organization's identity.

Let's explore this in depth by painting a picture of the key traits that define organizations with culturally integrated governance and then walking through the practical, detailed steps they take to get there.

Reflection: Is governance viewed positively across your organization, or is resistance still apparent?

Leadership as living examples of governance

In these organizations, leaders embody governance. Executives and managers do not see governance as a check box to tick off but as a strategic enabler that drives trust, innovation, and resilience. They're the first to articulate why governance matters, tying it to real-world outcomes like delivering ethical AI solutions and staying ahead of regulations. During company-wide meetings, you might hear a CEO share a story about how teams deployed a customer-facing AI-powered tool faster because compliance was baked in from the start. It's beyond lip service, it's a consistent narrative that governance is a competitive edge, and it's driven via every level of leadership and in some cases even a KPI.

But culture isn't just about the tone from the top – it's measurable. In organizations at this stage, we begin to see governance-related behaviors reflected in metrics like policy adherence rates, self-reported confidence in RAI practices, and reductions in incident volume. These signals confirm that governance isn't being enforced – it's being embraced.

To support cultural integration, many organizations at this level establish a governance ambassador model. This program identifies individuals from different business units, departments, or regions who act as local champions of governance and responsible AI.

These ambassadors are not full-time governance staff; they are trusted contributors embedded in their teams – product owners, data analysts, or business leads – who receive additional training and are empowered to

- Communicate updates to governance policies in accessible, role-relevant terms.
- Gather feedback on pain points or successes with governance tools and processes.
- Act as the first point of contact for responsible AI concerns or queries.
- Participate in governance councils or reviews to ensure representation from all corners of the business.

This model helps embed governance into the operational rhythm of the business. It also reduces resistance by making governance a peer-to-peer experience rather than something handed down by a distant central function.

Proactive Risk Management

Organizations focus heavily on proactively identifying and addressing risks before they materialize. Governance teams leverage advanced tools and actively monitor environments to foresee risks rather than simply responding to incidents. Advanced organizations like these implement comprehensive risk registers and utilize scenario planning to evaluate potential impact to platform availability. Regular governance audits further strengthen this proactive approach, enabling swift and decisive action when needed.

Some guidance to achieve this

- **Develop and maintain a comprehensive risk register** to clearly define and prioritize potential governance issues and assign clear ownership and accountability.
- **Employ automated monitoring and alerting systems** (such as Power BI dashboards integrated with audit logs and activity tracking) to detect and report unusual patterns or breaches proactively.
- Implement **regular scenario planning and risk assessment workshops**, engaging cross-functional teams to discuss potential risks, identify vulnerabilities in processes, and prepare mitigation strategies.
- **Establish defined protocols for rapid escalation and resolution**, ensuring teams are well-equipped and trained to respond swiftly to emerging threats.

Reflection: How predictive are your current risk management processes? Could more proactive measures reduce your exposure to unforeseen challenges?

Continuous Evolution

Governance frameworks are dynamic, designed to adapt as quickly as technology, markets, and legislation change. Policies are reviewed regularly, updated, and improved using insights and feedback from across the organization. Continuous evolution is driven by a structured feedback mechanism collecting feedback from across the organization. Actionable steps to ensure continuous evolution include the following:

- **Set up regular review cycles** for governance policies (e.g., quarterly or biannually), incorporating insights from operational data, user feedback, and external regulatory changes.
- **Leverage stakeholder feedback loops** to continuously gather and integrate perspectives from users, governance teams, and senior leadership, creating a responsive and adaptive framework.
- **Utilize benchmarking and industry best practices** to regularly update governance frameworks, ensuring your policies remain current and aligned with external standards and emerging trends.
- **Invest in knowledge-sharing and training programs** to ensure teams remain updated on best practices and evolving governance standards, enhancing organizational agility and responsiveness.

Reflection: Are your governance processes flexible enough to quickly adjust as new AI capabilities emerge?

Why scenario planning matters

Proactive evolution of your governance is critical, yet preparation needs to extend beyond incremental updates. You now recognize that anticipating future challenges and opportunities is just as essential as reacting swiftly to immediate issues.

Scenario planning workshops empower your organization not only to react quickly but also to proactively anticipate, plan for, and even influence future outcomes. They provide a structured approach to identifying potential risks and opportunities ahead of time, rather than dealing with crises reactively. By engaging in scenario planning, your organization builds resilience, minimizes disruptions, and positions itself strategically for ongoing success.

Let's look at organizing a scenario planning workshop.

Scenario Planning Workshops

Scenario planning workshops are powerful tools that help you anticipate, understand, and prepare for future uncertainties. Instead of reacting to disruptions as they arise, scenario planning encourages proactive consideration of multiple potential futures – both opportunities and challenges. However, conducting an effective scenario planning workshop requires thoughtful preparation and clear direction.

In this section, you'll learn how to carefully select relevant scenarios, set meaningful objectives aligned with your strategic goals, and organize a productive workshop that translates insights into actionable strategies. We'll explore how to identify and prioritize scenarios based on their potential impact and likelihood. By the end, you'll be equipped to plan and facilitate a workshop that both clarifies your organization's preparedness but empowers your teams to navigate uncertainty and confidently navigate uncertainty confidently and effectively.

Planning a Scenario Planning Workshop

The effectiveness of your scenario planning workshop depends heavily on selecting relevant, realistic, and impactful scenarios. Rather than guessing or choosing arbitrarily, scenario selection must be intentional, structured, and closely tied to your strategic goals.

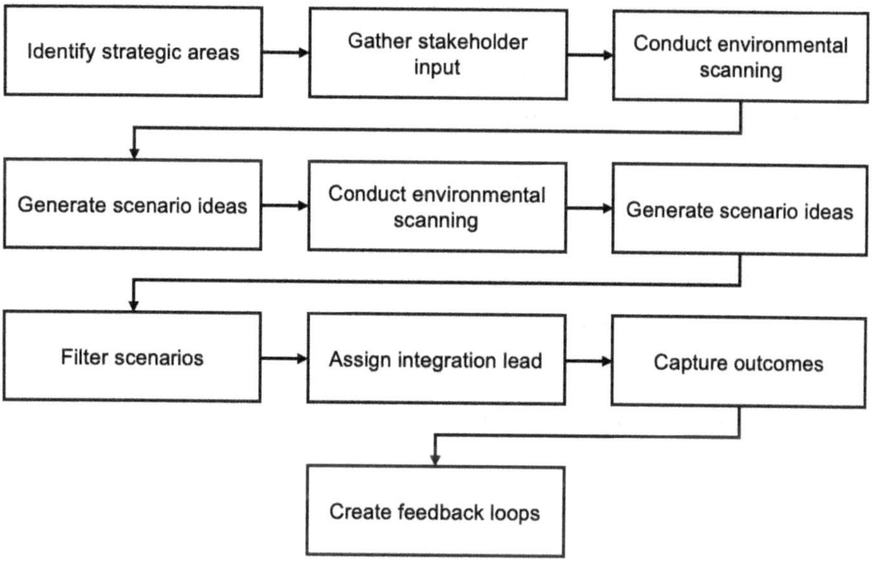

Figure 8-2. *Stages of a scenario planning workshop*

Here's how to approach scenario selection step-by-step:

Step 1: Identify Your Strategic Areas of Focus

Start by clarifying the broad strategic area you aim to explore. Scenarios might be based around a specific challenge or opportunity - such as responsible AI governance, customer data protections, or scaling Power Platform adoption.

Example:

> *"We want to understand scenarios that could impact trust in our AI-driven customer service applications."*

Step 2: Gather Input from Key Stakeholders

Engage stakeholders who have insight into potential risks and opportunities. This typically includes business unit leaders, governance and compliance teams, IT and security representatives, and senior executives. Consider posing questions like

- What future scenarios are you most concerned about?
- Where might our governance or RAI practices face challenges or disruptions?
- Are there upcoming regulatory changes or industry shifts we should anticipate?

Step 3: Conduct Environmental Scanning

Perform an external and internal environmental scan to uncover potential scenarios. This might include

- Industry research and reports
- Regulatory landscape shifts
- Competitor analysis
- Technological advancements or disruptions
- Internal incident reports, audits, or known vulnerabilities

Example:

"An internal audit revealed frequent use of undocumented data sources, posing significant data privacy risks. Additionally, an external trend indicates heightened regulatory scrutiny over AI-driven decisions."

Step 4: Generate Scenario Ideas

Based on your strategic focus, stakeholder inputs, and environmental scanning, draft multiple potential scenarios. Consider scenarios across a spectrum - from best-case to worse-case, ensuring they are plausible and varied.

Examples:

- **Best-case**: *"AI tools greatly enhance productivity and customer satisfaction, but governance and compliance mechanisms keep pace seamlessly."*
- **Worst-case**: *"A regulatory audit uncovers noncompliant AI tools causing reputational damage and financial penalties."*
- **Realistic compromise scenario**: *"AI adoption accelerates, exposing moderate compliance gaps that require quick mitigation."*

Step 5: Filter Scenarios by Relevance and Feasibility

Evaluate each scenario for its relevance to your strategic objectives and practical feasibility to manage within the scope of a workshop:

- Relevance
  - Does this scenario directly affect our critical strategic priorities?
  - Would proactive preparation for this scenario significantly benefit our organization?
- Feasibility
  - Can we realistically address and explore this scenario within a workshop format?
  - Do we have adequate information or data available to deeply understand and analyze this scenario?

A successful workshop is just the beginning – its real value lies in how effectively the insights are absorbed into your governance program. Many organizations struggle with this transition. To ensure scenario planning leads to measurable change, consider the following post-workshop integration approach:

1. **Assign an integration lead**: Designate a governance team member to track the implementation of all actions, maintain accountability, and coordinate follow-ups across stakeholders.
2. **Capture outcomes in structured documentation**: Use a shared space – like a governance portal to store

   The final set of prioritized scenarios

   Action owners, timeframes, and status updates
3. **Create lightweight feedback loops**: Set a recurring check-in (e.g., monthly) with stakeholders involved in the workshop to track progress, surface blockers, and update plans if new risks or shifts emerge.
4. **Convert key scenarios into testing templates**: Use the most valuable or likely scenarios as the basis for resilience testing or inject them into future incident response simulations. This helps train teams and refine protocols.
5. **Incorporate learnings into policy evolution**: Where a scenario reveals a gap in your current governance or RAI policy, document it and route it into your policy review cycle (e.g., quarterly review board), so insights translate into updated standards.

Step 6: Prioritize Scenarios Based on Impact and Likelihood

Using your shortlist, rate each scenario on two dimensions – potential impact and likelihood of occurrence. Aim for scenarios with a combination of high impact and sufficient probability to warrant preparation.

Example prioritization matrix:

Scenario	Impact	Likelihood	Priority
Regulatory audit exposes AI compliance gaps	High	Medium	High
Minor security breach due to unsecured/ undocumented data source	Medium	High	Medium

Step 7: Finalize Your Scenario Selection

Choose scenarios that offer a mix of immediate actionable insights and long-term strategic considerations. Typically, workshops handle two to four scenarios effectively, ensuring deep exploration without diluting focus.

Running the Scenario Planning Workshop: Step-by-Step

Once you have selected your scenario and prepared participants, the workshop itself moves through several key steps designed to explore each scenario comprehensively, assess its implications, and develop concrete actions to manage identified risks and opportunities.

Here's a practical outline you can follow for the scenario planning workshop:

Step 1: Workshop Opening and Context Setting (~30 Minutes)

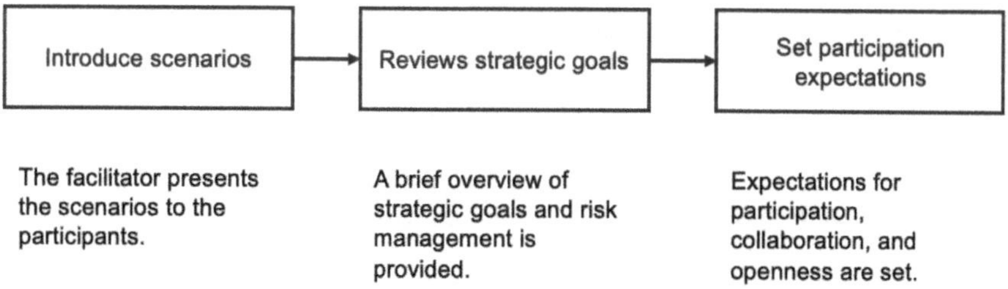

Figure 8-3. *Workshop flow (opening and context setting)*

- **Purpose:** Ensure all participants clearly understand the workshop's purpose, scope, and expected outcomes.
- **Activities**
    - Facilitator introduces the scenarios.
    - Briefly review strategic goals and the importance of proactive risk management.
    - Set expectations for participation, collaboration, and openness.
- **Outcomes:** Participants clearly understand the workshop's purpose, scenarios to be explored, and expectations for their involvement.

Step 2: Scenario Exploration (Breakout Groups – 60/90 Minutes)

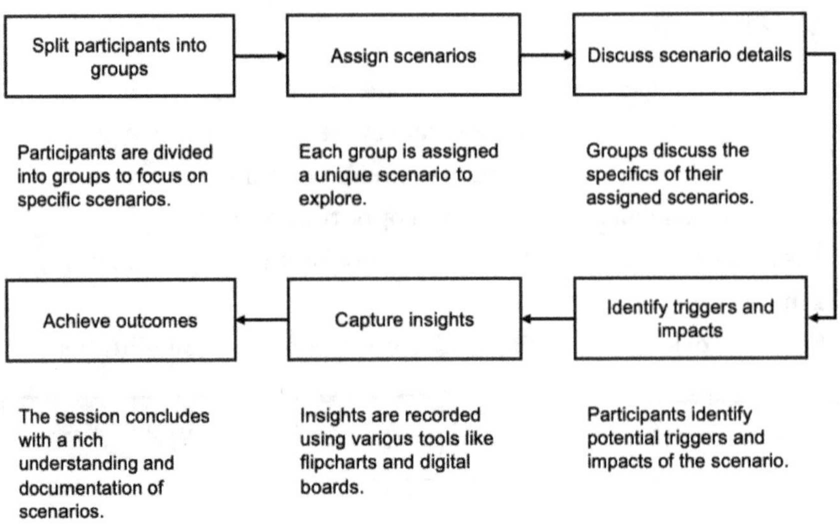

***Figure 8-4.** Scenario exploitation*

- **Purpose**: Explore scenarios in depth, understanding their implications for the organization.
- **Activities**
    - Split participants into groups, each group assigned to explore a specific scenario.
    - Participants discuss scenario details and provide prompts that drive critical questions:
        - What triggers might indicate this scenario is occurring?
        - How would this scenario impact operations, customers, employees, compliance, and reputation?
        - What are the early signs or indicators of this scenario emerging?
    - Capture insights using flip charts, sticky notes, or digital whiteboards.
- **Outcomes**: Rich understanding and documented insights about the scenario, highlighting potential impacts and early indicators.

Step 3: Scenario Impact Assessment and Prioritization (45–60 Minutes)

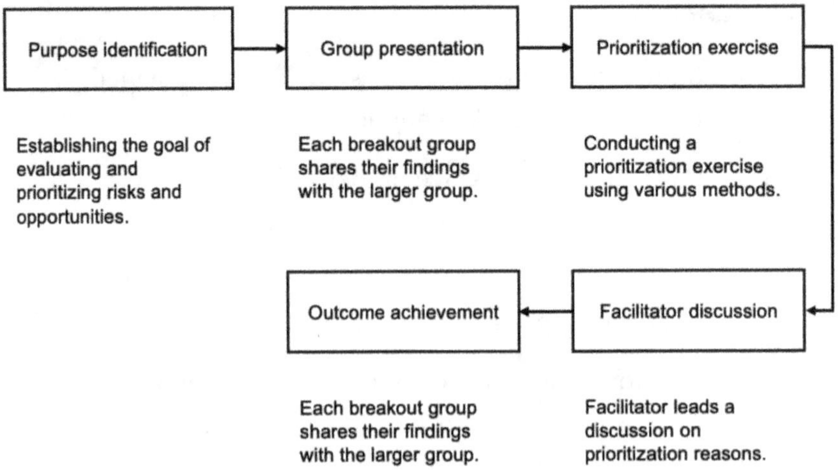

Figure 8-5. *Scenario impact and prioritization*

- **Purpose**: Evaluate and prioritize the identified risks and opportunities from each scenario.
- **Activities**
  - Each breakout group presents their findings briefly to the entire group.
  - Conduct a prioritization exercise (using voting, impact/likelihood matrices, or scoring methods) to highlight the most significant threats and opportunities.
  - Facilitator guides a discussion on why specific risks or opportunities have emerged as priorities.

- **Outcome**: Clear prioritization of risks and opportunities, establishing shared understanding and agreement on what the organization should address first.

Step 4: Developing Mitigation and Action Plans (60–90 Minutes)

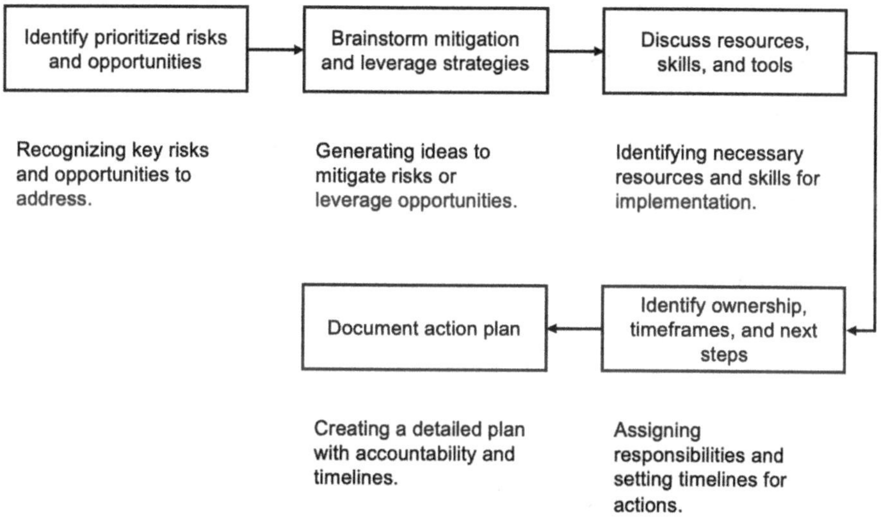

***Figure 8-6.** Developing mitigation and action plans*

- **Purpose**: Turn prioritized risks and opportunities into actionable strategies.
- **Activities**
  - Participants work in groups (either the original groups or reshuffled) to brainstorm and document actionable mitigation strategies for prioritized risks or leverage strategies for opportunities.
  - Discuss the resources, skills, and tools needed for implementation.

- Identify clear ownership, timeframes, and next steps for each action.

- **Outcome**: Clearly documented action plans for each prioritized risk/opportunity, complete with accountability, timelines, and initial resource requirements.

Step 5: Reporting Back and Validation (30–45 Minutes)

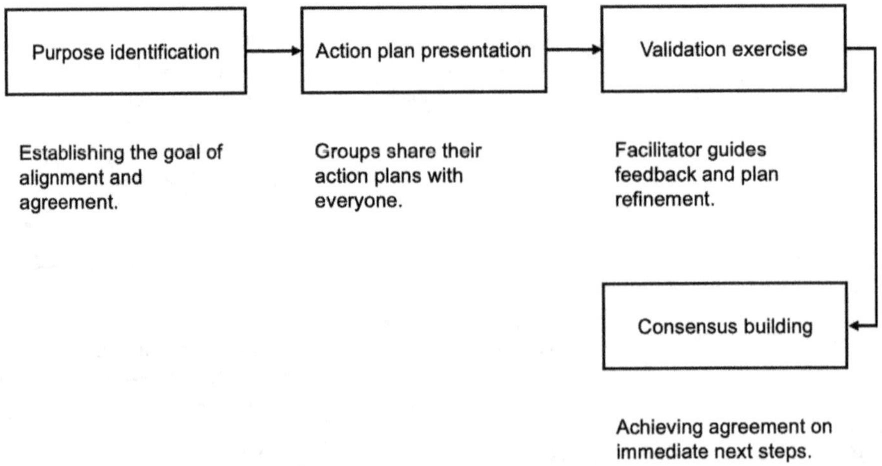

***Figure 8-7.** Reporting and validation planning*

- **Purpose**: Ensure alignment and agreement across the broader group about the identified actions and priorities.
- **Activities**
    - Each group presents their action plans back to the broader group.
    - Facilitator leads a validations exercise, allowing participants to offer feedback, refine plans, and build broader consensus on immediate next steps.

- **Outcomes**: Agreed-upon set of validated, refined action plans ready for post-workshop implementation.

Scenario planning prepares your organization by identifying risks and defining proactive mitigation strategies - but what happens when a risk becomes reality? Proactive preparation is crucial, but equally important is your organization's ability to react quickly, effectively, and decisively when incidents occur.

Incident response planning is your organization's practical playbook, detailing step-by-step actions, roles, and responsibilities that spring into action when a scenario materializes into an actual incident. Without a structured incident response plan, even minor incidents can escalate quickly, leading to reputational damage, compliance failures, or financial losses.

In the next section, we'll move from proactive planning into the immediate, responsive realm of incident management. You'll learn how to create clear actionable incident response plans, assign appropriate roles and responsibilities, and ensure your organization is ready to handle incidents with confidence, minimizing disruption and maintaining trust.

Incident Response Planning

Incident response plans represent your organization's readiness to respond swiftly and effectively when incidents occur. This is an area that is well covered by several international standards, frameworks, and other best practices that could apply depending on the scenario. Look at some of the more well-known standards and frameworks:

Standard/ framework	Purpose	Applicability	Key strengths
ISO/IEC 27035 (Information Security Incident Management)	Manage and respond to information security incidents	Security breaches, unauthorized data access, and system compromise	Clear, structured, internationally recognized methodology for incident life cycle
ISO 10002 (Customer Satisfaction and Complaints Handling)	Manage customer dissatisfaction and complaints effectively	Customer-facing incidents (e.g., inaccurate chatbot responses, poor user experiences)	Structured approach focused on customer satisfaction, root-cause analysis, and continuous improvement
ISO 223001 (Business Continuity Management)	Ensure continuity and minimize disruption following incidents	Incidents threatening operational continuity (e.g., major service outages, reputational crises)	Comprehensive continuity planning, business impact analysis, and recovery strategies
ISO 31000 (Risk Management)	Provide guidelines for managing all forms of risk strategically	Broad range of risks (operational, reputational, ethical)	Flexible framework for risk identification, assessment, and response at strategic level
ITIL V4 Incident Management	Best practices for IT service incident management and rapid service recovery	Service disruptions, performance incidents, and system errors affecting business or customer service	Widely adopted, practical guidance for rapid incident detection, categorization, communication, and resolution

How to blend standards and frameworks by scenario

While each of these standards provides robust guidance independently, combining specific elements from multiple frameworks creates a highly tailored and effective incident response. Here's an example:

Scenario example: Imagine it's a Tuesday morning, 9:00 AM. Your organization's customer support chatbot has unknowingly provided inaccurate product recommendations to several customers overnight. Complaints are starting to appear on social media, and customer satisfactions scores are dipping rapidly.

Here's how you might orchestrate an integrated response, using elements from each of the standards and frameworks previously outlined:

Time	Activity	Standard/framework
09:05 AM	Incidents detected and logged: A spike in customer complaints triggers immediate notifications. Service desk logs the incident in a central system and flags it as critical.	ITIL 4 incident management (incident identification, categorization, prioritization)
09:15 AM	Initial assessment and escalation: The incident manager assesses the severity, escalating to senior management and communications teams. The chatbot response is temporarily disabled.	ISO/IEC 27035 (incident assessment, immediate response actions)
09:30 AM	Customer engagement and apology: Customer service reaches out proactively to impacted customers, acknowledging the issue and providing accurate product advice. Public relations issue a clear, transparent apology on social media.	ISO 10002 (customer dissatisfaction management, complaint handling, clear communication)

(*continued*)

Time	Activity	Standard/framework
10:00 AM	Business impact assessment: Operational leaders assess the broader impact – brand reputation, regulatory risks, customer loyalty – and implement contingency measures to ensure minimal operational disruption.	ISO 22301 (business continuity assessment, recovery strategies, minimizing operational disruption)
11:30 AM	Root cause investigation: A cross-functional team (IT, governance, customer service) performs a root-cause analysis. They discover the chatbot was trained with outdated product information due to insufficient review process.	ISO 10002 and ISO/IEC 27035 (root-cause analysis, lessons learned, structured investigation)
02:00 PM	Risk analysis and strategic review: The team assesses long-term risks including regulatory compliance, trust erosion, and future incidents. Leadership decides to tighten AI governance policies, focusing on content review and data freshness.	ISO 31000 (strategic risk management, risk assessment, long-term mitigation planning)
04:00 PM	Corrective action and restoration: The chatbot is updated and retested thoroughly. Policies are amended to include mandatory accuracy checks. Service is gradually restored and carefully monitored.	ISO/IEC 27035 and ITL 4 (incident resolution, corrective action, service restoration)

(*continued*)

Time	Activity	Standard/framework
Next day 10:00 AM	Incident review and continuous improvement: Leadership reviews incident outcomes, approves improved procedures, and communicates lessons learned across the organization, embedding these into training and operational practices.	ISO/IEC 27035 and ISO 10002 (post-incident learning, continuous improvement, embedding improvements into culture and processes)

By structuring your incident response like this - drawing from multiple, established standards - you create clarity, accountability, and agility. Consider whether your current processes provide similar clarity and responsiveness. If gaps exist, reflect on how adapting elements from these frameworks could enhance your organization's resilience and responsiveness in future incidents.

In my experience, organizations at this stage shift from reactive governance toward predictive, intelligent governance models. They are no longer just responding; they attempt to anticipate incidents, prevent them, and continuously evolve their approach using advanced and data-driven methods.

Advanced Governance Policies

Advanced governance policies empower organizations to move beyond foundational compliance and reactive risk management. This next section explores how AI-driven governance, cross-platform integration, and automated policy evolution can reshape your governance landscape, positioning your organization both as a responsible adopter of AI and a leader in its strategic and ethical implementation.

Let's dive deeper into these advanced practices.

AI-Driven Governance

Leveraging AI itself to manage elements of governance allows organizations to scale effectively. Organizations look to agents to enforce policy intelligently. These specialized agents continuously monitor environments, detect policy deviations, and autonomously enforce governance actions, reducing manual oversight and ensuring compliance around the clock.

Example:

Consider a scenario where an app owner wishes to share a new Power App with their department. Instead of manually submitting detailed documentation and waiting days or weeks for a governance review, the app/flow/agent owner interacts directly with an autonomous governance agent.

The interaction might look like this:

1. **Information capture**

   The autonomous agent prompts the owner through Teams chat to gather essential details:

   a. **Data types**: Does the app/flow/agent handle personal, sensitive, or regulated data?

   b. **Connectors**: Which connectors external services will the app/flow/agent use?

   c. **User scope**: How many users or departments will access this app?

   d. **Impact assessment**: Would downtime or inaccuracies significantly disrupt business operations or customer interactions?

2. **Automated risk evaluation**

   The agent analyzes the provided information and dynamically evaluates risk factors:

   a. Identifies potentially sensitive data combinations

   b. Assesses connector-related risks based on predefined classifications

   c. Calculates potential business impact based on scope and criticality

3. **Automated next steps**

   The agent (if configured to do so) could autonomously

   a. Provision the correct environment.

   b. Automatically enforce recommended DLP policies.

   c. Notify relevant compliance or security teams for periodic reviews.

By leveraging autonomous agents for automated risk assessment, organizations gain immediate clarity and significantly accelerate governance decisions for app, flow, and agent owners. This intelligent automation substantially reduces workload for governance teams, allowing them to focus on strategic tasks rather than administrative oversight. Most importantly, automated assessments proactively enhance compliance and minimize risk exposure by consistently applying accurate governance measures. The result is a governance framework that empowers teams to innovate confidently, knowing oversight is both intelligent and always active.

For governance agents to operate effectively and responsibly, their design must be intentional and transparent. At this stage of maturity, organizations no longer create agents on a case-by-case basis - they build reusable, governed agent patterns that can be applied across departments and geographies.

Here are key design considerations for responsible governance agents:

1. **Purpose and scope must be clearly defined**: Each agent should have a focused charter - for example, “review app metadata for high-risk dimensions” or “monitor RAI compliance metrics weekly.” Avoid all-in-one agents that become brittle and opaque.
2. **Guardrails must be built-in**: Agents operating on sensitive actions - such as approving environment requests or initiating workflows - should include escalation logic (e.g., “if decision score = low confidence, escalate to CoE team”). Embed limits like
    a. Approval thresholds (e.g., no auto-approval if >50 users affected)
    b. Action logging (so decisions are traceable)
3. **Instruction tuning is critical**: These agents must be tuned using clearly defined instruction sets and prompts. Train them using real-world scenarios, including edge cases, and store decision criteria in a transparent repository (Dataverse).
4. **Oversight must be continuous**: Even intelligent governance agents require human oversight. Monitor decisions weekly, audit their logs monthly, and retrain them when false positives or unexpected gaps emerge.

5. **Make them explainable**: Governance agents should be able to justify their actions. Build in the ability to return rationale, like: "This flow exceeded the defined usage limits; therefore, I escalated the approval."

Reflection

- Are your governance mechanisms proactive enough to autonomously handle compliance risks as they emerge?
- How could autonomous agents help you embed responsible AI principles more deeply within your organization's processes?

Cross-Platform Integration

Governance, especially AI automated, cannot live in isolation - it must span diverse technologies and platforms seamlessly. Power Platform is often one of many platforms enabling AI-powered innovation within an organization. Cross-platform integration ensures consistent governance, compliance, and responsible AI practices across all these ecosystems.

Effective cross-platform integration involves

- **Unified governance frameworks**

  Developing cohesive policies and standards applicable across various tools (e.g., Power Platform, Azure AI Foundry, custom-developed AI solutions) to prevent fragmentation of compliance and ethical oversight

- **Centralized visibility and control**

  Implementing dashboards or governance consoles that aggregate data, insights, and compliance status from multiple platforms, providing administrators with holistic oversight

- **Seamless security and data compliance**

  Aligning policies on data access, storage, and usage across platforms – ensuring that sensitive information maintains consistent privacy and security protections, irrespective of the technology used

- **Cross-platform workflows and automations**

  Leveraging integrations that allow workflows (such as compliance checks, data classifications, or ethical risk assessment) initiated in one platform to automatically trigger governance processes or controls in another

- **Shared risk management and incident response**

  Establishing standardized incident response procedures that apply uniformly across platforms, ensuring coordinated and rapid action during incidents or compliance breaches

As AI capabilities scale across Power Platform, Azure, third-party APIs, and proprietary systems, regulatory compliance must scale with them. Mature organizations at the integrated excellence stage ensure that legal interpretations and governance policies don't vary wildly by platform or technology.

Otherwise, a Copilot in Power Platform may be subject to one set of rules, while a similar tool built on Azure ML follows another – creating compliance gaps and inconsistencies.

To address this:

- **Centralize legal interpretation**: Create a cross-functional working group - legal, IT, and data protection officers - that defines how global standards (like the EU AI Act) apply across platforms.
- **Publish platform-neutral RAI guidelines**: These serve as a unified playbook for AI projects, regardless of the technology used. For example, a fairness checklist used for both Copilot Studio and Azure AI Services.
- **Audit and reconcile policy gaps**: Use a quarterly cycle to review governance implementations across platforms. Where policy enforcement differs, document why, and identify risks or remediation plans.
- **Enable cross-system alerts**: When governance violations occur in one platform, surface them in shared dashboards or incident management tools - this ensures issues don't remain siloed.

Coordination is critical

Integration isn't just ethical - its organizational. As AI governance extends across teams and tools, coordination becomes the linchpin. Mature organizations establish

- Interoperability standards (e.g., tagging AI systems with metadata across environments for shared dashboards)
- Unified governance terminology, ensuring that terms like "risk tier" and "RAI sensitivity level" mean the same thing everywhere
- Cross-domain escalation paths, where incidents discovered in one domain (say Azure) can trigger investigation or updates in another (e.g., Power Platform)

Reflection

- Is your current governance model consistent across platforms, or are there gaps and blind spots?
- How could unified, cross-platform governance streamline compliance and enhance organizational agility?

An integrated approach strengthens overall governance, ensuring AI-driven innovation remains secure, ethical, and consistently compliant across your entire technological landscape.

Automated Policy Evolution

Organizations use intelligent systems to continuously monitor feedback from end users, scan for regulatory changes, and suggest policy adjustments proactively.

Rather than manually reviewing every piece of feedback or every legislative update, these organizations employ advanced agents that sift through vast amounts of data – internal surveys, internal forum comments, compliance reports, and regulatory bulletins. These agents categorize feedback, flag urgent issues such as updates to regulations, and generate clear, actionable recommendations for policy adjustments.

Governance teams and policy committees become overseers rather than manual processors, reviewing concise insights, approving policy changes, and focusing on strategic oversight. The result is significantly reduced administrative overhead, rapid policy responsiveness, and enhanced compliance across the organization.

Structuring Feedback Loops for Policy Evolution

To achieve continuous improvement, integrated excellence organizations design structured feedback loops that transform internal signals into actionable policy updates. This means going beyond sporadic user comments or audit logs and building a system that routinely gathers, classifies, and escalates insights.

Here's how such a system might be constructed:

- **Input capture**: Collect feedback from multiple sources:
    - End-user forms (e.g., "Was this AI-generated result useful?")
    - Post-incident reviews and audit outcomes
    - Changes in regulatory policy surfaced by legal teams
- **Categorization logic**: Use Power Automate or Azure Logic Apps to
    - Tag feedback by RAI principle (e.g., privacy, fairness).
    - Score sentiment or urgency (low, medium, high).
    - Flag recurring themes.
- **Policy adjustment proposals**: An autonomous agent or designated policy lead prepares a short proposal outlining
    - The policy area impacted
    - Suggested changes
    - Risk implications if left unaddressed

- **Approval and implementation**
    - Use a digital record to log every policy update and rationale, with a timestamp.
    - Use versioning and governance portals to notify relevant stakeholders of the changes.

Conclusion: Establishing Trust and Governance

Reaching the integrated excellence stage is a significant achievement - not only have you operationalized governance, but you've also embedded it deeply into the culture, processes, and strategy of your organization. Governance has evolved into a dynamic enabler of responsible, agile innovation. Advanced policies powered by autonomous agents not only minimize risk but also actively support and accelerate business goals.

However, achieving this level of maturity isn't the end of the journey. Integrated excellence demands continuous evolution, proactive management, and a commitment to staying ahead of technological, regulatory, and market changes. It requires an ongoing investment in cultural transformation, AI-driven tooling, and robust feedback loops that sustain agility and responsiveness.

As you reflect on your organization's journey through each maturity level - from foundational to integrated excellence - recognize that governance, trust, and responsible AI are powerful drivers of sustained success. Embrace the lessons learned, remain adaptable, and continue to embed ethical decision-making into every aspect of innovation. With this foundation, your organization will not only navigate future challenges but also lead with integrity, confidence, and genuine excellence.

This concludes this part of the book where we went deep into governance. The next part of the book, while closely related to governance, needs to stand on its own. In the next part we'll dive deeply into **data readiness and security**, exploring strategies to safeguard your most critical assets, maintain compliance, and position your organization to leverage data as a strategic advantage.

Integrated Excellence: Progression Checklist

The checklist that follows distills the core achievements and behaviors that define Integrated Excellence. It's both a reflection point and a validation tool – something your governance team, leadership, and even your broader community can use to check alignment and momentum. These indicators are signals of readiness for what comes next.

Progression checklist

- ☐ Governance is embedded into organizational culture, with leadership actively championing the Platform and responsible AI principles.
- ☐ Governance is consistently applied across platforms, with centralized visibility and cross-platform risk monitoring.
- ☐ Risk management is predictive and proactive, supported by automated tools and scenario planning workshops.
- ☐ Policies are dynamically adjusted through data-driven insights, reducing manual overhead and increasing agility.
- ☐ Governance policies are continuously updated in response to internal feedback, external changes, and performance insights.
- ☐ Training is role-specific, up-to-date, and continuously refined based on feedback, adoption metrics and observed behavior.
- ☐ Incident response plans are integrated, cross-functional, and draw from recognized standards such as ISO/IEC 27035 or ITIL 4.
- ☐ Governance dashboards offer real-time insights into compliance, exceptions, trends and RAI adherence across all environments.
- ☐ AI-powered autonomous agents support real-time governance enforcement (e.g., assessing risk, triggering policy responses).
- ☐ The CoE has become a strategic advisor and ethical enabler across all AI-powered platforms and initiatives.

PART III

Data Readiness and Security

Throughout our exploration of Power Platform, AI, and responsible AI, we've emphasized the strategic importance of leadership, culture, governance, and trust. But when we get down to the essentials, AI innovation rests squarely on the quality and security of your data. No matter how advanced your AI capabilities are, their effectiveness, reliability, and ethical integrity will ultimately reflect the readiness and security of your underlying data.

Organizations today manage vast oceans of structured and unstructured data – ranging from neatly organized databases, spreadsheets, and operational systems to more complex documents, emails, images, and audio files. Each type of data presents distinct challenges and unique opportunities for unlocking AI's potential. To achieve genuine data readiness, your organization must ensure data accuracy, availability, and relevance – and equally important: commit to responsible management, ethical use, and robust protection.

Yet, data readiness is only part of the story. Security and compliance complexities significantly increase the stakes. Organizations often operate across multiple regions, each with unique regulatory frameworks, privacy standards, and compliance mandates. These geographic and regulatory factors, coupled with the accelerating pace of technological change, make effective data governance and security more challenging than ever.

In this final and arguably most critical part of our journey, we'll establish a clear, practical foundation for data readiness and security. We'll delve deeply into how organizations can

- **Establish basic visibility and ownership of their data**: Moving beyond assumptions and guesswork and building a clear understanding of what data exists, where it lives, and who is responsible for it
- **Strengthen data quality and readiness**: Embedding simple, repeatable practices that improve accuracy, consistency, and fitness for purpose while building early habits of responsible data use across teams
- **Operationalize governance and security**: Automating controls, embedding risk management into development workflows, and treating data as a living asset that requires ongoing stewardship and oversight
- **Elevate data to a strategic enabler of trusted AI**: Unifying governance across platforms, using predictive tools to manage risk, and embedding fairness, transparency, and accountability into the very fabric of your AI and data life cycle

To further ground your understanding, we'll introduce concrete examples, actionable checklists, and strategies drawn from industry-proven frameworks.

By the end of this part, you will have the practical tools, clear guidance, and proven approaches to confidently position your data as a strategic advantage.

CHAPTER 9

Foundational: Data Readiness and Security

Reactive | Fragmented | Experimental

In the world of AI, the importance of data quality and readiness cannot be overstated. AI systems are trained on and respond based on data, so the old saying "garbage in, garbage out" still applies – poor data will inevitably produce poor model outcomes. High-quality, well-prepared data is the bedrock of any successful AI project. Whether you're a business user configuring a Power Platform AI tool or a data scientist building a custom model, data is the fuel that powers AI.

A model trained on reliable data can reveal powerful insights, but the same model trained on inaccurate or biased data may lead to faulty or even harmful results. In short, **getting your data right is the first, foundational step toward AI success**. This chapter explores how to achieve data readiness – from understanding your data landscape and ensuring data quality, to addressing fairness, transparency, accountability,

S. Jeffery, *The Low-Code AI Maturity Model*, https://doi.org/10.1007/979-8-8688-1730-4_9

and security. By laying this foundation, you set your AI initiatives up for success before a single model is used. Typically, an organization operating at the foundational level displays the following characteristics:

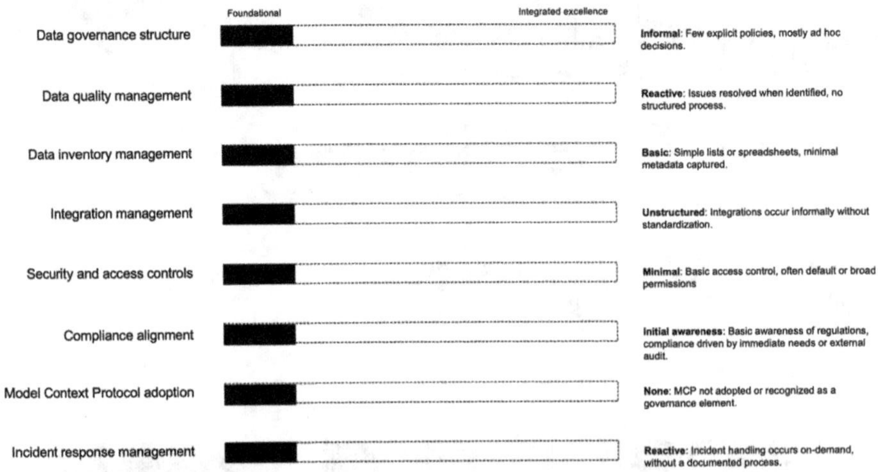

Figure 9-1. *Operational run state of an organization at the foundational level*

Responsible AI Begins with Data

As you know by now, I am a firm believer that responsible AI isn't something you layer on top of AI systems once they're built - it starts here, with the decisions you make about data collection, quality, governance, and fairness.

As your organization matures, these foundational efforts will evolve into more advanced risk management and compliance strategies - but their roots begin here.

Here's how the guidance in this chapter aligns with responsible AI expectations:

- **Fairness** begins with data quality. Accurate, complete, representative data reduces the risk of bias before a model is ever trained. Techniques like bias analysis, demographic balancing, and fairness-aware tooling (such as Fairlearn) help enforce this principle early.
- **Transparency** starts with visibility. A clear data inventory, documentation of data usage, and early adoption of model cards ensure you can trace how data flows into your AI solutions - and explain those decisions later.
- **Accountability** is enabled by clarity. Defining data ownership, stewardship, and governance roles from the outset means someone is always responsible for maintaining integrity and managing risk.
- **Privacy and security** must be designed in from the start. Even in early AI experimentation, applying data sensitivity labels, enforcing access controls, and minimizing unnecessary data collection builds good habits that scale later.
- **Reliability and safety** depend on disciplined preparation. Validating structured and unstructured data for consistency, timeliness, and readiness ensures your AI systems perform as expected - not just technically but ethically.
- **Inclusiveness** is supported through reusability and discoverability. A well-maintained data inventory enables diverse teams to access high-quality, trusted datasets - reducing silos and supporting equitable AI development across the organization.

Understanding Your Data Landscape

Every organization's data landscape includes a variety of data, generally falling into two broad categories: structured and unstructured data. Understanding these categories and how to manage them is crucial for preparing data for AI.

- **Structured data** is highly organized and formatted so that it's easily searchable and analyzable. This is the kind of data you find in tables or spreadsheets – for example, customer records with defined fields (name, age, purchase history) or an HR database with employee IDs, titles, and salaries.

  Structured data has an **established schema** (rows and columns, or defined data types) that makes it straightforward to query. For instance, an HR agent might draw on a structured database of employees, allowing it to quickly retrieve an employee's department or tenure. The clear organization of structured data simplifies tasks like filtering, sorting, and joining datasets, which are operations often needed in AI model training.

- **Unstructured data** lacks a predefined schema or consistent format. It includes free-form text, images, audio, video, and other formats that aren't neatly organized in tables. Examples abound in every business: support ticket descriptions, chat transcripts, customer product reviews, emails, or social media posts. For a customer support app, the conversation logs and customer feedback are unstructured text; for an HR virtual assistant, resumes or interview transcripts are unstructured.

This data is rich in information but harder to process because it doesn't fit into a simple row-column structure. You can't query an image or a paragraph of text as easily as a database table. Thus, leveraging unstructured data for AI often requires techniques like natural language processing for text or computer vision for images to extract meaningful results.

Most real-world AI solutions in Power Platform will blend both structured and unstructured data. For example: an HR customer service agent might use structured data (like employee records, payroll, etc.) alongside unstructured data (like HR policy documents).

Understanding what data you have and its type is the first step in preparing it for AI. It allows you to choose the right tools and techniques: you might use SQL Server or Dataverse for structured data and cognitive services or AI Builder text analysis for unstructured data. In all cases, a clear view of your data landscape - knowing where your data resides, in what format, its quality, and how it can be accessed - sets the stage for effective data preparation.

Why a Deeper Inventory Matters for AI

AI solutions don't just consume data - they're shaped by it. This means your ability to trace how data flows into models, agents, or applications isn't just a compliance check box - it's the basis for trust, explainability, and risk mitigation.

Beyond the basics, your inventory should begin to answer key questions like

- What level of RAI sensitivity is associated with each dataset?
- Which assets are feeding AI models directly (training data) vs. acting as knowledge sources?

- What access controls and protections are currently in place?
- Has the data been validated for quality and fairness?
- When was it last reviewed and by whom?

This level of visibility allows you to make informed decisions about

- Which data sources are safe to reuse across AI scenarios
- Where gaps in coverage might affect model fairness or completeness
- Which datasets require additional oversight or privacy controls

Mapping What Matters

Creating an inventory is only valuable if it tells you something about your assets.

A list of data sources without context is just another spreadsheet. What turns an inventory into a governance tool is the metadata – the descriptive information about each asset that guides how it can be used, shared, or protected.

This doesn't need to be overly technical or complex. The goal is to establish a minimum standard for what gets captured about every dataset, API, connector, or source being registered.

Think of this like a "governance starter pack" – providing enough clarity for responsible reuse, ownership, and risk awareness without overburdening teams.

Example: Metadata fields for a basic data inventory

Metadata field	Purpose	Example value	Governance benefit
Asset name	Human-friendly identifier	CustomerOrderData	Easy to search and reference
Description	Summary of the asset contents	Order details by customer ID	Clarifies scope and reference
Data owner	Person or team accountable	DataOps	Enables stewardship and escalation
Intended use	Approved use case(s)	Reporting, AI model training	Prevents misuse in sensitive scenarios
Sensitivity level	Data classification	Internal only/regulatory sensitive	Guides access control decisions
Last verified	Most recent validation date	February 1, 2025	Signals data freshness and quality
Integration method	How this asset is accessed	Custom connector, Dataverse[1]	Helps assess integration risk
Reuse count	Optional metrics for value tracking	Seven solutions connected	Highlights valuable, high-impact assets
Type	Structured or unstructured data, AI model	Prediction model	Signals the type of data asset

[1] Dataverse may store data natively or expose it virtually from external systems.

Where to Source Metadata for Your Inventory

A list of data sources without context is just another spreadsheet. What turns an inventory into a governance tool is the metadata – the descriptive information about each asset that guides how it can be used, shared, or protected.

You've already seen an example metadata table (included above). Now let's explore where that metadata can come from in practice.

Metadata field	Where to get it from
Asset name	Populated based on asset type (model, structured data source, unstructured knowledge source)
Description	Provided by data asset owner
Data owner	Confirmation from asset creator
Intended use	Provided by data asset owner
Sensitivity level	Established in RAI sensitivity evaluation
Last verified	Provided by data asset owner
Integration method	Based on asset type
Reuse count	Automated count field
Type	Populated during detection

As you can see in the table above, some of the metadata you need already exists at the environment level (asset name, created date, created by, type). You just need to know where to look and how to extract it consistently.

Here's a breakdown of the key places you'll find usable metadata across Power Platform environments – especially for AI-related solutions.

Key Metadata Sources in Power Platform

1. **AI model table (Dataverse)**

   This table provides information on AI Builder models created in an environment.

   Look for

   a. Name (this is the name of the model specified by the creator)

   b. Template (e.g., TextTranslation, DocumentScanning), this specifies the model "template" used. Note, this also lists custom GPT prompts and "bring your own model" for models imported from Azure Machine Learning

   c. Created on

   d. Created by

   This is important for identifying any custom models that might be interacting with sensitive or regulated data.

2. **Copilot components table (Dataverse)**

   This table tracks components used in Microsoft Copilot Studio, including

   a. Knowledge sources (e.g., SharePoint site links, public URLs)

   b. **Bot file attachment**: Uploaded documents used in retrieval-augmented generation (RAG) scenarios

c. **ParentBotID**: Where the component is used

d. Created on

e. Created by

These system tables give you the "what" – the foundational metadata elements baked into every AI model, copilot component, or data-powered solution in your environment. But capturing what already exists is only part of the journey. To turn these fragments into a usable inventory, you now need to operationalize the "how."

How do you take this raw metadata and enrich it with accountability, ownership, purpose, and risk context? How do you ensure that the assets you discover aren't just catalogued – but validated, trusted, and ready for responsible reuse?

In the next section, we'll walk through exactly how to do that: step-by-step, with adaptive cards, review flows, and simple governance loops that scale.

How to Source Metadata for Your Inventory

Capturing metadata is really about establishing a process – one that gradually enriches each data asset with clarity, accountability, and trustworthiness. At this stage of maturity, we are aiming to make the process systematic, repeatable, and governance-aligned.

Let's walk through that process.

Step 1: Identify the Data Asset and Its Creator

Your initial metadata collection often comes directly from Power Platform environments.

For each environment, you can retrieve

- **Copilot agent assets**

  Use the Copilot Component table to identify assets like

  - KnowledgeSource (e.g., URLs or SharePoint sites)
  - botFileAttachment (e.g., documents)

- **AI models**

  Stored in the AI models table with fields for

  - **Template**: For example, textTranslation, ImageDescription

- **Power Apps with data sources**

  Each app exposes which AI components it uses in a collection called dataSources collection (returned from the Get App As Admin action). This can include prompts, tables, or AI components that suggest AI interaction or sensitive data use.

  This metadata lets you determine

  - What type of asset is used
  - Where it's used
  - Who created it

But creation ≠ ownership. So now comes the most important human-in-the-loop step.

Step 2: Confirming Data Asset Ownership

(RAI principle: Accountability)

Automatically assign the creator of the data asset as tentative owners – then verify ownership with an adaptive card. The concept is simple:

"Are you responsible for this asset?"

If yes – record them as the data owner.

If no – ask them to assign ownership to someone else.

Once confirmed, this person becomes the **primary steward for metadata enrichment** and future reuse decisions.

Step 3: Capture Additional Metadata

(RAI principle: Transparency, Privacy and Security, and Reliability and Safety)

Next, use a second adaptive card – or guided form – to capture important descriptive metadata.

This might include

- **Intended use**

  For example, "Used for product feedback analysis in monthly dashboards"

- **Sensitivity level**

  For example, "Internal only, confidential, regulatory, or RAI-sensitive"

- **Integration method**

  For example, "Used in Copilot Studio as a knowledge source"

- **Last verified**

  For example, Automatically timestamped or manually entered

- **Data format/type**

  Structured, unstructured, custom model, prompt, etc.

All of this gets added to the inventory - enriching the asset with context that supports governance automation, reuse logic, and RAI compliance.

If the data asset is a **custom AI model** - for example, a model created in AI Builder or a custom prompt-based solution - your inventory process might include a **lightweight model card** step.

A model card[2] is a structured way to describe the model's intent, inputs, risks, and evaluation history. It promotes transparency and traceability - and aligns directly with RAI principles such as **Transparency, Fairness**, and **Accountability**.

You don't need it to be overly complex. The key is to capture intent and risks early, so that stakeholders understand what the model is for and what its limitations are. You can collect this metadata using an **adaptive card** or form triggered during inventory registration - just like you would for sensitivity or ownership.

Example: Model card metadata fields

Field	Description
Model name	Human-readable name (e.g., "Invoice classifier v2")
Model purpose	What the model is intended to do
Type/template	AI Builder template or GPT prompt type
Known limitations	Bias risks, sensitivity flags, known blind spots
Last evaluation date	When it was last assessed or tested
Performance metric	Basic success metric (e.g., accuracy, precision) – doesn't need to be lab-level
Responsible party	Data steward or model owner for accountability

Even this minimal model card adds value. It means that when someone reuses or integrates the model in a copilot or app, they're working with transparent expectations - not assumptions.

[2] `https://arxiv.org/pdf/1810.03993`

Step 4: (If Applicable) RAI Sensitivity Evaluation

(RAI principles: Fairness, Transparency, and Reliability and Safety)

Not all data assets are equal. If the asset is to be used in

- Training an AI model
- Powering a Copilot
- Exposed in a public-facing application

it should undergo a **RAI sensitivity evaluation**.

This evaluation (covered in detail earlier) asks

- Does this data contain personally identifiable information (PII)?
- Are fairness risks known or suspected?
- If it is used in decision-making, scoring, or predictive logic?

The result:

- Flag the asset if it's high-risk.
- Route it into further governance workflows.
- Apply restrictions on reuse or external integration.

The goal isn't to block innovation – it's to ensure visibility, traceability, and safety.

Step 5: Perform a Data Quality Review

(RAI principle: Reliability and Safety)

Even a perfectly governed asset loses value if the data is outdated, inconsistent, or incomplete.

You don't need an enterprise-grade data profiling engine yet, but you do need a repeatable method to check

- **Accuracy**: Are values plausible?
- **Completeness**: Are all key fields populated?

- **Consistency**: Do values align across datasets?
- **Timeliness**: When was this last updated?

You might automate basic checks or use another adaptive card for the data steward to validate quality manually.

Tip Add a "quality verified" flag to your inventory – along with the reviewer's name and date.

Step 6: Data Readiness Complete - Now Ready for Reuse

When ownership is confirmed, metadata is complete, RAI risks are assessed, and quality is reviewed - the asset is now inventory-grade.

This means:

- It can be discovered via search/agent.
- It can be reused confidently by other teams.
- It can be included in compliance reporting.
- It can be programmatically assessed for fairness or risk.

Most importantly: this process transforms inventory from a catalog into a value enabler - surfacing high-quality, trusted data that fuels safe and responsible AI.

Designing Data for Reuse and Responsibility

Most organizations treat data inventories as administrative overhead - a one-time mapping exercise to tick a compliance box or support a specific project. This mindset is limiting.

In reality, the data inventory has the potential to become one of the most valuable assets in your AI strategy. Why? Because discoverability and reusability sit at the heart of scalable, responsible AI. If every project starts from scratch – hunting for data sources, duplicating effort, or guessing about data quality – you're not just wasting time; you're increasing risk.

Mature organizations treat their data inventory like an internal product – something that is continuously maintained, enriched, and made accessible to **accelerate future solutions**. This shift unlocks benefits far beyond operational efficiency:

- It promotes trust by increasing transparency around data use.
- It reduces risk by centralizing ownership and quality standards.
- And most importantly, it creates a culture of reuse – where knowledge flows freely, not in silos.

Let's look at some practical ways to move your inventory from a static record into a dynamic, reusable capability – one that support every AI project that follows.

- **Connect it to AI development workflows**: Use the inventory as a reference during solution planning. Do this by adding an agent to the inventory. If someone proposes a new agent or model, the inventory helps identify which datasets exist, whether they meet quality standards, and whether the data owner has approved reuse.
- **Tag data assets by AI usage type**: Distinguish between datasets used for training, prediction, conversational search, or human-in-the-loop workflows. This enables more precise oversight and risk analysis.

- **Include RAI-specific metadata**: For each asset, include fields such as
    - Fairness considerations (e.g., Does it contain sensitive attributes?)
    - Sensitivity classification (internal, confidential, regulatory-sensitive)
    - Prior bias reviews or mitigation actions taken
    - Data lineage (where it originated, how it's transformed)

Use case spotlight: Inventory as AI risk radar

Imagine you have this well-maintained inventory of datasets used across Power Platform solutions. Now, picture taking it a step further – connecting a conversational agent, powered by Microsoft Copilot Studio, directly to that inventory.

This transforms the inventory into a strategic gateway to discovery, governance, and responsible reuse.

Scenario: HR wants to build a new conversational agent

The HR team wants to build an agent that helps employees check their bonus eligibility and vacation entitlement. Before they start collecting new data or reaching out to IT, they query the "Data advisor" agent:

> "Do we already have any datasets with bonus eligibility information?"
>
> "Who owns the employee leave records?"
>
> "Have these datasets been reviewed for use with AI?"
>
> "Was bias ever assessed on the salary data we used last year?"

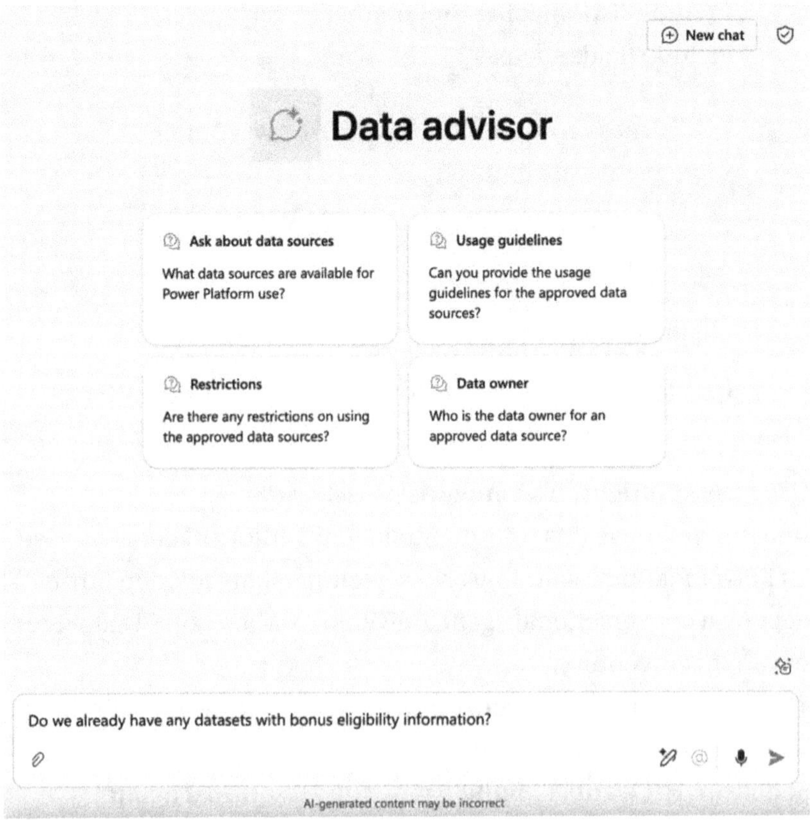

Figure 9-2. *Data advisor agent querying data inventory*

The agent, drawing from the inventory, returns immediate and context-rich answers:

- "Yes, employee leave records exist and are owned by the HR Data Steward."
- "The data is structured, confidential, and was used in a previous HR chatbot."
- "Bonus eligibility fields are partially complete in the existing HR compensation table."

Benefits of a connected Copilot

This one interaction triggers a cascade of value:

1. **Promotes responsible reuse**

    a. The team reuses an existing, vetted data asset instead of duplicating effort.

    b. They build upon previous risk assessments rather than starting from scratch.

    c. The agent can leverage data already formatted and reviewed for compatibility.

2. **Reduces AI risk**

    a. The agent reveals that the dataset was marked with "High RAI sensitivity," flagging the need for fairness review before use.

    b. It informs HR that sensitive data (e.g., salary, tenure) requires tighter access controls in any new solution.

3. **Accelerates development**

    a. Developers and teams no longer rely on email chains; Teams chats or hallway conversations to locate data.

    b. Developers get instant insight into schema, owners, and usage policies - cutting days or weeks off their timelines.

4. **Improves collaboration and accountability**

    a. Data owners are clearly identified and automatically notified when their data is being requested for reuse.

   b. This ensures human oversight remains in the loop, supporting governance policies established earlier in the maturity journey.

5. **Supports data discoverability across use cases**

   a. A single dataset might be used in a virtual assistant, a custom predictive model, and a Power BI dashboard.

   b. The inventory - accessible via natural language - helps different teams' surface relevant datasets, reducing silos and reinforcing consistency across AI solutions.

Why this matters

In fast-moving AI environment, reuse with context becomes a competitive advantage. It reduces duplication, aligns teams, and ensures responsible design patterns are embedded from the start. By turning your inventory into a queryable knowledge base, you shift from reactive governance to proactive enablement. In short, an agent connected to your data inventory does more than answer questions - it builds a bridge between governance, opportunity, and trust.

Purview vs. Data Inventory: Seeing vs. Governing

Tools like Microsoft Purview Data Map provide automated discovery and classification of your data estate - a vital capability in large, complex environments. They help organizations answer critical questions like "Where is our sensitive data?" or "Who has access to customer records?"

But the data inventory solves a very different problem: governance of AI-powered solutions.

The inventory isn't about scanning the environment - it's about enabling solution teams to make responsible decisions, align to responsible AI principles, and ensure that data use is ethical, fair, and transparent.

In mature environments, both capabilities work together:

- Purview provides discovery, classification, and technical controls.
- The data inventory adds human governance, oversight, and responsible AI context.

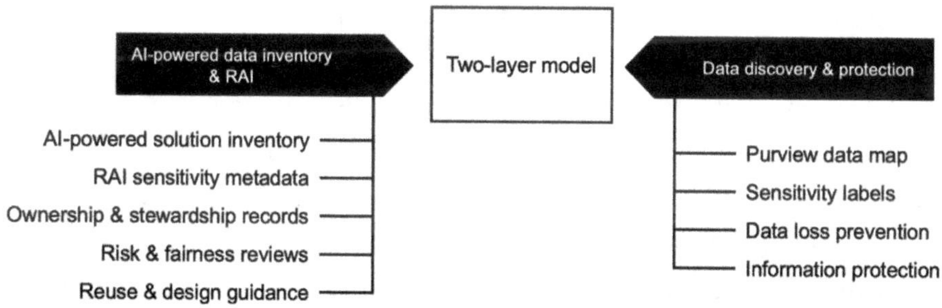

***Figure 9-3.** A dual-layered approach using Purview and an AI-powered data inventory*

As your organization progresses from foundational practices toward operationalizing AI at scale, new protocols are emerging to standardize how AI solutions access and use data. One such innovation is the **Model Context Protocol (MCP)** - an open source framework designed to provide a universal "plug and play" standard for AI systems to access data, tools, and services securely.

Think of MCP like a USB-C port - creating a consistent, reusable way for AI agents and models to access validated, governed data assets. While MCP adoption may sit beyond the foundational stage for most organizations, it's important to understand the direction the industry is moving.

For organizations serious about scalable, governed AI adoption, MCP represents the natural evolution of what starts today with a simple, well-maintained inventory.

Preparing Data with Purpose

Structured and unstructured data not only look different – they often require different approaches to prepare for AI. Both types are valuable, but each comes with its own challenges and best practices for quality assurance and readiness. Let's look at how to handle each, using an example of a customer support app and a HR assistant to illustrate.

With **structured data**, the emphasis is on leveraging its organized nature. Since it's already in databases or spreadsheets, you can use querying and automated checks to enforce quality. Much of the guidance below is closely related to the kinds of task that a DBA (Database Administrator) would undertake, but for clarity:

In an HR agent scenario with structured employee records, you might implement *data validation rules*: ensure that salary fields are numeric and within reasonable bounds or that dates of employment are not in the future. Duplicate records are another issue to watch for – if an employee is entered twice in a table, an AI that counts employees could double-count. *Deduplication processes* or *unique constraints* (e.g., no two records with the same employee ID) help maintain integrity.

Consistency checks are also easier in structured data: you can run scripts to find if a field contains unexpected values (e.g., a country field containing "UK" vs. "United Kingdom" vs. "U.K.") can be standardized to one format. For structured customer data in a support app, you might use Power Platform dataflows to merge data from a CRM or billing system, ensuring that customer ID's match across sources and consolidating all info for each customer.

The key is to exploit the schema: use it to your advantage by defining and understanding relationships, mandatory fields, and valid ranges. This way, much of the data quality enforcement can be automated by the database or data import process itself. Before feeding structured data into an AI model, it's common to do a final sanity check. The point here is to have a process defined, which we'll cover in the next section.

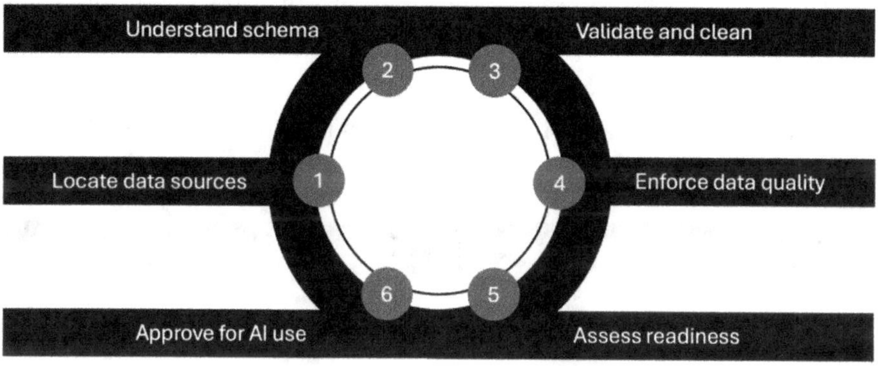

Figure 9-4. *High level process for structured data readiness*

Preparing **unstructured data** is a different beast. Since this data doesn't have a neat schema, preparation (depending entirely on the use case and technological approach) will vary greatly. Let's take an example of an HR agent that answers HR policy-related queries. HR policies are usually kept as PDF or DOCX files, so that employees can easily print, search, and download copies. Microsoft Copilot Studio can leverage unstructured text, but, if it is provided with poor-quality or disorganized data, then the results will be equally poor. The key to success is preparing and organizing these unstructured knowledge assets so that Copilot Studio can use them effectively as knowledge sources, all while maintaining security and accuracy.

If you're preparing documents to be used as knowledge sources in Copilot Studio, whether creating new content or evaluating existing materials, it's important to consider how well they're structured for AI consumption.

Formatting documents consistently is essential. Use *clear headings and subheadings* to define content structure, and *leverage bullet lists or tables* to simplify the extraction of key information. Avoid embedding critical information in images and keep the language clear, concise, and standardized to ensure accurate search results.

Ensuring that data is ready for AI involves more than simply confirming that it's neatly stored or formatted; it requires building a robust and repeatable process for validating and maintaining data quality. In the next section, we'll take a closer look at how to practically manage structured data, step-by-step, so you can trust the outputs of your AI-powered solutions from the outset.

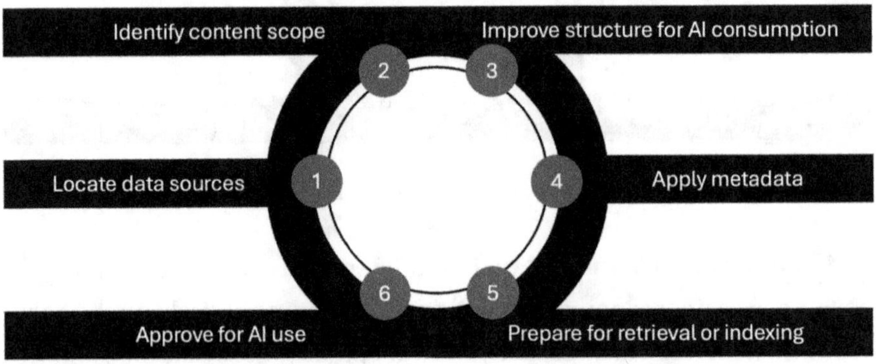

Figure 9-5. *High-level process for unstructured data readiness*

Establishing Trust in Structured Data

High-quality data means your AI models can learn the right patterns and produce reliable results. Data quality has several dimensions, but four key aspects to focus on are **accuracy, completeness, consistency, and timeliness**. By systematically improving these dimensions, you enhance your data's suitability for AI.

Data quality isn't a one-time event but an ongoing discipline that needs clear expectations and active communication across the organization. Establishing data quality standards involves defining what quality means in practical, operational terms. These standards serve as benchmarks guiding data preparation, ongoing management, and assessment. Clear, easy-to-understand quality standards help teams recognize the importance of data integrity, ensuring everyone from technical teams to business stakeholders has a shared understanding and commitment.

Accuracy

What it means

Accuracy ensures your data correctly reflects the reality it's intended to represent. Inaccurate data leads directly to flawed insights and poor decision-making and can significantly damage the credibility of AI solutions. By starting to define accuracy requirements for data used in AI-powered solutions now, you are setting yourself a clear bar for quality moving forward. The guidance below is designed to provide you with a foundational process that you can integrate into governance:

Step 1: Identify Key Data Points

- Focus on the data that matters most. For example, customer names, addresses, financial information, or product details. You don't have to check everything at once – prioritize the data fields most critical to your business operations.

Step 2: Basic Spot Checks

- Conduct simple reviews of a small sample (e.g., 10–20 records) from your most important data sets.
- Does the data look correct? (e.g., are dates realistic, are amounts sensible).

- Check against trusted sources, if possible, such as cross-checking addresses with customer invoices or comparing sales figures with accounting records.

Step 3: Engage Data Owners and Users

- People who regularly use or collect data (like salespeople, customer support agents, or administrative staff) often have the best sense of what's correct.
- Ask these individuals to periodically review data accuracy, report common issues, and suggest improvements. As part of this process, encourage them to identify which systems or sources they trust most - and why. This insight helps prioritize data quality efforts and strengthens understanding of which systems are perceived as reliable across different roles.

Step 4: Simple, Clear Guidelines

- Provide straightforward rules for data entry where applicable, and ensure consistency in how formats such as dates, currencies, or units are stored across systems. While users may interact in local formats, the underlying data should follow a standardized structure to support integration and downstream accuracy.
- Even simple guidelines help maintain accuracy, making sure everyone is on the same page.

Step 5: Encourage Feedback

- Create a simple way for team members to flag data that looks incorrect or questionable - this could be as easy as a shared spreadsheet or Microsoft Form.

- Review these flags regularly to spot and resolve recurring accuracy issues.

Reflection

- Have you clearly identified which data fields are most critical?
- Do teams feel empowered to flag accuracy issues easily?

Establishing clear policies around accuracy ensures everyone involved knows what's expected before an AI-powered solution goes live. At this foundational stage, a straightforward accuracy policy doesn't need to be complex; rather, it should clearly outline responsibilities and basic actions required before launch.

Example accuracy policy:

> *"Before any system or application using AI capabilities moves into production (i.e., goes live to end users), an accuracy review must be conducted and documented. The review involves the developer and the data owner jointly verifying that key data points used by the solution accurately represent real-world information."*

Here's a clear breakdown of how you can implement this policy:

Step 1: Define Roles Clearly

- **Developer**: Responsible for verifying technical accuracy, such as data sources correctly integrated and outputs aligned with expectations.
- **Data owner**: Typically, a business user or subject matter expert responsible for verifying that the data content accurately reflects real-world scenarios or operations.

Step 2: Create a Simple Checklist

Provide a straightforward checklist to guide the accuracy check. Here's a practical example:

Accuracy checklist	Verified (Y/N)	Comments
Data entries reflect reality (spot-check)		
Data formats meet organizational standards		
Sample outputs reviewed by data owner		
Any flagged inaccuracies resolved		

Step 3: Require Sign-Off

Both the developer and data owner should sign off (either physically or digitally) to formally acknowledge that accuracy has been reviewed.

Step 4: Store and Maintain Documentation

- Keep a centralized record of these accuracy reviews (a simple shared folder, SharePoint site, or document library will suffice).
- Ensure these records are accessible to future reference, audits, or troubleshooting.

Reflection

- *Do teams clearly understand their roles regarding data accuracy?*
- *Is your current accuracy documentation easy to use and maintain?*

Implementing this type of straightforward accuracy policy early sets a strong foundation, helping teams understand their accountability and build good data governance habits as AI use grows in your organization.

Completeness

Completeness refers to how thoroughly your data captures all necessary information required by your AI-powered solutions. Think about completeness like the pieces of a puzzle – missing pieces mean an incomplete picture, leading to ineffective or misleading insights. At the foundational stage, establishing basic completeness means making sure your key data fields are consistently filled out with no critical gaps.

Here's how you can practically address completeness at this early stage:

Step 1: Clearly Identify Required Data Fields

- Start by clarifying the minimum essential data fields required for each AI use case.

  Example: If building a customer-support conversational agent, essential fields could include "Customer ID," "Product," "Issue," "Description," and "Resolution status."

  Action: Document these critical fields clearly in simple language accessible to everyone.

Step 2: Quickly Audit Existing Data

Review a representative sample (e.g., 10–20% of records) of the data. Check for blanks or gaps in key fields.

Example audit approach:

Data field	Records checked	Missing data? (Y/N)	% complete
Customer ID	Y	N	100%
Product	Y	Y	90%
Issue description	Y	Y	75%
Resolution status	Y	Y	40%

Action: Focus improvement efforts on fields with significant gaps, especially if critical to the AI use case.

Step 3: Define Simple Completeness Rules

Create easy-to-follow rules for data collection:

- Clearly identify mandatory fields (highlighted in forms or apps).
- Introduce validation checks to prevent incomplete submissions.

Step 4: Enforce Minimal Data Standards Before Moving to Production

Establish a policy that requires a basic completeness check before solutions go live:

> "Before deploying an AI-enabled system, ensure all required data fields have at least a 90% completion rate based on sampled data."

Step 5: Implement Basic Reminders or Nudges

Encourage completeness through user-friendly reminders or nudges:

- Simple alerts or notifications prompting users when critical fields are left blank.
- User-friendly error messages that explain why certain fields are mandatory.

Summary of Actionable Recommendations

- Clearly define the minimum fields needed for your AI solutions.
- Periodically audit your critical datasets to identify incomplete records.

- Create clear, concise completeness standards, and communicate them to all stakeholders.
- Implement straightforward pre-deployment checks.
- Use reminders or nudges to encourage adherence to data completeness guidelines.

You've now defined essential accuracy checks and completeness standards. But reliable AI solutions require consistency. Think of consistency as speaking the same language across all data sources: it ensures that your data fields have uniformity, making it easier to trust and integrate from multiple places without confusion or conflict. Without consistency, you risk misinterpreting data, encountering errors, or losing valuable insights.

Consistency

Consistency matters, but it doesn't always mean full standardization at the source. Similar data - regardless of where it originates - should ideally be defined clearly and formatted in predictable ways, even if it's captured through different systems or regions. Where that's not possible, tools like dataflows, AI translation models, or metadata mapping can help bridge gaps to support reliable AI outcomes.

Why does consistency matter practically?

- **Clear communication**: Teams spend less time resolving misunderstandings caused by conflicting data.
- **Reliable analytics**: AI systems can easily process data without misinterpretation, improving insights and outcomes.
- **Reduced complexity**: Easier integration of data across multiple systems, teams, or regions.

Here's a straightforward process to set up clear guidelines and policies around data consistency:

Step 1: Identify and Define Critical Fields Clearly

You've already done some of this in previous dimensions (accuracy, completeness), but you begin by creating simple, universally understood definitions for critical data fields, ensuring everyone uses a common language.

Example:

Data field	Agreed definition	Formatting rule	Example
Date	Date a sale or event occurred	DD-MM-YYYY (e.g., 01-05-2025)	05-05-2025
Currency	Currency used in transactions	ISO Currency Codes (e.g., USD, GBP)	USD
Region	Geographic sales region	Predefined regions (e.g., NA, EU, APAC)	EU

Step 2: Create a Simple Consistency Guide or Cheat Sheet

Develop a concise document that captures key definitions and formatting rules. Ensure this guide is available to everyone – consider hosting it on intranets, SharePoint, or other collaborative spaces.

Example policy statement:

> *"All teams must reference the consistency guide for defining, formatting, and recording critical data fields to ensure uniformity across systems."*

Step 3: Conduct Periodic Consistency Audits

Like accuracy and completeness, periodically sample your data across various sources to ensure uniformity.

A straightforward audit process

- Randomly select 10% of records (monthly, quarterly).
- Check against your consistency definitions.
- Note discrepancies and assign responsibility for correction.

Data field	Sample checked	Inconsistencies (Y/N)	Action
Currency	100 records	Yes (15% incorrect codes)	Retrain teams on currency codes
Date	100 records	No	None

Step 4: Implement Entry-Level Data Validation

Where feasible, introduce basic validation measures in Power Apps to encourage consistency. Consider creating a template app that makers can use to get started. Consider quick onboarding sessions, email reminders, or short training videos to help reinforce coding standards.

Examples

- Drop-down controls for currency selection instead of free text
- Date pickers enforcing correct formats

Reflection

- *Have you identified the most critical fields requiring consistency in your data sets?*
- *Are your current communication and training approaches effectively embedding consistent data practices?*

Establishing consistency now makes your AI initiatives more reliable and scalable.

Timeliness

Imagine your organization created a conversational agent designed to answer customer questions. It has access to accurate, complete, and consistent data, but it's only updated once per quarter. What happens when the product information changes or customers' needs shift rapidly? The conversational agents answer quickly becomes outdated, diminishing usefulness, or worse, it quotes an outdated price, leaving the company liable to fulfill the order at a loss.

This is where timeliness comes into play.

Timeliness means your data is accessible and up-to-date exactly when it's needed to support effective decision-making, operations, and customer interactions. Timely data directly impacts the reliability and relevance of AI-driven solutions, making it an essential component of your foundational data strategy.

How to ensure timeliness at the foundational level

Here's a straightforward and actionable approach you can follow to ensure your data is timely and supports effective AI solutions:

Step 1: Identify Data Update Frequency Based on Business Needs

Not all data needs real-time updates. Determine how often each data type should be refreshed based on operational requirements.

Example guideline:

Data category	Recommended update frequency	Justification
Product inventory levels	Real-time to hourly	Immediate customer inquiries, ecommerce accuracy
Customer complaints	Daily	Quick response to customer issues
Sales figures	Weekly	Frequent operational decisions, forecasting needs
Long-term performance metrics	Quarterly	Strategic planning, less frequent operational need

Step 2: Define Clear Timeliness Policies

Document how quickly critical data must be updated, and ensure teams clearly understand their responsibilities.

Sample policy statement:

> *"Product availability data must be refreshed hourly to ensure accurate information for customers interacting with our online platforms and AI-driven customer service tools."*

Step 3: Establish Practical Processes for Timely Data Updates

Implement manageable processes for updating data at the recommended frequency, keeping complexity low at this stage.

Simple ways to ensure timely data

- **Automate simple data updates**: Use Power Automate flows, or schedule jobs to automate hourly or daily data refreshes (dataflows, PowerShell scripts, etc.).

- **Clear roles and responsibilities**: Identify data owners clearly, and ensure they understand their role in maintaining timeliness.

Data type	Responsible team	Update method/tool	Frequency
Inventory availability	Operations	Automated via Power Automate	Hourly
Customer feedback	Customer service	Daily manual update	Daily

Step 5: Communicate Importance and Benefits

At this stage, as you've seen in the governance part of this book, establishing clear and frequent communications is vital to establish control. Help teams understand the value of timely data through short engaging communication - highlighting examples of how timeliness directly affects business outcomes.

While structured data provides a clear path to data quality, your organization's data landscape extends way beyond organized databases and spreadsheets. Let's now turn our attention to unstructured data, exploring how to effectively manage and utilize these vital but often overlooked assets.

Making Knowledge Work for AI

Unstructured data - emails, documents, images, customer chats, audio files - often accounts for most of an organization's data. When preparing your organization for AI adoption, especially in the early stages, its common to initially overlook unstructured data due to its complexity. However, harnessing the value of unstructured data is often precisely where AI solutions can shine.

Imagine an internal conversational agent built with Microsoft Copilot Studio, designed to answer employee questions about HR policies. The accuracy and usefulness of this agent are directly tied to the quality of the underlying data - those very HR policy documents.

Because unstructured data lacks a predefined schema, ensuring quality can feel challenging - but a clear approach makes it achievable. You don't need to manage every piece of unstructured data within your organization. Instead, focus only on the specific sets of data explicitly identified as part of your AI initiatives.

Here's how to apply foundational data quality principles - accuracy, completeness, consistency, and timeliness - to your unstructured data.

Accuracy: Capturing Real-World Context Clearly

Accuracy in unstructured data means capturing the context, content, and meaning clearly and correctly. This is especially critical for AI solutions like conversational agents, where inaccurate information can lead to incorrect or misleading responses.

At this foundational stage, the focus should be primarily on reliable sourcing and clear capture practices.

Practical guidance

- **Reliable source validation**

  Use authoritative sources for your knowledge base. For instance, use HR policy documents from your designated internal HR portal or validated knowledge repository - rather than outdated or unofficial copies. In complex organizations, it's helpful to maintain a central register of trusted sources to clarify which version is considered authoritative.

- **Clear labeling and metadata**

  Assign clear metadata to each document describing its source, purpose, and last update.

Example foundational policy:

"All unstructured data integrated into conversational agents or AI-driven systems must come directly from authoritative, validated sources. Each data asset should include clear metadata detailing the purpose, version, source, and last revision date."

Completeness: Providing Comprehensive Context

Completeness ensures the AI system has all necessary information to deliver accurate and reliable responses. Incomplete data can lead to misunderstandings or partial answers by your conversational agent.

Practical guidance

- **Checklists for knowledge assets**

  Establish simple checklists to ensure every document includes critical details. For instance, every HR policy document should be well organized. A key feature here is the use of structured headings, such as H1 for main sections and H2 for subsections, which create a logical hierarchy.

- **Regular gap checks**

  Before integrating documents into AI systems, perform basic reviews with subject matter experts to confirm that no essential sections are missing.

Example policy:

"Before being used in AI solutions, all unstructured data must be reviewed against a standard completeness check to ensure critical context is fully captured."

Consistency

While conversational agents can begin delivering value even with loosely structured data, consistency in structure and format greatly improves accuracy, reduces ambiguity, and supports better responses at scale. It allows AI systems to more reliably interpret context, extract meaning, and respond to user queries with confidence - especially when dealing with complex or regulated content.

Practical guidance

- **Standard templates**

  Implement simple, standardized document templates for similar content types, such as policies or FAQ documents.

- **Naming conventions and file formats**

  Standardize naming conventions and file formats (such as DOCX or PDF for text-based documents) to enable seamless integration with AI systems.

Example policy:

"Unstructured data intended for use with AI systems must adhere to defined formatting standards, including a standardized naming convention and approved document templates."

Timeliness: Maintaining Current and Accessible Information

Timeliness means ensuring your conversational agents rely on the latest available information. Outdated documents can lead to incorrect responses or noncompliance issues.

Practical guidance

- **Schedule reviews and updates**

  Ensure that document owners are responsible for reviewing and updating critical content on a defined cadence (e.g., quarterly or annually), based on the sensitivity or regulatory importance of the information.

  AI governance teams should focus on confirming that content in use reflects the most recent version, not necessarily managing the update cycle itself. Encourage owners to clearly mark the revision date on each document to support transparency and downstream trust.

- **Version control and archiving**

  Use clear version control practices to ensure that AI solutions access only the most current documents.

Example policy:

"Documents used in AI-powered conversational or autonomous agents must be reviewed and updated at a defined frequency (e.g., quarterly), clearly marked with version and revision dates."

Summary

You've taken crucial first steps toward understanding how to manage and maintain the quality of structured and unstructured data used by AI-powered solutions.

However, reliability is just one part of a larger picture. As AI increasingly shapes decisions in business, healthcare, and beyond, it's not enough for these systems to simply work well – they must also be ethical and trustworthy.

In the next section, we'll dive deeper into responsible AI through the lens of data readiness and security, exploring how to put these principles into practice in your AI projects.

Using real-world examples, such as the Adult Census Dataset model, we'll examine how to address fairness, enhance transparency, and uphold other responsible AI standards. By embracing these practices, you can develop AI systems that are not only powerful and reliable but also ethical, secure, and inclusive.

Where Trust Begins: Responsible Data Foundations

This is perhaps the most critical section of the book – and for good reason. Earlier chapters introduced awareness-building strategies, forming AI councils, and navigating the "softer" aspects of responsible AI adoption. However, when it comes to data, the stakes rise dramatically. This is the point where trust is either built or shattered. It's where transformative value and innovation can be discovered or where missteps can bring initiatives to an abrupt halt.

Unlike other dimensions of adoption – where foundational stages typically involve exploration and experimentation – the foundational stage for data demands immediate seriousness, rigor, and deliberate effort.

Here, the decisions you make set a trajectory that affects everything that follows. Your goal is not just to raise awareness but to embed initial checks, balances, and clear accountability into your AI development workflows from the outset.

Data fairness, transparency, explainability, security, and reliability must be explicit considerations from the very beginning - well before your solutions scale or become widely adopted. And that's exactly where we'll begin.

Organizations often begin using tools such as AI Builder - a low-code tool within Power Platform that allows you to quickly create custom AI models like prediction or classification models - and Azure Machine Learning, particularly Azure Machine Learning Studio, for building and deploying custom models. Both platforms provide an opportunity to embed fairness into your AI practices right from the start.

Model Fairness: A Brief Primer

In machine learning, fairness means ensuring the model's predictions or decisions do not disproportionately favor or disadvantage any group. Fairness issues typically arise from biases embedded within the training data itself rather than the algorithms.

These biases, if not managed, propagate into the model's decisions, leading to discriminatory or unfair outcomes.

Addressing fairness early, prior to training your model, is crucial. Why? Because correcting unfairness post-training is often complex, costly, and might result in substantial rework.

Fairness Before Training the Model

Ensuring fairness begins even before the model is trained, through careful inspection and preparation of your training data. Applying fairness standards at the data preparation stage helps to identify and reduce biases before they impact your model's performance. Techniques include

- Reviewing data to identify and mitigate **underrepresentation of groups**
- Ensuring balanced datasets across **sensitive attributes** (age, gender, ethnicity, etc.)
- Utilizing tools such as Microsoft's Fairlearn toolkit for **early detection of biases or imbalances within datasets**

Example scenario: AI Builder prediction model and fairness

Scenario: Imagine a public health organization leveraging a custom AI model to forecast healthcare access for various demographic groups. The model identifies individuals who might face barriers to medical services.

Historical data used to train the model reveals a potential issue. It primarily represents younger, urban residents with stable employment, while older males, rural residents, and part-time or unemployed workers are underrepresented.

- **Potential bias**: When trained on imbalanced data, the model learns primarily from the dominant demographic. As a result, when presented with loan applications from older individuals or rural applicants, the model is likely to unrealistically predict lower approval probabilities, simply due to the limited exposure to those groups in the training dataset.
- **Impact of bias**: The biased predictions mean individuals from underrepresented demographics may be unfairly denied loans based not on their true creditworthiness but on an inherent data imbalance.

How to avoid this issue

1. **Analyze and balance your data before training**

   Evaluate your dataset to identify demographic imbalances. Ensure the training data represents all relevant groups sufficiently.

2. **Apply fairness-aware techniques before training**

   Utilize tools such as Microsoft's Fairlearn to measure data fairness and then implement strategies (such as reweighting) to rebalance the dataset or adversarial debiasing to actively reduce the potential biases before training.

3. **Document and establish standards**

   Clearly document your standards for acceptable demographic representation in training data. For example: "All training datasets for custom models must demonstrate balanced representation across sensitive attributes (gender, age, ethnicity). Any imbalance must be explicitly documented and addressed prior to model training."

Having established the importance of addressing fairness before training, the next question becomes practical: How do you turn these principles into real-world action? This is where the right tools - used in the right way - can accelerate your efforts and prevent costly rework.

Tooling isn't the hard part - knowing what to look for and having clear quality standards is. But once you know the risks, these tools accelerate your ability to enforce quality, fairness, and responsible data use at scale.

In practice, there is no single tool that can prepare every dataset or solve every fairness challenge. Instead, Microsoft provides a family of tools designed to help you prepare, clean, classify, and assess your data - each one focused on a specific part of the challenge. For structured data, Power Query, Dataverse validation rules, and Power Automate help enforce consistency and integrity. For unstructured content, tools like SharePoint Syntex and Microsoft Purview help classify and manage documents effectively. And for fairness and bias analysis, Microsoft Fairlearn and the responsible AI dashboard enable you to detect and mitigate risk before your model is every deployed.

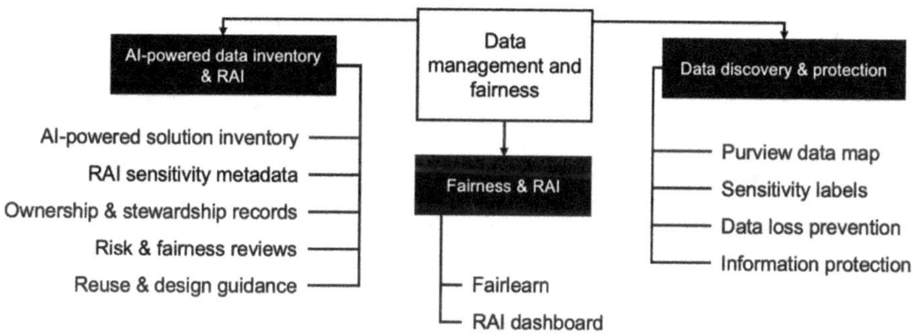

Figure 9-6. *Relationship of tooling and approaches to data management and fairness*

Microsoft Fairlearn provides a robust, yet approachable, toolset specifically designed to assist organizations in analyzing and improving the fairness of datasets and models. By incorporating Fairlearn into your workflow at this foundational stage, you can identify biases early, avoid costly rework later, and confidently scale your AI-powered solutions.

Let's explore precisely how you can leverage Microsoft Fairlearn to assess and mitigate fairness issues within your dataset, step-by-step.

Using Microsoft Fairlearn for Data Fairness Analysis

When you're adopting AI responsibly at scale, ensuring fairness within your AI models becomes critical. Microsoft Fairlearn is an open source toolkit designed specifically to help organizations identify, measure, and mitigate biases and unfairness within machine learning models and datasets.

At this stage, Fairlearn can serve as a robust and approachable tool to ensure your training data is fair, balanced, and aligned with ethical principles.

What is Microsoft Fairlearn?

Fairlearn is a Python-based open source toolkit developed by Microsoft that

- Helps identify fairness-related issues in datasets and AI models
- Measures disparities across sensitive attributes (like age, gender, ethnicity, etc.)
- Offers tools to mitigate unfairness and bias through algorithmic techniques

Fairlearn aligns directly with Microsoft's responsible AI principles, helping organizations build AI systems that are equitable and transparent.

The toolkit was initiated in 2018 by Miro Dudik from Microsoft Research, evolving from a research paper on fair classification to a comprehensive Python package. It supports various fairness definitions, such as demographic parity, and is designed to navigate trade-offs between fairness and model performance.

Use Case Scenario: Predicting High Earners for Preapproval Loans

Imagine we work for a financial organization aiming to make its loan application process faster and more efficient. Typically, deciding whether a customer qualifies for a loan involves a detailed review of their financial history, credit score, and income - a process that can take days. To speed things up, the bank decides to use the Adult Census Dataset model to predict whether loan applicants are high earners (defined as earning over $50,000 annually) based on features like education, occupation, and hours worked. This prediction would support a "decision in principle" (DIP).

In this scenario, the organization plans to use the Adult Census Dataset model to streamline its DIP process for loans. Predicting high earners helps the organization make faster decisions about who's likely to qualify while tailoring offers to match applicants' financial profiles. Of course, this is just the starting point. To make this work in the real world, the organization would also need to ensure the model is fair (not biased against certain groups) and compliant with regulations. But as a use case, it shows why predicting high earners matters and how the model can help. In the following section, we'll go through, step-by-step, how we can ensure the model is fair and transparent in its decision-making process.

Step-by-Step Guide: Using Microsoft Fairlearn for Data Fairness

This is (probably) the only part of this book where you'll find code. I've purposely tried to avoid it, but here it is unavoidable. Below is a practical guide showing you how to use Microsoft Fairlearn to measure and improve the fairness of your data sets. You might wonder why we're already deep into the weeds while at the foundational level. I've mentioned many times in this book that data readiness is one of the most important aspects, along with responsible AI - so ensuring that, even if this feels a little overkill, you are prepared for the future.

Installation process

To use Fairlearn, you need to install it along with its dependencies. The installation is platform-agnostic, working on both macOS and Windows. Here are the steps:

- **Using pip**: Run the following command in your terminal (macOS) or command prompt (Windows):

  ```
  pip install fairlearn
  ```

- **Ensure dependencies**: Install additional libraries required for the example:

  ```
  pip install scikit-learn pandas matplotlib
  ```

Using the adult census dataset

Let's now walk through a detailed process using the Adult Census Dataset,[3] which is accessible via Fairlearn's `fetch_adult` function.[4] This dataset is ideal for demonstrating fairness issues, as it includes sensitive features like "sex" and "race" and is commonly used in fairness research.

Step 1: Import Necessary Modules

Begin by importing the required libraries. These are standard Python packages, ensuring compatibility across platforms:

```
import matplotlib.pyplot as plt
import pandas as pd
from sklearn import metrics as skm
from sklearn.linear_model import LogisticRegression
from sklearn.model_selection import train_test_split
```

[3] Frances Ding, Moritz Hardt, John Miller, and Ludwig Schimdt. Retiring adult: new datasets for fair machine learning. Advances in neural information preprocessing systems, 34:6478–6490, 2021. URL: https://proceedings.neurips.cc/paper_files/paper/2021/file/32e54441e6382a7fbacbbbaf3c450059-Paper.pdf

[4] URL: https://fairlearn.org/main/user_guide/datasets/adult_data.html

```
from sklearn.preprocessing import LabelEncoder, StandardScaler
from fairlearn.datasets import fetch_adult
from fairlearn.metrics import MetricFrame, count, selection_
rate, selection_rate_difference
from fairlearn.reductions import DemographicParity, GridSearch,
ErrorRate
```

Think of libraries as toolboxes containing special tools (functions) we can use. For example, `matplotlib.pyplot` helps create plots, and `Fairlearn.datasets` helps load our dataset.

Step 2: Load and Preprocess the Data

Load the Adult Census Dataset and preprocesses it for analysis. The dataset includes features like education, occupation, and income, with a binary target indicating whether income exceeds $50,000:

```
# Load the dataset - step 2
dataset = fetch_adult(as_frame=True)
features_raw = dataset.data
target = (dataset.target == ">50K") * 1
sensitive_feature = features_raw["sex"]
features = features_raw.drop(labels=["sex"], axis=1)
features = pd.get_dummies(features)

# Scale numerical features for consistency
scaler = StandardScaler()
features_scaled = scaler.fit_transform(features)
features_scaled = pd.DataFrame(features_scaled,
columns=features.columns)

# Encode target variable
label_encoder = LabelEncoder()
target_encoded = label_encoder.fit_transform(target)
```

```
# Split into training and test sets, ensuring stratification
X_train, X_test, y_train, y_test, A_train, A_test = train_
test_split(
    features_scaled, target_encoded, sensitive_feature, test_
    size=0.4, random_state=0, stratify=target_encoded
)

# Reset indices for consistency across splits
X_train = X_train.reset_index(drop=True)
A_train = A_train.reset_index(drop=True)
X_test = X_test.reset_index(drop=True)
A_test = A_test.reset_index(drop=True)
```

Preprocessing is like preparing ingredients before cooking, converting text to numbers, and scaling for consistency. This step ensures the data is ready for modeling, with features scaled and split into training (60%) and test (40%) sets, maintaining the distribution of the target variable.

Explanation

When we say "*maintaining the distribution of the target variable*" while splitting a dataset into training and test sets, it's like trying to divide a pizza fairly between two friends. Imagine the pizza has two types of toppings: pepperoni (representing people with income > \$50K) and mushrooms (representing people with income ≤ \$50K). If you just randomly slice the pizza, one friend might end up with all the pepperoni and the other with just mushrooms. That's unfair, and it's also not a good way to test how well your pizza-sharing strategy works for everyone.

In machine learning, the "pizza" is your dataset, and the "toppings" are the different categories of your target variable (in this case, income levels). If you don't maintain the distribution, your training set might have all the "pepperoni" (high-income earners), and your test set might be all "mushrooms" (low-income earners). Then, when you train your model on the training set, it's like teaching it to love pepperoni but never letting it taste mushrooms. When you test it on the test set, it's going to be confused – like a kid who's only ever eaten pizza with one topping and suddenly must deal with something completely different.

So, **stratification**[5] is like carefully slicing the pizza to ensure both friends get the same proportion of pepperoni and mushrooms. That way, your model learns from a balanced diet and can handle whatever slice of life (or pizza) comes its way!

Step 3: Train a Fairness-Aware Predictor

Train a logistic regression model without considering fairness to establish a baseline:

```
# Train the logistic regression model - step 3
baseline_model = LogisticRegression(solver="liblinear", fit_
intercept=True)
baseline_model.fit(X_train, y_train)
```

Explanation

We're building a basic model to predict income without worrying about fairness first. It's like making the pizza we just mentioned but without considering dietary needs – it might work for some but could exclude others. This step shows us the starting point, so we can see how fairness changes later.

Step 4: Assess Fairness of the Unmitigated Model

We use Fairlearn's metrics to assess potential bias. We focus on the selection rate (proportion of positive predictions) by group, defined by the sensitive feature "sex":

```
# Predict on the test set with the baseline model -  step 4
y_pred_baseline = baseline_model.predict(X_test)
```

[5] https://en.wikipedia.org/wiki/Stratified_sampling

```
# Calculate selection rate by group for the baseline model
baseline_metric_frame = MetricFrame(
    metrics={
        "accuracy": skm.accuracy_score,
        "selection_rate": selection_rate,
        "count": count,
    },
    sensitive_features=A_test,
    y_true=y_test,
    y_pred=y_pred_baseline,
)
print(baseline_metric_frame.overall)
print(baseline_metric_frame.by_group)
baseline_metric_frame.by_group.plot.bar(
    subplots=True,
    layout=[3, 1],
    legend=False,
    figsize=[12, 8],
    title="Accuracy and selection rate by group",
)
```

Explanation

We're checking if our model is fair, by seeing how often it predicts high income for men vs. women. If one group gets predicted more often, it might be biased. This helps us spot unfairness.

You should expect to see results like below:

```
accuracy              0.851820
selection_rate        0.194707
count             19537.000000
```

Or in bar chart form:

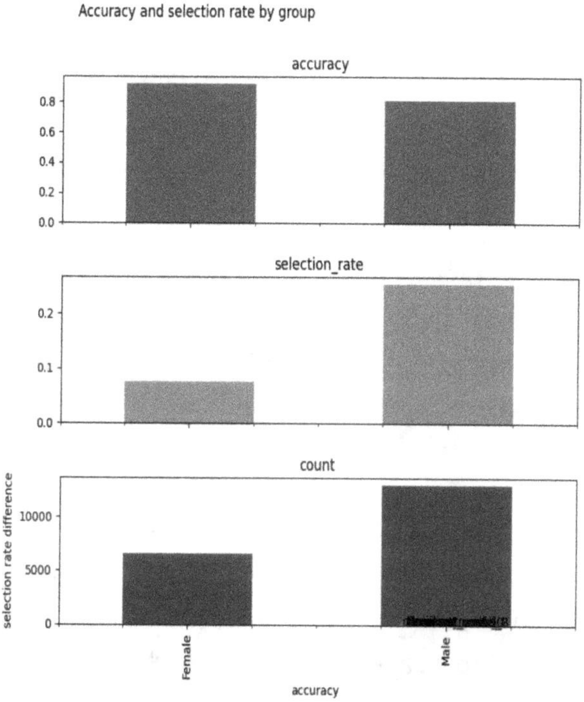

***Figure 9-7.** Output from Fairlearn: unmitigated model*

Examining the differences in accuracy, males show roughly triple the rate compared to females (count). Even more striking is the gap in loan approvals (selection_rate), where males are approved three times more often than females.

Even though we excluded the sex feature from the training data, the model still shows bias toward males. This highlights that just leaving out a sensitive detail doesn't usually fix fairness, as other related factors in the data can still cause unequal outcomes. This is known as **proxy bias** – where seemingly neutral features in the data still reflect patterns tied to sensitive attributes like sex. In other words, the model learns indirect signals, even if the direct one is removed. Fortunately, tools like Fairlearn's mitigation algorithms (such as *GridSearch* or *ExponentiatedGradient*) can help reduce these disparities by explicitly optimizing for fairness during model training.

Step 5: Mitigate Fairness Issues Using GridSearch

To reduce unfairness, we use Fairlearn's **GridSearch** with the **DemographicParity** constraint, which aims for similar selection rates across groups:

```
# Mitigate with GridSearch
grid_search = GridSearch(
    LogisticRegression(solver="liblinear", fit_intercept=True),
    constraints=DemographicParity(),
grid_size=31,
)
grid_search.fit(X_train, y_train, sensitive_features=A_train)
# Predict on the test set with the mitigated model - step 5
y_pred_mitigated = grid_search.predict(X_test)
# Calculate selection rate by group for the mitigated model
mitigated_metric_frame = MetricFrame(
metrics=selection_rate,
y_true=y_test,
y_pred=y_pred_mitigated,
sensitive_features=A_test
)
# Display selection rates for the mitigated model
print("Selection rate by sex (mitigated):")
print(mitigated_metric_frame.by_group)
# Evaluate all predictors from GridSearch
predictors = grid_search.predictors_
errors, disparities = [], []
for predictor in predictors:
```

```
    def classifier(X):
        return predictor.predict(X)
    error_metric = ErrorRate()
    error_metric.load_data(X_train, pd.Series(y_train),
    sensitive_features=A_train)
    disparity_metric = DemographicParity()
    disparity_metric.load_data(X_train, pd.Series(y_train),
    sensitive_features=A_train)
    errors.append(error_metric.gamma(classifier).iloc[0])
    disparities.append(disparity_metric.
    gamma(classifier).max())
results_df = pd.DataFrame({"predictor": predictors, "error":
errors, "disparity": disparities})
non_dominated_models = []
for row in results_df.itertuples():
    errors_for_lower_disparity = results_df["error"]
    [results_df["disparity"] <= row.disparity]
if row.error <= errors_for_lower_disparity.min():
        non_dominated_models.append(row.predictor)
predictions = {"baseline": baseline_model.predict(X_test)}
metric_frames = {"baseline": baseline_metric_frame}
for i, model in enumerate(non_dominated_models):
    key = f"dominant_model_{i}"
    predictions[key] = model.predict(X_test)
metric_frames[key] = MetricFrame(
    metrics={
        "accuracy": skm.accuracy_score,
        "selection_rate": selection_rate,
        "count": count,
    },
```

```
    sensitive_features=A_test,
    y_true=y_test,
    y_pred=predictions[key],
)
x_vals = [frame.overall["accuracy"] for frame in metric_frames.
values()]
y_vals = [frame.difference()["selection_rate"] for frame in
metric_frames.values()]
keys = list(metric_frames.keys())
plt.scatter(x_vals, y_vals)
for i in range(len(x_vals)):
    plt.annotate(keys[i], (x_vals[i] + 0.0003, y_vals[i]))
plt.xlabel("accuracy")
plt.ylabel("selection rate difference")
plt.show()
```

Explanation: If we found bias, we're now adjusting the model to be fairer. Think of it like tweaking a recipe to make it inclusive for everyone, ensuring men and women have equal chances of being predicted high income. GridSearch tries different versions of the model to find the best balance.

Look at the results now.

```
accuracy 0.851820
selection_rate 0.194707
count 19537.000000

sex
Female 0.925960 0.076794 6537.0
Male 0.814538 0.254000 13000.0

Selection rate by sex (mitigated):
sex
```

```
Female 0.150069
Male 0.152846
Name: selection_rate
```

What Is This Output Telling Us?

You've built a basic model and then used Fairlearn to make it fairer, especially when it comes to men and women. The output shows how well this works - and it's quite telling.

Step 1: The Starting Point (or Before Mitigation)

The first set of numbers is like a report card from the original model:

```
accuracy 0.851820
selection_rate 0.194707
count 19537.000000

sex
Female 0.925960 0.076794 6537.0
Male 0.814538 0.254000 13000.0
```

Overall performance

- **Accuracy (0.851820)**: The model gets it right about 85% of the time, which is pretty good. It's like guessing correctly who's a high earner 85 times out of 100.
- **Selection rate (0.194707):** It predicts "high earner" for about 19.5% of people overall. Think of this as the percentage of applicants the model says "yes" to.
- **Count (19537):** This is the total number of people it tested - 19,537 - in this test set.

By sex

- **Females**
    - **Accuracy (0.925960)**: For women (6537 total), it's right 92.6% of the time - very accurate.
    - **Selection rate (0.076794)**: Despite being highly accurate for women (92.6%), the model only predicts high income for 7.7% of them - suggesting that women are selected for positive outcomes far less frequently, even when the model is correct.
- **Males**
    - **Accuracy (0.814538):** For men (13,000 total), its right 81.5% of the time, less accurate than for women.
    - **Selection rate (0.254000):** Yet it says "yes" to 25.4% of men, over three times more often than women!

What's happening?

The model is biased - even though we didn't tell it about sex (we dropped this as a sensitive property), it's picking up bias from other areas, like occupation, hours worked, etc., that correlate with sex. This is bias sneaking in through the back door!

Step 2: Fixing It with Fairlearn (After Mitigation)

Next, we used Fairlearn's `GridSearch` with `DemographicParity` to tweak the model, aiming for equal "yes" rates across sexes. Here's the result:

```
Selection rate by sex (mitigated):
sex
Female 0.150069
Male 0.152846
Name: selection_rate, dtype: float64
```

- **Mitigation selection rates**
  - **Females (0.150069):** Now women get a "yes" 15% of the time - that's up from 7.7%.
  - **Males (0.152846):** Now 15.3% of men get a "yes" - that's down from 25.4%; it's almost even.

What happened?

Fairlearn adjusted the model so that is says "yes" to men and women at nearly the same rate. The gap shrank from **17.7** percentage points (25.4–7.7%) to just **0.3** percentage points (15.3–15%). That is fair.

The trade-off

There is a catch: while fairness improved, we don't see the accuracy of the mitigated model in this output. Accuracy might drop a bit, say, from 85% to 82% - because the model is now prioritizing equal treatment over perfect predictions.

How GridSearch and DemographicParity fixed the bias

In step 1, the original model predicted who earns over $50,000 using the Adult Census Dataset; it was unfair - saying "yes" to 25.4% of men, but only 7.7% of women. We used Fairlearn's GridSearch and DemographicParity to fix this. But what did they do under the covers?

1: GridSearch: Trying lots of recipes

Imaging your original model is like a chef preparing a large batch of soup to serve high earners. The soup is delicious (85% accurate), but there's a fairness problem: some diners get full bowls, while others barely get a taste. That's the selection bias - the model favors one group far more often, even if the soup itself is good.

GridSearch steps in like a kitchen manager running experiments to fix this. It doesn't change the soup - it tries 31 different serving methods to figure out how to portion it more fairly. Some methods prioritize equal servings; others stick close to the original proportions. The goal? Balance fairness without ruining the flavor.

- **What it does:** GridSearch takes your base model (a logistic regression, which guesses based on patterns like education and hours worked) and creates 31 versions of it. Each version tweaks how much weight it gives to individual ingredients - like how much "hours worked" or "occupation" influences the "yes" decision.
- **How it tweaks:** It adjusts something called the "lagrange multipliers"[6] (think fancy knobs on the stove). These knobs control how much the model cares about fairness vs. accuracy. Some versions might lean hard into fairness, others less so.
- **Behind the scenes:** For each of the 31 attempts, GridSearch trains a new model on your training data (that 60% chunk that we split), testing different knob settings. It's like experimenting with salt and pepper until we get the taste right.

Right here in the code:

```
# Mitigate with GridSearch
grid_search = GridSearch(
    LogisticRegression(solver="liblinear", fit_intercept=True),
    constraints=DemographicParity(),
    grid_size=31,)
grid_search.fit(X_train, y_train, sensitive_features=A_train)
```

In our case, GridSearch ran these 31 experiments to find models that could close that 17.7% gap between men and women.

[6] https://en.wikipedia.org/wiki/Lagrange_multiplier

Step 2: DemographicParity - Setting the fairness rule

While GridSearch is busy doing the cooking, DemographicParity is the rulebook it follows. It's like a judge saying, "every group - men and women - should get the same-sized service of soup, no matter what."

- **What it means:** DemographicParity demands that the selection rate (the percentage of "yes" predictions) be equal across groups defined by a sensitive feature - in our case "sex." So, if men get a "yes" 25.4% of the time, then women should too, or at least close to it.
- **How it enforces this:** It adds a penalty to the model's math. Normally logistic regression tries to maximize accuracy by fitting the data as closely as possible. DemographicParity steps in and says, "hold on, if your 'yes' rate for men is higher than for women, I'm adding a penalty to your score." This forces the model to adjust its predictions.
- **Under the covers**: The models equation gets an extra term - a fairness constraint. It's like telling the chef, "You can't just pile soup on the men's trays; spread it evenly, even if it means a few bowls aren't perfectly seasoned." Mathematically, it's minimizing the difference in selection rates (e.g., 25.4–7.7%) while still trying to keep accuracy decent.

Why Transparency Matters

Transparency means that the decisions made by AI systems are clear and understandable to users and stakeholders. Without transparency, AI can feel like a mysterious "black box," leading to mistrust.

Transparency ensures users know why decisions are made, fostering confidence in AI tools, especially in critical areas like hiring, lending, or customer service.

For many solutions built in Power Platform, it won't be necessary to apply the same level of rigor that we are diving to in this book. However, understanding how AI models can affect fairness - and knowing which tools can help you evaluate and adjust them - is essential as your use of AI becomes more advanced. The examples in this book are designed to build that awareness early, so you're equipped to grow responsibly as your solutions mature in scale and complexity.

Model Cards

Imagine you are buying a box of cereal. You flip it over, check the nutrition label, and instantly know what's inside - calories, ingredients, maybe a warning about allergens. Now, picture doing the same for an AI model. That's what a model card does - it's a concise, standardized way to document an AI model's purpose, performance, and limitations. Think of it as a "transparency passport" that travels with your model, giving users the info they need to trust and use it responsibly.

Model cards were first introduced in 2019 by Margaret Mitchell and her team at Google, in a groundbreaking paper titled "Model Cards for Model Reporting."[7] Mitchell, a pioneer in AI ethics, saw that as AI systems grew more complex and widespread, people needed a simple way to peek under the hood. Her solution? A framework that's easy to create, easy to read, and packed with critical details. Since then, model cards have become a cornerstone of responsible AI, adopted by companies, researchers, and even hobbyists who want to build trust in their models.

[7] https://dl.acm.org/doi/10.1145/3287560.3287596

For our Adult Census Dataset story – where we trained a fairness-aware model to predict income levels – creating a model card is the perfect next step. It's our chance to show not just what we did to reduce bias but how we're making it transparent. Let's break this down step-by-step, using our model as a hands-on example, so readers can follow along and create their own.

Step-by-step: Building a model card for transparency

Here's how we'll craft a model card for our fairness-aware income prediction model. Each section builds transparency, and I'll explain what to do so you can replicate this with confidence:

- **Model details: who, what and how**

  Start with the basics – this is your model's ID card.

- **What to include**
  - **Creator**: Who built the model?
  - **Version**: What's the current version?
  - **Type**: What kind of model is it (e.g., logistic regression)?
  - **License**: How can others use it (if sharing broadly)?

- **Our example**
  - **Creator**: Your name/organization – let's say Steve Jeffery.
  - **Version**: 1.0 – our first iteration.
  - **Type**: Logistic regression with fairness constraints – specifically, we used demographic parity to balance outcomes.
  - **License**: MIT license – open and free for others to use or adapt.

- **Intended use: What's it for?**

  Define the model's purpose and boundaries – think of this as its job description.

  - **What to include**
    - What does the model do?
    - Who's it for?
    - What's not in scope?
  - **Our example**
    - **Purpose**: Predicts if someone earns over $50,000 annually using census features like age, education, and occupation.
    - **Intended users**: Researchers and policymakers studying income inequality and fairness in AI.
    - **Out-of-scope**: Not for individual financial decisions or real-time predictions – it's a research tool, not a personal finance app.
- **Factors: What influences it?**

  Highlight the key variables and fairness considerations – think of this as the model's lens on the world.

  - **What to include**
    - What features does it use?
    - Any sensitive attributes (e.g., gender, race) you've focused on?

- **Our example**
    - **Features**: Age, education level, hours worked per week, occupation.
    - **Fairness focus**: Gender - we ensured the model treats men and women equitably in decisions.

- **Metrics: How good is it?**

  Share performance numbers - think of this as the model's report card.

    - **What to include**
        - Accuracy or other core metrics
        - Fairness metrics (e.g., selection rate differences)
        - Before-and-after mitigation results, if applicable
    - **Our example**
        - **Accuracy**: 85%
        - **Selection rate difference**:

          Before mitigation: 17.7% (males: 25.4%, females: 7.7%)

          After mitigation: 0.3% (males: 15.3%, females: 15%)
        - **Fairness metric**: Demographic parity - selection rates are now nearly equal across genders.

- **Evaluation data: How was it tested?**

  Describe the test data – think of this as the model's proving ground.

  - **What to include**
    - What data did you use?
    - How representative is it?
  - **Our example**
    - **Data**: Test split of the Adult Consensus Dataset (1994 US Census data).
    - **Representativeness**: Mirrors the US population in 1994 but may not reflect today's demographics.
- **Training data: What shaped it?**

  Explain the training data – think of this as the model's education.

  - **What to include**
    - What data was used?
    - Any known biases?
  - **Our example**
    - **Data**: Training split of the Adult Census Dataset
    - **Biases**: Underrepresentation of some groups (e.g., racial minorities, older adults) – a common issue with 1994 data

- **Quantitative analyses: Does it treat everyone fairly?**

  Break down performance across groups – think of this as the fairness audit.

  - **What to include**
    - How does it perform for different groups?
    - Any disparities?
  - **Our example**
    - Before mitigation

      Males: 25.4% selected (predicted >$50k)

      Females: 7.7% selected

      Gap: 17.7% – unfairly favoring males
    - After mitigation

      Males: 15.3% selected

      Females: 15% selected

      Gap: 0.3% – equitable
- **Ethical considerations: What could go wrong?**

  Address risks and mitigations – think of this as the model's conscience.

  - **What to include**
    - Potential ethical issues
    - How you've handled them

- **Our example**
  - **Risks**: Could reinforce historical bias if ever applied to new data without checks.
  - **Mitigation**: Reduced gender bias, but other biases (e.g., race, age) need monitoring.

- **Caveats and recommendations: How to use it wisely**

  Flag limitations and tips – think of this as the use manual.

  - **What to include**
    - What are the limits?
    - How can users avoid pitfalls?
  - **Our example**
    - **Limitations**: May not generalize beyond 1994 US data.
    - **Recommendations**:

      Monitor fairness metrics over time.

      Retrain with fresh, diverse data for modern use.

Creating a model card is a powerful act of accountability. For our Adult Census model, it shows we've reduced bias and lays out everything users need to know in plain terms. Readers can take this exact process – thanks to Margaret Mitchell's brilliant framework – and apply it to their own projects.

Let's stick with our Adult Census prediction model story. So far, we've used Fairlearn to ensure fairness, achieving a balanced selection rate for both men and women in predicting income levels. But now, we're shifting gears. Imagine, for compliance or regulatory reasons, we needed to provide full transparency into how our model makes its decisions - a level of detail that goes beyond what a basic model card can offer.

The Need for Deeper Transparency

While model cards provide a solid high-level overview, they don't dig into the nitty-gritty details of how individual features influence predictions. This is where tools like SHAP (SHapely Additive exPlanations) and LIME (Local Interpretable Model-agnostic Explanations) come in, offering the deeper, feature-level insights we need. Let's continue our story and explore when and why you might need these tools, using our Adult Census Dataset model as the example.

When Might You Need SHAP or LIME?

Before we apply these tools, let's clarify when they're most useful. SHAP and LIME shine in situations where:

- **Regulatory compliance**: Laws or policies demand detailed explanations of AI decisions. In our case, regulators might require us to prove that features like education or occupation aren't unfairly skewing income predictions.
- **Stakeholder understanding**: Auditors, legal teams, or business leaders need to see not just what the model predicts but why. A feature-level breakdown builds trust and accountability.
- **Model debugging**: You suspect biases or errors - like overreliance on a single feature - and need to pinpoint where they're coming from.

- **User trust**: Transparency reassures end users that the model is fair and reliable, especially in sensitive areas like income prediction.

For our Adult Census model, compliance is the driver. Since income prediction can impact people's lives - and fairness is a hot issue - we need to show exactly how the model works, down to the individual feature level.

Choosing between SHAP and LIME

Both SHAP and LIME can give us the depth we need, but they're suited to slightly different purposes:

- **SHAP**: Offers a mathematically rigorous, consistent way to explain feature contributions. It works for both individual predictions (e.g., why this person was predicted as a high earner) and global trends (e.g., how education impacts predictions overall). The downside? It's computationally heavy.
- **LIME**: Focuses on explaining individual predictions by approximating the model locally. It's faster and more intuitive - great for quick, digestible insights - but less consistent for predictions.

Given our compliance focus, SHAP is the better pick. Its consistency and theoretical grounding make it ideal for regulatory scrutiny, where precision and reliability matter. That said, LIME could still play a supporting role for simpler, stakeholder-friendly explanations. Let's start with SHAP and see how it transforms our transparency.

Applying SHAP to the Adult Census Model

Our model is a logistic regression[8] trained on the Adult Census Dataset. We've got our training data (X_train) and test data (X_test), with features like age, education, and hours worked. Here's how we'd use SHAP to peel back the curtain:

Step 1: Install SHAP and compute SHAP values

We need to install the SHAP module:

```
pip install SHAP
```

And then in our code file, we'll calculate how each feature contributes to predictions across the test set:

```
import shap
explainer = shap.LinearExplainer(baseline_model, X_train)
shap_values = explainer.shap_values(X_test)
```

Step 2: Global insights

To show regulators the big picture, we create a SHAP summary plot:

```
shap.summary_plot(shap_values, X_test, feature_names=X_test.
columns)
```

This plot reveals how features impact predictions across all individuals. For example, it might show that "hours worked" consistently pushes predictions toward "high earner," while "age" has a more mixed effect. This helps prove the model isn't overly dependent on any one feature - a potential bias red flag.

[8] https://en.wikipedia.org/wiki/Logistic_regression

Here is what is generated:

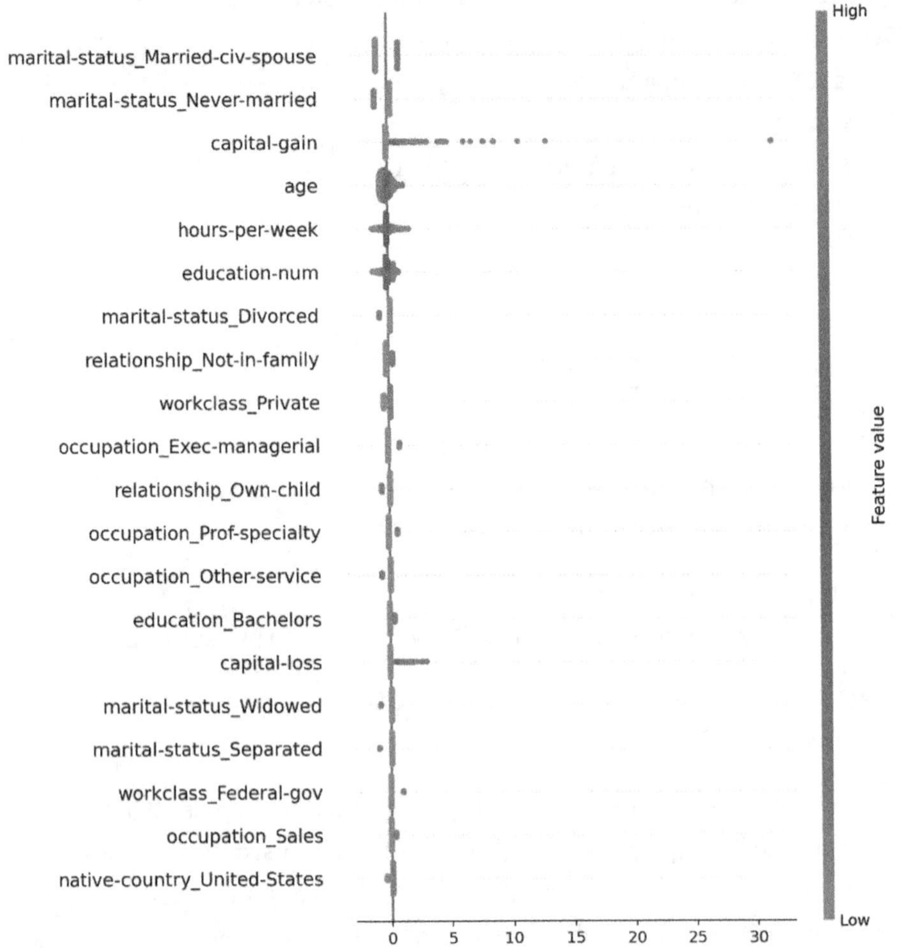

***Figure 9-8.** SHAP explanations show how features impact predictions*

This visual breaks it down: maybe "occupation" or "education" boosted the likelihood of "high earner," while "hours worked" (hours worked per week) nudged it down slightly. For compliance, this level of detail lets auditors verify that no sensitive features (like gender or race, if included) are unfairly driving decisions.

Meeting regulatory requirements

With SHAP, we can satisfy regulators by

- **Documenting feature contribution**: Generate a report with SHAP values for a sample of predictions, showing how each feature weighs in.
- **Proving fairness**: If sensitive features are part of the model, SHAP can demonstrate their impact is minimal or balanced across groups, aligning with our Fairlearn results.
- **Interactive exploration**: Export SHAP insights to a Power BI dashboard to let stakeholders explore feature contributions themselves.

Adding LIME for perspective

While SHAP is our go-to for this story, LIME offers a quick alternative for explaining a single prediction:

```
pip install lime
import lime
from lime.lime_tabular import LimeTabularExplainer
from sklearn.linear_model import LogisticRegression
from IPython.display import display, HTML

# Train the model on a NumPy
baseline_model = LogisticRegression()
baseline_model.fit(X_train.values, y_train)

# Set up LIME explainer
lime_explainer = LimeTabularExplainer(X_train.values, feature_
names=X_train.columns, class_names=['Low Earner', 'High
Earner'])
```

```
# Select an instance and generate explanation
instance = X_test.iloc[0].to_frame().T
lime_explanation = lime_explainer.explain_instance(instance.
values[0], baseline_model.predict_proba, num_features=5)

# Display as HTML
import os
# Set the path to the desktop
desktop_path = os.path.join(os.path.expanduser("~"), "Desktop")
# Save the LIME explanation to the desktop
output_file = os.path.join(desktop_path, 'lime_
explanation.html')
lime_explanation.save_to_file(output_file)
```

This produces a simple chart highlighting the top five features for that prediction - perfect for nontechnical stakeholders who need a clear, concise story:

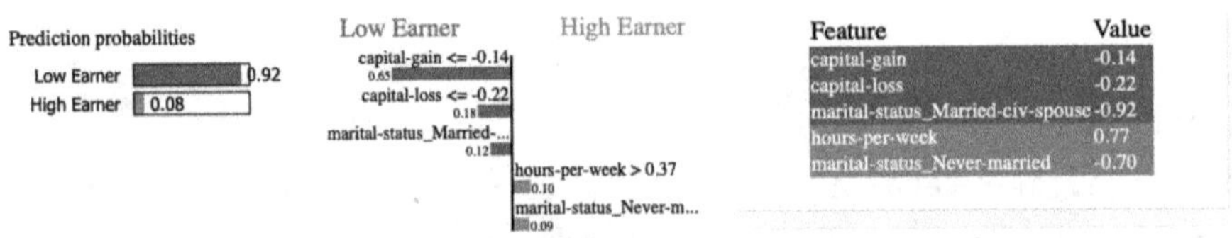

***Figure 9-9.** Output from LIME displaying features used in single decision-making*

Summary

In this section, we dove into the world of fairness in AI using Microsoft Fairlearn and the Adult Census Dataset, a real-world collection of income and demographic data. We started with a basic model that predicted who earns over $50,000, only to find it was biased - favoring men with a 25.4% "yes" rate while giving women just 7.7%. Even though we hid the "sex" feature (column), the model sniffed out patterns in other data, proving bias isn't that easy to dodge.

Fairlearn's GridSearch and DemographicParity: GridSearch tried 31 different model variations, tweaking how much weight it gave to different factors, while DemographicParity enforced a fairness rule - equal "yes" rates for men and women. This resulted in a fairer model with selection rates of 15% for women and 15.3% for men, shrinking the gap from 17.7% to a mere 0.3%. Accuracy stayed strong (85% before, likely dipping slightly after), showing Fairlearn balanced fairness and performance.

We then took our story even further, into the world of compliance and regulation. Using custom models like the one we created may not represent a large percentage of the solutions you build in Power Platform, but it doesn't mean that you are not on the hook for any solutions that do, whether this is now or further down the line. Being prepared is key and is the reason why we've gone deep into fairness and transparency - these are two key principles that, if you are not prepared for them, will erode trust in AI capabilities very quickly.

We looked at model cards - high-level documentation that provides all the information you might need about a model in a single document.

We then looked at how using SHAP and LIME we were able to produce reports that provide next-level transparency - to the feature level. This not only satisfies regulators that models are both fair and transparent but also keeps key stakeholders informed about how decisions are being made.

While fairness and transparency are critical at the foundational stage, responsible AI encompasses broader dimensions you'll encounter as you mature your AI practices. As you advance, you'll need to consider additional aspects such as

- **Accountability**: Defining clear roles and responsibilities to determine who is accountable if an AI solution causes unintended outcomes
- **Privacy and security**: Extending beyond basic data protection to ensure AI solutions preserve user privacy and uphold trust

- **Human oversight**: Maintaining human supervision and control, ensuring AI solutions support - not replace - human judgement and accountability

Right now, the primary task is awareness - knowing these considerations exist and starting to think about how your organization might approach them. Deeper exploration and practical guidance will follow as your journey progresses to the emerging, operationalized, and integrated excellence stages.

Accountability

Having explored fairness and transparency, we now turn to a principle that ensures these efforts remain effective over time: accountability. Accountability in responsible AI means taking responsibility for how AI systems perform, ensuring they stay fair, transparent, and aligned with ethical standards. It involves clear oversight, defined roles and processes to monitor and address issues as they arise. For our Adult Census Dataset model - used to predict high earners for loan approvals - accountability ensures that fairness and transparency are not just one-time achievements but ongoing commitments.

Accountability in AI is about ensuring that someone is responsible for the system's actions and outcomes. It must include

- **Clear roles**: Assigning people to oversee the AI's performance and ethics
- **Monitoring:** Regularly checking that the system meets its goals and doesn't drift into bias or error
- **Corrective actions:** Having plans to fix issues, like retraining the model or adjusting policies

For our scenario, accountability means that if new data causes the selection rates to become unequal again, there's a plan - and a team - to step in and correct it.

Key Components of Accountability

Here's how to implement accountability in your AI projects:

1. **Define roles and responsibilities**

   Clear ownership is the foundation of accountability. Assign roles to ensure someone is always answerable for the AI's performance.

   a. **AI ethics lead:** Oversees ethical use, ensuring fairness, transparency, and compliance.

   b. **Data steward:** Manages data quality and security, keeping training data accurate and representative.

   c. **Model owner:** Handles technical aspects, like development, updates, and performance monitoring.

2. **Monitor performance and fairness**

   Regular evaluation is critical.

   a. **Performance monitoring:** Track accuracy, fairness, and other KPIs over time.

   b. **Fairness audits:** Use tools like Fairlearn to recalculate selection rates or other fairness metrics.

   c. **Automated alerts:** Set thresholds (e.g., selection rate difference > 3%) to trigger reviews or retraining.

3. **Document decisions and processes**

   Transparency supports accountability by making actions traceable. Keep clear records of

   a. **Model documentation:** Note the model's purpose, training data, and fairness adjustments (model cards).

   b. **Decision logs:** Record key choices, like why certain features were included or excluded.

   c. **Explainability tools:** Use SHAP or LIME to clarify how the model makes decisions.

4. **Establish corrective actions**

   When issues arise – like bias creeping back in or ethical concerns – have a plan to resolve them.

   a. **Retraining protocols:** Define when and how to update the model with new data.

   b. **Escalation paths:** Set up a process to alert the AI ethics lead or governance board if problems are detected.

   c. **Human oversight:** Include human review for high-stakes decisions, like loan approvals, to catch and correct errors.

Accountability is a continuous effort – starting at the foundational level builds trust with stakeholders, customers, and users. As data and contexts evolve, so must your oversight. By integrating accountability into your AI life cycle, you ensure fairness, transparency, and ethical use remain priorities, creating systems that are both powerful and responsible.

While accountability ensures responsible management of AI systems, security guarantees that these systems remain robust and trustworthy by defending against threats that could undermine their purpose.

Why Security Matters in Responsible AI

The goal of security is no longer simply to protect - it is to enable trusted innovation at scale. You've already built the foundations: DLP policies are active, environments are segregated, and sensitive data has owners, classifications, and controls. But as your organization steps into a leadership role in AI adoption, security becomes more than a checklist. It is an architectural commitment - visible, explainable, and verifiable by customers, regulators, and peers.

Rather than focusing on edge-case risks, or theoretical threats, the emphasis here is on demonstrating that governance and security are fully aligned, not bolted on. With Microsoft Copilot Studio and Power Platform, that alignment is increasingly native - security is embedded directly into the product. For many organizations, this is a powerful differentiator - enabling them to scale responsibly without reinventing protective mechanisms. From encrypted storage and role-based access controls to DLP enforcement, connector governance, and tenant isolation, these are not optional extras. They're engineered into the product, and they represent a baseline of protection that should be expected from any platform used at enterprise scale.

It's important to recognize that product capabilities alone aren't enough. The principle of RAI demands more: it requires clarity around risk, proportional safeguards, and clearer accountability when those safeguards are exceeded or bypassed. This is where organizational maturity comes in.

Governance scales with confidence

Security is an architectural feature of your governance model. You know which apps, flows, and agents are accessing sensitive data. You've set policies for which connectors can be used, where, and by whom. Your environments reflect business intent, and your inventory tells you not just what's in use but whether it meets your standards for safe use, integration, and AI-powered decision-making. You can prove that

- Data does not cross boundaries unintentionally.
- Agents and other assets are restricted to least-privileged access.
- Runtime behavior is logged and auditable.
- Makers operate within clear and enforced limits.
- Risks are contextual - and so are the controls.

This is what it means to implement the security principle of responsible AI: not just protecting against harm but enabling innovation in a way that is explainable, trustworthy, and appropriate to the use case.

Avoiding overengineering

It's important to recognize that most solutions do not rely on custom models or exotic integration patterns. They use secure, pre-integrated components with well-defined data access and storage paths. Overengineering controls in these scenarios may create friction without meaningful gains in protection.

It's important to know when to apply high-assurance scrutiny - and when to let trusted, managed defaults do their job. For instance:

- If your agent only queries internal knowledge sources and uses Microsoft Graph connectors, your DLP and connector classifications may already provide sufficient coverage.

- If your organization begins exploring custom prompts, embeddings, or external APIs, that's where enhanced validation and governance practices should come into play.

The principle is proportionality: secure what needs securing, but don't let fear create unnecessary roadblocks to value.

Operationalizing Governance Through Your Data Inventory

Data governance is often misunderstood. It's seen as a necessary overhead - a set of controls to minimize risk or avoid regulatory penalties. However, in an AI-driven organization, good governance does something far more powerful: it creates value.

Your data inventory - if treated correctly - becomes the engine of that value.

Governance starts with visibility - but matures through reuse.

Earlier in this book, we introduced the concept of building a data inventory: mapping your data assets, understanding where they live, who owns them, and how sensitive they are. This is a vital first step - but it's not the end goal.

Inventories are not just for control. They are for *discovery*, for *reuse*, and for *compounding the value of your data assets over time.*

Think of the inventory less like a ledger - and more like a product catalogue. It's the single source of truth that enables every future AI initiative to go faster, safer, and with less waste.

Without seeing or using it in this way, every new solution starts from scratch:

- Searching for data owners
- Duplicating data sets

- Introducing avoidable risks
- Wasting time and energy

With it, your governance function shifts from enforcing rules to accelerating opportunity.

Why Reusable Data Assets Create Compounding Value

Data assets multiply their impact every time they're reused. This is a crucial mindset shift. Reusable data assets save effort, accelerate development, reduce risk, and enable faster innovation – all great tangible value benefits and not just once but every time they are applied to a new problem or solution.

In effect, the more a data asset is used, the more valuable and resilient it becomes – both technically and organizationally. It works like this:

- The first solution bears the highest cost – finding the data, preparing it, ensuring quality, and establishing governance.
- Every subsequent solution starts faster, builds on known patterns, and leverages existing approvals.
- Feedback loops improve data accuracy, completeness, and documentation.
- Trust spreads – both in the data and in the governance process itself.

This creates a flywheel effect:

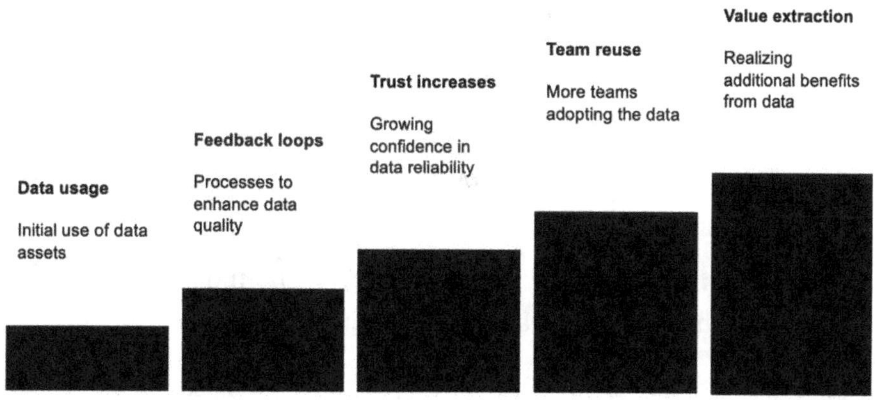

Figure 9-10. *Compounding effect of reusable data assets*

This is why modern governance strategies increasingly view value (tracking and communicating) as an extension of their key responsibilities. However, if data reuse drives compounding value – how do you measure it?

Capturing Value Signals from Data Reuse

Here are some early, foundational signals that your data inventory is generating value:

1. **Reuse frequency**

   Track how often specific datasets are referenced in solutions, Power Apps, AI models, Copilot Studio agents, or dashboards.

   a. **Metric**: # of solutions per data asset.

   b. **Signal**: High reuse indicates trusted, well-prepared assets.

2. **Cross-team adoption**

   Measure how many distinct teams or departments are using the same dataset.

   a. **Metric:** # of distinct business units reusing a data asset.

   b. **Signal:** Reuse across organizational boundaries shows data portability and strategic relevance.

3. **Time-to-value reduction**

   Estimate how much faster solutions reach production because they used an existing, governed dataset rather than building from scratch.

   a. **Metric:** Estimated effort saved per solution (hours or cost).

   b. **Signal:** Demonstrates tangible productivity and agility benefits.

4. **Governance load reduction**

   Measure how often RAI checks, sensitivity labels, or approvals were already in place because a dataset was reused.

   a. **Metric:** # of solutions using preapproved data assets.

   b. **Signal:** Governance as an accelerator, not a blocker.

5. **Feedback and improvement loop**

   Track the number of quality issues surfaced, resolved, or enhancements made to a dataset after reuse.

a. **Metric**: # of feedback loops per dataset.

b. **Signal**: A living, improving data asset - not a static one.

Practical guidance: Start with simple signals

This doesn't need to be complex in the early stages. Even manually tagging datasets in your inventory with basic metrics (e.g., reuse count, teams involved) creates visibility.

Over time, mature organizations automate this by integrating their data inventory with platform analytics tools - like the CoE Starter Kit, Azure Application Insights, etc.

Why this matters

Data governance should create value as well as protect it. When you measure reuse, you create new conversations about data value. Value extends beyond tangible value benefits as data stewards can prioritize maintaining the most valuable data assets. Data stewards are one of the key roles in ensuring that your data assets remain trusted, accurate, and ready for reuse. Let's explore these roles - and wider responsibilities - more closely.

Defining Core Data Roles and Responsibilities

Most organizations already have people managing data in some capacity - whether through formal data teams or informal ownership within business units. This section is about making existing responsibilities explicit and visible within the context of AI readiness.

Clear accountability doesn't just help prevent mistakes - it gives people confidence. It makes the obvious who to ask, who approves reuse, and who is ultimately responsible for the quality and integrity of the data being used.

For now, these roles don't need to be complex - but they *do* need to be clear.

Key roles typically include the following:

Role	Responsibility	Example in action
Governance lead	Coordinates meetings, policy creation, and ensures alignment across teams	Schedules monthly data council reviews to assess AI use of sensitive datasets
Data owner	Accountable for specific data sets (e.g., HR data, customer data), making sure the data stays accurate, complete, and secure	Signs off on salary dataset before it's used in an AI-driven compensation analysis
Data steward	Supports data owners, enforces day-to-day data policies, and helps colleagues resolve basic data questions	Flags a data quality issue in Power BI and coordinates corrections with the team

These roles don't need to be new full-time positions – they may align with existing business owners or IT roles. What matters is clarity, and as we'll explore next – clarity is also what allows governance habits to take root.

Embedding Lightweight Governance Habits

At the foundational stage, governance should be focusing less on controls and more on establishing consistent, repeatable habits.

These habits become the scaffolding for more advanced governance later. They teach teams that good data management isn't something done all in one go – it's part of how AI solutions are built, deployed, and improved.

Four habits to operationalize early

These are simple, repeatable actions I recommend embedding across teams from the very beginning:

1. **Ownership is recorded, visible, and enforced**

   This aligns with the RAI principle of accountability. Every dataset or data source should have an identified owner - and that ownership shouldn't live in a spreadsheet buried away somewhere.

   Practical approaches

   a. Include a *mandatory data owner* field in your inventory.

   b. **Build reuse prompts into processes**: When someone requests access to data, the owner is automatically notified or included.

   Outcome: Ownership isn't theoretical - it's embedded into daily tools and visible to makers, developers, and governance teams.

2. **Every dataset has a documented purpose**

   This aligns with the RAI principle of transparency. A dataset without a clear intended use increases risk. People start to copy it, repurpose it, or feed it into models it was never meant to support.

   Practical approaches

   a. Maintain a *purpose* field in the inventory, linked to AI usage types - training data, knowledge source, conversational search, reporting only, etc.

   b. Encourage teams to validate the intended purpose before reuse or integration.

   Outcome: You reduce ambiguity and avoid data being misapplied in sensitive scenarios.

3. **Sensitivity and risk markers guide decision-making**

   This aligns with the RAI principles of fairness and reliability and safety. Foundational governance is about giving teams the context to make better choices.

   Practical approaches

   a. **Use a lightweight classification model**: Internal only, confidential, regulatory-sensitive, RAI-sensitive

   b. **Encourage teams to document known risks or fairness considerations against datasets**: Bias potential, underrepresentation, outdated data

   Outcomes: People can move fast – but they're moving with their eyes open.

4. **Data quality checks are part of solution delivery**

   This aligns with reliability and safety and accountability principles. If an asset depends on specific datasets – there should be a defined expectation for verifying accuracy, completeness, consistency, and timeliness (the four dimensions of quality).

   Practical approaches

   a. Integrate a simple data readiness checklist into your asset deployment process.

   b. Encourage or mandate a review between developer/maker and data owner prior to release.

c. Store this review in your inventory record - a living history of readiness actions.

Outcomes: Data quality isn't assumed - it's verified in a lightweight, scalable way.

Framing the Habit Shift

These habits turn the inventory from a passive record into an active part of your AI solution life cycle. They build muscle memory, so when governance matures into automated policies, advanced controls, and auditability, your teams are already aligned in mindset and behavior.

This is how trust scales - not because a policy says so but because the habit was built from day one.

Establishing Basic Data Security: Building Guardrails, Not Roadblocks

Data governance without security is like locking the front door but leaving the windows wide open. Earlier in this chapter, we focused on creating clarity - understanding what data you have, who owns it, and how to prepare it for AI solutions. Clarity without control, however, is fragile.

Security at the foundational stage is about building the minimum viable guardrails that enable safe innovation - not blocking progress but ensuring that every AI-powered asset is developed with trust built in from the start. This is not a world of zero risk; it's a world of well-managed risk.

Why Security Is a Data Readiness Concern

In responsible AI, we talk a lot about privacy, fairness, and accountability. Security underpins all of them.

- *Fairness* means data is used appropriately for its intended purpose.
- *Transparency* means users understand what data is collected and why.
- *Privacy and security* ensure data is protected from misuse, leakage, or exploitation.
- *Accountability* ensures there are clear controls in place for who can access what.

Without security habits in place, even the best AI solution can collapse under the weight of a data breach or misuse.

Four Practical Habits for Foundational Data Security

1. **Classify sensitive data early**

   Start simple. Use broad categories like

   a. Public/internal use/confidential/regulatory-sensitive/RAI-sensitive

   Add a field for classification in your data inventory. Make this visible in your AI solution life cycle.

Use Microsoft Purview to automatically discover, classify, and label sensitive data at scale as you mature. Purview's data map[9] provides a centralized view across services - Dataverse, SharePoint, SQL, Azure Data Lake.

Outcome: Classification drives better decisions and sets the stage for automated enforcement later.

2. **Apply role-based access control (RBAC) as a default**

   In Power Platform, it's easy to underestimate access risk - because users are empowered to build fast. But sensitive data demands clear access boundaries.

   Practical steps

   a. Use Dataverse security roles for data assets used in apps, flows, or agents.

   b. Limit read/write access to sensitive tables or records.

   c. Ensure your environment strategy in Power Platform separates development from production - preventing accidental exposure.

   Outcome: Access control shifts from manual decision-making to embedded guardrails.

[9] https://learn.microsoft.com/en-us/purview/data-map

3. **Protect data at rest and in transit**

   Even within trusted environments, encrypting sensitive data is a nonnegotiable standard for responsible AI.

   Practical steps

   a. **For Dataverse**: Encryption at rest is enabled by default. For higher sensitivity (financial data, healthcare), consider Customer Managed Keys (CMK)[10] to control your own encryption keys.

   b. **For AI knowledge sources (e.g., HR policy documents in SharePoint)**: Use sensitivity labels (Microsoft Purview) to apply encryption and access restrictions consistently.

   Outcome: Data remains secure even if copied, moved, or exported.

4. **Monitor usage and anomalies**

   Security is not a one-time configuration – it's an ongoing practice of detection and response.

   Practical steps

   a. Use Power Platform's built-in Data Loss Prevention (DLP) policies to block risky connector combinations (e.g., Dataverse to social media).

[10] https://learn.microsoft.com/en-us/power-platform/admin/customer-managed-key

b. As maturity increases, use Purview to apply data access policies at scale – ensuring regulatory-sensitive data is only accessible by preapproved roles.

Outcome: Monitoring creates a signal loop – detecting risk early and reinforcing accountability.

Establishing Data Principles: Setting a Foundation for Trust and Culture

Governance processes guide what we *do* with data. Data principles guide *how* we think about it.

At the foundational stage, you've already begun to build the scaffolding of operational governance – inventories, ownership, oversight, and basic security. But these structures only become meaningful when teams share a consistent mindset about data: why it matters, how it should be used, and the standards it must uphold.

That's the role of data principles.

Why Establish Data Principles Early?

Data principles are not policies. They aren't technical controls. They are short, memorable, organization-wide statements that shape behavior (like building AI solutions) and everyday decisions (like sharing a dataset of designing a form).

Critically, well-designed data principles scale. They don't need rewriting as your organization matures – they grow in relevance as your use of AI accelerates.

What Makes a Good Data Principle?

At this stage, your principles should

- Be clear, brief, and easy to remember.
- Balance value creation with protection.
- Align closely with responsible AI principles.
- Create confidence - not fear - around using data well.

Example foundational data principles

Here's a set of example principles that could form the basis of a simple data charter:

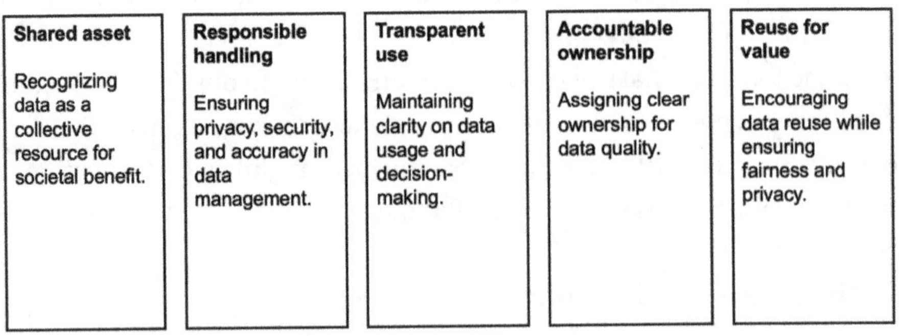

***Figure 9-11.** Example data charter aligned with RAI principles*

1. **Data is a shared asset**

   "Our data is created by people, for people. It creates more value when it's discoverable, reusable, and maintained for the benefit of all."

   This aligns with the responsible AI principle of "inclusiveness."

2. **Data must be handled responsibly**

   "Every person has a role in protecting the privacy, security, and accuracy of the data they work with."

   This aligns with "reliability and safety" and "privacy and security" responsible AI principles.

3. **Data use must be transparent**

   "People should understand what data is being used and for what purpose and have visibility into decisions driven by data."

   This aligns with the responsible AI principle of "transparency."

4. **Data owners are accountable for its quality**

   "Every dataset has a defined owner who is responsible for ensuring accuracy, completeness, and fitness for use."

   This aligns with the responsible AI principle of "accountability."

5. **Reuse increases value**

   "Data reuse is encouraged and celebrated - but must be governed, with appropriate checks for fairness, bias, and privacy."

   This aligns with the responsible AI principle of accountability and fairness.

Putting Principles into Practice

Principles, not unlike governance policies, have little impact if they live in a PDF or a SharePoint folder nobody reads.

Here are some practical ways to embed them into your AI readiness journey:

Practice	Example	Outcome
Reinforce in solution design	Ask solution teams to document how their AI solution aligns with the data charter.	Shift focus from technical build to ethical impact.
Incorporate into governance rituals	Open data council meetings by revisiting a principle and discussing recent examples.	Keeps principles active, not passive.
Use in onboarding and training	Include data principles as part of AI and Power Platform maker training	Builds a shared language and expectations early.

Principles don't replace governance processes – they are an enabler for them.

A well-run inventory is valuable because people understand their responsibility for the data it contains. Security controls are effective because people respect the boundaries they enforce. Ownership works because teams understand why stewardship matters.

As your organization matures in its use of AI, these principles will become the thread connecting technical governance, operational processes, and ethical leadership.

Conclusion: Data Readiness and Security – Foundational Stage

Data readiness isn't glamorous work. It doesn't create the loudest headlines in AI adoption, and often it happens quietly - in the background - while organizations race toward innovation.

But this is precisely what makes it so powerful.

When you invest in foundational data readiness, you create conditions where AI solutions can emerge faster, scale more safely, and earn greater trust from the people who use them. You move from "data as a risk" to "data as a renewable resource" - something that improves with each reuse, becomes more valuable with every quality check, and builds organizational confidence over time.

At this stage, your focus hasn't been perfection. It's been clarity, visibility, and consistency.

You've laid the groundwork for

- Governance that operates through habits, not bureaucracy
- Inventories that drive reuse, not just record-keeping
- Data stewardship that turns responsibility into value creation
- Security practices that protect without stifling innovation
- A shared set of principles that will grow with your organization

In the chapters ahead, this foundation will allow you to scale more ambitious AI initiatives. You'll expand your governance reach, embed responsible AI practices more deeply, and create environments where AI can thrive safely and ethically.

But none of that happens without data readiness.

This is the quiet (but busy) work that makes everything else possible.

Foundational Data Readiness Progression Checklist

This checklist summarizes the outcomes, practices, and behaviors that signal you've established a solid foundation for data readiness and security.

Governance habits

- [ ] Lightweight, repeatable governance habits are embedded across teams.
- [ ] Ownership is recorded and enforced. Every dataset has a documented purpose and AI usage classification.
- [ ] Sensitive features have been identified and communicated.
- [ ] Data quality checks (accuracy, completeness, timeliness) are part of solution delivery.

Data security

- [ ] Sensitivity labels or classifications are applied to all key datasets.
- [ ] Basic security controls are enforced based on classification.
- [ ] Integration with Microsoft Purview (or equivalent tooling) for metadata management and data governance.
- [ ] Data Loss Prevention (DLP) policies are active and aligned to sensitivity classifications.

Data principles

- [ ] A short, memorable set of data principles or a charter is defined.
- [ ] Principles reflect responsible AI values (e.g., transparency, accountability, privacy, reliability).
- [ ] Principles are communicated and reinforced through onboarding processes, solution design and governance discussions.

Reuse as value creation

- [ ] Reuse of data assets is tracked, encouraged, or at least recognized.
- [ ] Frequently reused datasets are prioritized for quality improvements, governance oversight and stewardship attention.
- [ ] The organization understands and communicates that reuse increases the value of data assets – creating compounding benefits over time.

CHAPTER 10

Emerging: Data Readiness and Security

Exploratory | Structured | Evolving

By this stage in your journey, a quiet but powerful transformation has already begun. You've moved from experimentation to intentionality. From isolated projects to emerging, visible practices. From building things to building trust.

In the earliest chapters of this book, we explored the cultural foundations of AI readiness – leadership, vision, responsible principles, and the habits that help organizations navigate uncertainty. We've talked about governance as a means of creating clarity, not control. We've embedded responsible AI thinking, not as an afterthought but as a design principle. In the last chapter, we tackled data readiness at its most fundamental level – understanding what you have, securing it, and preparing it for safe and responsible use.

But now, that landscape shifts.

This is the point in the maturity journey where data governance stops being quiet work – and becomes a strategic advantage.

S. Jeffery, *The Low-Code AI Maturity Model*, https://doi.org/10.1007/979-8-8688-1730-4_10

Data is no longer just being managed – it's being integrated. It's no longer just being protected – it's being activated. And most importantly, it's no longer just supporting isolated AI solutions – it's becoming the connective tissue across platforms, processes, and business decisions.

Emerging data readiness is where your organization begins to operate differently – because the value of connected, trusted, reusable data is becoming undeniable.

This is also the point where the concept of data products starts to take hold – treating key datasets not as hidden infrastructure but as discoverable, reusable, governed assets that fuel innovation across the enterprise.

And just as importantly – this is where responsible AI practices start to scale within your data ecosystem.

Fairness becomes something you audit systematically – not just something you check once. Transparency becomes embedded in your data life cycle – not retrofitted after the fact. Security becomes a continuous discipline – not a gate at the end of the process.

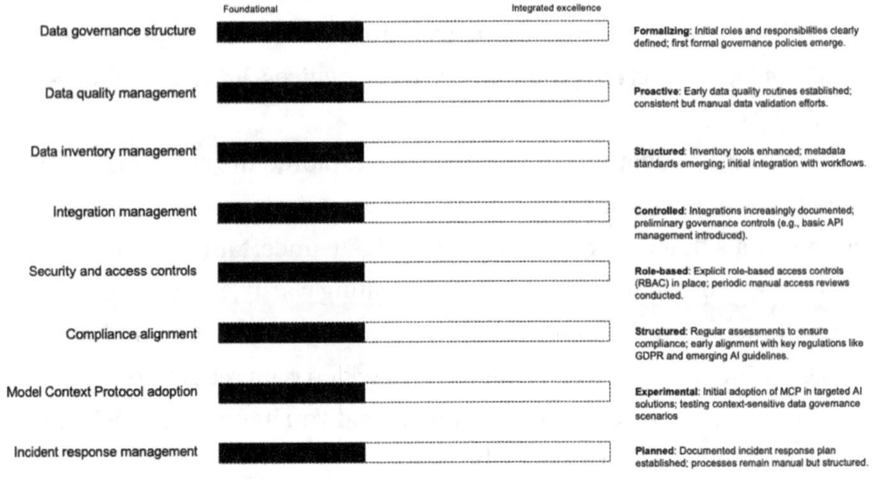

***Figure 10-1.** Typical characteristics of an organization at the emerging stage*

This chapter is about moving from stewardship to strategy. It's about building a data ecosystem that is not only technically integrated but ethically grounded, resilient, and positioned to power the next wave of AI innovation.

Let's explore what that looks like.

Strengthening Internal Data Governance

At the emerging stage, the shift is from inventorying what you have to how you use it - operationalizing ownership, quality, and control as living parts of your AI life cycle.

This section builds on your foundational work and provides clear guidance for expanding governance across your internal data landscape.

Automating Ownership Validation

Ownership clarity isn't scalable if it depends on manual upkeep. You need to automate ownership identification as well as ongoing validation. The goal here is to create self-maintaining data ownership models that reduce manual governance overhead.

Here are some recommended practical practices:

- Require ownership fields to be mandatory in the inventory. This should not be negotiable.
- **Automate regular validation**: Use Power Automate to send quarterly check-in messages to data owners:
    - "Are you still the correct owner of [Dataset X]?"
    - "Have any changes been made to this dataset's source of structure?"

- **Implement a default escalation:** If no owner responds after a defined period (e.g., 30 days), escalate to a governance lead or CoE teams' channel.
- Build automated notifications to notify owners when their dataset is used in a new AI solution.

Signals of maturity

- Ownership confirmation rates exceed 95% per review cycle.
- No "orphaned" data assets exist in the inventory.
- Reuse requests are routed automatically to the correct owner for approval.
- Data council review includes ownership engagement metrics.

Why this matters

Clear ownership underpins RAI principles of accountability and transparency - without it, every other governance practice weakens.

Formalizing Data Quality Policies

Data quality is now a defined expectation across solution teams. The goal with this is to establish documented and enforceable standards for the four dimensions of data quality:

- Accuracy
- Completeness
- Consistency
- Timeliness

Establish clear organizational quality policies that

- Publish data quality templates for structured and unstructured data.

  What this means in practice

  Data quality templates provide teams with a standardized starting point - defining the minimum expectations for how data should be assessed, prepared, and maintained before it's used in AI-powered solutions.

  What a structured data template might include:

Section	Example content
Critical fields	Define mandatory fields (e.g., CustomerID, ProductID, Email)
Sensitive fields (e.g., fields that require special permission to view)	Salary, passport number, bank account
Validation rules	Numeric range checks, date formats, uniqueness constraints
Completeness threshold	Minimum % of non-null values for key fields (e.g., 95%)
Accuracy controls	Cross-check required before use
Ownership	Named data owner with contact details

  Provide these templates as downloadable word docs, Microsoft forms, a model-driven app, etc. - make sure that it's easy to find, located in your CoE hub site or equivalent.

What an unstructured data template might include:

(Aimed at SharePoint content, policy documents, knowledge bases, etc.)

Section	Example content
Source verification	Document must originate from trusted system or owner (HR portal, legal site).
Metadata requirements	Author, last updated date, version number, intended audience.
Structure expectations	Use of H1/H2 headings, clear summaries, bullet lists where appropriate.
Review cadence	Quarterly or annual content review for accuracy and relevance.
Classification	Sensitivity label applied (public/internal/confidential, etc.).

Consider providing deeper guidance about how to configure (for example) a SharePoint document library with preconfigured metadata (content types), automated review workflows (information management policies, Purview risk, and compliance solutions[1]), and a sample word template.

- Implement checklists for AI solution development, enforced via CoE Starter Kit components or solution templates.

  Where should checklists live?

[1] https://support.microsoft.com/en-gb/office/create-and-apply-information-management-policies-eb501fe9-2ef6-4150-945a-65a6451ee9e9

Embed them inside your existing solution development life cycle, or create an AI solution design template/form.

Example: AI solution readiness checklist (structured data)

Check	Owner	Verified (Y/N)
Data owner identified	Solution developer	
Accuracy spot check performed	Data owner	
Completeness threshold met	Data owner	
Consistency validation completed	Developer	
Sensitivity classification applied	Developer	
Data purpose documented	Developer	

Example: AI knowledge readiness checklist (unstructured data)

Check	Owner	Verified (Y/N)
Document source verified	Solution developer	
Metadata applied (owner, review date)	Developer	
Content structure validated	Developer	
Latest version published	Data owner	
Sensitivity label applied	Developer	

- Integrate Power Automate flows to enforce minimum data completeness prior to deployment (e.g., block deployment if key fields are missing values).

- Define review and verification cadence:
    - Quarterly for critical operational datasets
    - Annually for knowledge assets or reference data

Signals of maturity

- Data quality exceptions are captured and reviewed via Power BI dashboards.
- Critical datasets have documented validation routines and success criteria.
- Business users are equipped with simplified guides for maintaining data quality.

Operational guidance

- Incorporate data quality acceptance criteria into solution delivery pipelines.
- Provide solution templates (Power Apps, Power Automate) with built-in validation rules for common data entities.
- Consider enabling Power BI Data Quality Metrics where appropriate for reporting.

Why this matters

Quality is not just about internal trust - poor-quality data creates downstream RAI risks, such as unfair outcomes or incorrect recommendations.

Designing and Operationalizing Data Validation Routines

A validation routine is a repeatable process designed to check whether a dataset meets specific quality standards before it is used in a solution, published, or shared for AI development.

What should validation routines check for?

Most validation routines at the emerging stage focus on the four foundational dimensions of data quality:

- **Accuracy**: Spot-checks against source systems; lookup verification (e.g., Customer ID exists in CRM).
- **Completeness**: Checking that mandatory field data is populated; % fill rate thresholds per field.
- **Consistency**: Standard formats are enforced (e.g., currency codes, dates, region names).
- **Timeliness**: Date-based checks to ensure records are not older than expected.

The obvious challenge here is automating/configuring platforms for scale. Here are a few suggestions:

Example: Structured data validation routine

Validation step	Method	Frequency
Field completion check	Power Automate flow to run weekly and flag records with null values in critical fields – note: consider throttling/performing batch operations for a large corpus.	Weekly
Duplicate detection	Dataverse duplicate detection rules or Power Automate.	Continuous/ real-time
Lookup verification	Power Automate calling API to verify (for example) Customer ID exists in CRM.	Daily
Data freshness check	Scheduled flow to alert if records older than 30 days.	Daily

Example: Unstructured data validation routine (knowledge sources)

Validation step	Method	Frequency
Metadata completeness	Mandatory fields in SharePoint	At upload
Formatting validation	Power Automate enforces suitability checking for headers or structure	At upload
Content review workflow	Data owner approval step in document library before inclusion in knowledge based	Quarterly

Signals of maturity

- Critical datasets have automated validation routines designed and documented.
- Data quality acceptance criteria are explicitly listed in solution deployment pipelines.
- Data stewards maintain these routines as part of their role - supported by tooling.
- AI solution development cannot proceed to production without passing a defined quality gate.

Data quality policies give you confidence in the integrity of your data - but quality alone isn't enough. Where data lives - and how environments are structured - has a direct impact on how safely, responsibly, and effectively it can be used.

A dataset that's perfectly accurate and validated might still be at risk if it's stored in a poorly governed environment, shared too widely, or exposed to unintended users.

This is why the next step in emerging data governance focuses on establishing environment-specific policies that create clear boundaries between experimentation, development, and production environments. These policies help teams move fast without compromising security, privacy, or responsible AI principles. Just like we introduced governance dimensions to assess and manage solution risk, your data estate deserves the same structured approach.

Establishing Environment-Specific Data Policies

One of the critical enablers of responsible data readiness is the deliberate separation of environments for different types of work.

AI development brings new risks, particularly when experimenting with sensitive or production-grade data. It is not enough to rely on general platform-level policies - your environments themselves should carry explicit, embedded expectations around how data is handled, protected, and used within them.

Your environments act as risk boundaries for your data. The goal here is to apply proportionate, risk-aware controls that balance agility with responsibility:

Proposed data governance dimensions

Dimension	Description	Example risk factors
Sensitivity	Confidentiality and regulatory impact	Customer PII, salary data, health records
Criticality	Business impact if unavailable or corrupted	Financial transactions, operational systems
Volatility	Frequency of data change	Dynamic datasets vs. static reference data
Usage scope	Breadth of consumption across teams	Enterprise-wide vs. local team datasets
AI sensitivity	Direct use in AI models or decisioning	Training datasets, model inputs, knowledge bases

Risk Classification Guidance

Using these dimensions, data owners and stewards can classify datasets within your inventory. This enables environment-specific controls, as follows:

Environment type	Recommended policies based on data dimensions
Sandbox/development	No sensitive or critical data. Use mock or masked data where possible.
Test	May include higher-sensitivity data, but with strict access controls and monitoring.
Production	Full governance enforcement based on sensitivity and criticality. Logging, DLP policies, and sensitivity labels applied.

How to operationalize data dimensions and risk classification

1. Define the dimensions (start with the five above).
2. Publish simple scoring criteria (low, medium, high for each).
3. Add fields to your data inventory to capture these scores.
4. Build automated reporting that flags
    a. Sensitive data in non-production environments
    b. Critical data without ownership assigned
    c. AI-sensitive data without quality validation logged
5. Use your governance council or CoE to review outliers quarterly.

Practical Assessment: Example Matrix

For each dataset, establish a simple scoring system, during your inventory enrichment or review:

	Low risk	Medium risk	High risk
Sensitivity	Public or non-confidential	Internal only	Confidential/ regulated/ personal data
Criticality	Non-operational	Operational reporting	Business-critical systems depend on data availability
Volatility	Static/ rarely changes	Periodically updated	Highly dynamic, real-time feeds
Usage scope	Single-user/ ad-hoc	Team / department	Organization-wide or AI-discoverable
RAI sensitivity	Reporting only	Informing agent responses	Feeding predictive or decision-making models

Figure 10-2. *Dataset risk assessment matrix*

Policy and Environment Application: Translating the Matrix into Controls

Once assessed, the matrix directly informs your environment strategy and policy application:

Risk profile	Recommended environment	Security policies to apply
Low risk	Developer or sandbox	Minimal restrictions
Medium risk	Team or department scoped	DLP policies, monitored access controls, automated ownership validation
High risk	Production environment	Mandatory data readiness checklist, approvals for AI reuse, sensitivity labels, data encryption, Purview integration

You see, it's not about locking everything down equally - your ability to apply precision controls where they matter most is what builds trust and encourages data reuse and even further innovation and value.

The reality of modern solution development is that data doesn't stay confined to one platform, one system, or one environment. **AI amplifies this challenge**. Models, copilots, and agents increasingly rely on data drawn from a web of interconnected systems - many of which live outside your direct control.

It's one thing to classify and protect data you own. It's another to govern the movement of that data across systems, vendors, and platforms - without losing visibility or increasing risk.

That's the challenge we'll tackle next.

Governing Data Integrations Responsibly

Every integration is a decision point.

When you connect a system of record, expose an API, or query an external data source, you're not just solving a technical problem - you're opening a trust boundary. You're extending your data governance principles into places you might not fully control.

At the emerging stage of data readiness, this becomes one of the most important transitions your organization will make. Until now, we've focused on strengthening internal governance - building habits, classifying assets, and driving clarity around data ownership. But the moment you integrate with external data or expose your own data for consumption, your risk profile changes dramatically.

Integration governance is where responsible AI becomes **operational**.

- *Fairness* is impacted by data lineage - are you inheriting unknown bias from external data?
- *Transparency* is impacted by integration clarity - do teams know where the data originates?

- *Privacy and security* are challenged by open APIs and hybrid architectures.
- *Accountability* becomes complex when ownership spans multiple vendors or systems.

Why Integration Governance Is a Maturity Signal

Organizations often find themselves overwhelmed by integration sprawl:

- API connections built without documentation
- Flow and automations querying sensitive data without review
- Power Platform solutions connecting to SaaS platforms with unclear data handling practices
- AI agents surfacing answers based on poorly governed external knowledge bases

Integration governance is where you can enable these connections responsibly. The most mature organizations view integrations as a value stream - a deliberate, managed capability that accelerates innovation without sacrificing control.

This is the mindset we'll adopt next.

We'll explore

- Common integration scenarios you'll need to govern
- Practical governance patterns and controls for responsible data integration
- Emerging industry practices like the Model Context Protocol (MCP) - shaping how AI agents consume data responsibly

- The risks of unmanaged integration sprawl
- And – most importantly – how to turn your integration capability into a driver of business value

Integration Scenarios

Data integration is where governance maturity gets tested in the real world.

Inside your own environments, governance habits feel achievable – your teams control the systems, the data, the people. However, as soon as your AI-powered solutions need to connect outside of those environments – to SaaS platforms, legacy systems, on-premises data, or public APIs – complexity multiplies.

Integration is no longer just a technical design decision.

It's a governance boundary decision.

It's a risk management decision.

It's a value realization decision.

This section is about making sure you don't treat integration as an afterthought – or a convenience layer – but as a strategic capability that can either amplify trust or erode it.

Common Integration Scenarios Requiring Governance

The emerging stage of data readiness is often where integration patterns start to diversify rapidly. Your governance approach needs to anticipate four broad categories:

1. **Internal systems of record**

   These are the foundational systems inside your organization – the single source of truth for key data entities:

a. HR systems (Workday, Dynamics 365 HR)

b. Finance and ERP (SAP, Oracle)

c. CRM (Dynamics 365 Sales, Salesforce)

Governance considerations

- Are you connecting directly to the system of record or to a derived dataset?
- What update frequency is appropriate - real-time vs. batch?
- Are the integration points documented, secured, and monitored?
- Which team is governing this system?

2. **First-party and third-party SaaS platforms**

These are vendor-managed services your teams depend on:

a. ServiceNow

b. SAP SuccessFactors

c. Salesforce

d. Industry-specific platforms

Governance considerations

- Does the SaaS provider have acceptable security, privacy, and RAI practices?
- Are you using approved connectors or building custom API integrations?
- Is data residency or regulatory compliance a factor?

3. **On-premises and hybrid data sources**

   Legacy systems won't disappear overnight. They need to be integrated securely - often requiring hybrid architectures like

   a. Data gateways (Power Platform data gateway)

   b. API proxies

   c. VPN-secured connectivity

   Governance considerations

   - Are integration patterns documented for resilience?
   - Are credentials and secrets managed securely?
   - Is data flowing through trusted infrastructure?

4. **External APIs and public data sources**

   AI solutions often rely on enrichment data - language models, industry datasets, or third-party knowledge sources - like

   a. Open data APIs

   b. Public knowledge graphs

   c. AI language APIs

   Governance considerations

   - Is the source reputable, licensed, and maintained?
   - Are rate limits, SLAs, and usage costs understood?
   - Is data being cached appropriately or calling APIs on demand?

This mapping exercise matters as every integration scenario changes your governance responsibilities:

Scenario	Primary governance focus
Internal systems	Ownership, accuracy, stewardship
SaaS platforms	Vendor compliance, API security
Hybrid/on-prem	Infrastructure trust, connectivity security
Public APIs	Source validation, usage control

Governance Controls and Patterns for Integration

Good integration is invisible to end users – but highly visible to governance teams. At the emerging stage of data readiness, the focus is no longer just enabling integration – it's governing it responsibly, consistently, and transparently.

Think of integration governance as having three layers of control:

1. **Control what can integrate: Define integration entry points**

   Responsible integration starts with defining trusted entry points to your data estate. Not every API or connector should have direct access to sensitive or critical systems.

   Implementation guidance

   a. Standardize on approved connectors; make them visible to all in your CoE site or repository.

   b. Use environment-specific integration zones – for example:

i. **Sandbox environments**: Open experimentation with limited external connectivity.

ii. **Production environments**: Restricted to validated, approved connectors only.

c. Use Azure API management (or similar services) to expose internal systems securely, applying authentication and monitoring.

2. **Control how data is consumed: Enforce sensitivity and risk-aware integration**

Your data inventory already classifies sensitivity, risk, and usage restrictions - not extend those classifications into integration patterns.

Implementation guidance

a. Require metadata on integration endpoints to declare

i. Sensitivity level of data being shared

ii. Intended use case (e.g., reporting vs. training an AI model)

iii. Limitations or downstream data handling requirements

b. Apply sensitivity-based restrictions:

i. Confidential data cannot be exposed publicly.

ii. Regulatory-sensitive data requires additional approval steps.

Example: AI-powered customer feedback analysis

Let's say you're integrating a customer feedback system with an agent that detects sentiment and emerging issues.

- The data includes customer emails and complaint transcripts - flagged in your inventory as RAI-sensitive due to identifiable personal information and emotional tone.
- A maker wants to pull this data into an agent using sentiment analysis in AI Builder using Power Automate.

Here's how responsible integration governance might apply:

- The integration is configured to expose only pseudonymized data fields (e.g., removing name, email).
- The integration record is tracked in your solution inventory, with a 90-day review cadence to revalidate privacy alignment.

This ensures the AI use case moves forward - but with safeguards appropriate to the sensitivity of the data.

3. **Control who can integrate: Govern integration ownership and lifecycle**

   Every integration should have an owner - just like every dataset. Treat integrations as first-class assets within your governance framework.

   Implementation guidance

   a. Record integrations in your data inventory or solution inventory - including

      i. Connected system

      ii. Owner and technical contact

iii. Data sources touched

iv. Risk level

v. Life cycle (creation date, review frequency)

b. Automate periodic reviews of active integrations:

i. Is the integration still in use?

ii. Has the risk profile changed?

iii. Are data classifications still valid?

Integration governance is about increasing organizational clarity. It ensures that when new AI-powered solutions emerge - pulling from diverse sources - there's already a shared understanding of how data flows, what risks are present, and who owns what. This approach scales innovation safely.

Traditional integration governance focuses on what data is accessed, by whom, and how. That's important. But as AI systems increasingly consume, reason over, and act on data - we also need to govern what they understand about that data.

This is where the Model Context Protocol (MCP) comes in.

Introduction to Model Context Protocol (MCP)

As AI agents and copilots become more autonomous, they need more than raw data - they need context. The Model Context Protocol was created so solve exactly this problem.

Originally developed by researchers and engineers at Anthropic - one of the leading AI safety research organizations - MCP is now a growing open standard, stewarded by the wider technology community via an open source initiative at modelcontextprotocol.io. Microsoft, through Azure AI and Copilot Studio, is one of the early adopters of MCP in a productionized, enterprise-ready way.

Why Was MCP Invented?

Anthropic introduced MCP[2] because they recognized a fundamental limitation in how large language models (LLMs) interact with external systems. While LLMs are powerful, they are not databases - and they should not be treated as a substitute for trusted system-of-record data.

MCP separates reasoning from retrieval.

It provides a structured way for models to request specific, governed context - ensuring the model reasons only over information that is

- Trusted
- Scoped
- Explainable
- Governed

As Anthropic put it:

> "MCP is a protocol for transmitting structured context to language models in a transparent, auditable way - allowing organizations to retain control over data while enabling models to perform complex reasoning."

This has far-reaching implications for enterprise AI - and makes MCP a natural fit for organizations embracing responsible AI practices.

Let's look at an example: a conversational agent without MCP.

Imagine a customer-facing AI agent built in haste, without MCP or structured context patterns.

A customer asks:

> "What is my latest invoice status?"

[2] https://www.anthropic.com/news/model-context-protocol

Without MCP, the agent might query a back-end API and return raw JSON - including sensitive internal metadata, irrelevant fields, or even data from the wrong customer record if access controls aren't enforced properly.

The agent is functioning - but without clear context boundaries, transparency, or governance controls.

Example: The same agent with MCP

Now imagine the same agent built with MCP principles:

- The API returns a structured context object: InvoiceStatusContext.
- This object includes only permitted fields: invoice number, status, due date.
- Metadata defines usage constraints: valid for this user only, expires in five minutes.
- The agent uses this context to generate its response - clean, accurate, and fully auditable.

Later, during a governance review, the CoE can see exactly what context was provided, for which users, from which system, and at what time - all without exposing sensitive back-end logic.

This shift is what makes MCP so valuable - it's about controlling how data interacts with language models in a way that's transparent, auditable, and safe.

What is MCP in practice?

At the technical level, the Model Context Protocol is not a service or database. It's a standardized JSON schema - an open protocol - used to pass structured, typed, and auditable data to a model. It sits between the system of record and the LLM, ensuring the context provided is

- **Deliberate**: Not an arbitrary dump of system data.
- **Typed**: Using defined schemas (e.g., InvoiceStatusContext, OrderDetailsContext).

- **Ephemeral**: Context expires after use or within a set time limit.
- **Governed**: Visible, reviewable, and policy-controlled.

It is probably best understood as a contract that bridges your structured data systems and your AI model - defining what the model sees and how it can use it.

Is MCP another layer in the stack?

MCP introduces a clear architectural model designed to sit between AI agents (such as Copilot Studio agents) and the systems or data sources they consume. At its core, MCP defines a standardized, machine-readable way to describe what data or actions are available - and under what rules. For example, rather than exposing a full employee record from Dataverse or an HR system, the developer could define a tightly scoped EmployeeContactContext like this:

JSON:

```
{
    "type:": "EmployeeContactContext",
    "name":"Steve Jeffery",
    "email": "steve.jeffery@contoso.org",
    "title": "HR Advisor",
    "department": "Human Resources",
    "expiresIn": 300
}
```

This context object becomes part of the agent's reasoning scope - passed into the prompt pipeline when relevant - but automatically expires after five minutes.

Core MCP components:

Component	Role in architecture	Governance consideration
MCP server	Publishes a manifest that describes available data, actions, usage policies, and context for an AI agent to consume. This is the sources of truth the agent queries.	Governed like any API endpoint – secured, version-controlled, with auditability over manifest changes.
MCP client	An agent built in Copilot Studio, acting as the consumer. It reads the manifest, interprets rules, and makes requests accordingly.	Clients should validate manifest signatures and ensure compatibility with approved schemas.
Manifest file	A machine-readable file that defines what resources (data/actions) exist, access requirements, sensitivity, and additional metadata.	This is a governance asset – subject to approval, sensitivity review, and life cycle management.
GitHub repository	Microsoft's Copilot Studio allows MCP manifests to be hosted in GitHub repositories for transparent versioning and governance.	GitHub repository governance – branch protection, code owners, audit logs, and review processes apply here.

Architectural flow (simplified)

1. Data owners or solution architects define a manifest file describing the available API or data resource.
2. The MCP server hosts this manifest - it could be a dedicated endpoint, static file, or GitHub repo.
3. The MCP client (your agent) queries the manifest at runtime.

4. Based on manifest-defined rules (auth, rate limits, sensitivity), the client makes requests to back-end systems.

MCP in Copilot Studio

Microsoft's adoption of MCP within Copilot Studio is a major step forward in aligning low-code AI agents with best practices for responsible AI.

In a typical conversational agent, integrating structured data from multiple systems has always been possible - but until now, there hasn't been a standard, governable way to define what data is passed to the model, how it's structured, how long it remains valid, and whether it should even be used for reasoning.

MCP provides this structure, but how does it work behind the scenes? What's happening under the hood?

1. You create an MCP manifest

   The .mcp.json file defines

   a. The action name and description

   b. Expected inputs and outputs

   c. API endpoint details

   d. Any constraints, expiration, or access logic

2. You host the manifest on an MCP server

   I recommend deploying[3] the official MCP server in an Azure Web App.

   a. This server reads MCP files (from GitHub.

   b. It exposes a REST API that Copilot Studio can connect to via a custom connector.

[3] Refer to the Microsoft Copilot Studio MCP lab for detailed walk-through: `https://github.com/microsoft/mcsmcp`

3. You create a custom connector in Power Platform

   a. This connector points to the Azure Web App (your MCP server).

This architecture doesn't just simplify integration - it gives governance teams a clear point of control, making it easier to review, approve, and monitor what agents are allowed to access and under what terms. Let's explore this more deeply.

Governance Implications for MCP

This model provides a governance lever that didn't exist in many traditional integrations.

Governance opportunities

- Approval workflows for manifest publication or updates
- Sensitivity labeling embedded in the manifest[4]
- GitHub repo governance (code owners, mandatory reviews, protected branches)
- Monitoring and logging of manifest consumption (access logs, usage telemetry)

MCP represents more than just a technical pattern - it's a serious governance enabler. Where traditional integrations often bypass governance controls (hard-coded API calls, undocumented endpoints, inconsistent usage), MCP forces design clarity:

- What data is exposed
- Who owns it

[4] Sensitivity labels are not yet standardized, but the MCP specification allows for custom metadata fields, which means you could include sensitivity information.

- How it should be used
- What constraints apply

This is where AI readiness and governance intersect. In an organization where autonomous agents and generative AI thrive, integrations can no longer be silent, undocumented handshakes in code.

Instead, MCP offers a blueprint for responsible integration - making every connection visible, governed, and adaptable over time.

Without governance patterns like MCP, integrations can sprawl unchecked, introduce shadow IT, leak sensitive data, or create fragile AI systems that are impossible to trust or audit.

Turning Integrations into a Value Stream

Every integration you build expands your organizations reach - into data, systems, and customer experiences that weren't possible before. The risk is real but so is the opportunity. If integrations are visible, well-managed, and built with reuse in mind, they become a compounding asset - not a liability.

Why Integration Capability Is a Value Stream

Integrations create value in multiple ways:

Value lever	Description
Data availability	Teams can access trusted, well-governed data without duplication.
Speed to market	Reusable integrations accelerate delivery of new solutions or AI agents.
Resilience and reliability	Centralized monitoring detects failures before they impact users.
AI readiness	Context-aware integrations (e.g., MCP) enable safe AI consumption of data.

The mindset shift at the emerging stage is realizing that integrations are not tactical glue – they are strategic assets. A well-governed integration becomes the foundation for faster solution development, reduced maintenance effort, and lower operational risk. Instead of dozens of hidden point-to-point flows, the organization invests in shared integration patterns and reusable data assets, each governed, discoverable, and monitored.

Capturing integration value

To move beyond assumptions of value, organizations need explicit metrics that quantify reuse, performance, and acceleration. Here's how:

- **Track reuse**: For every shared integration, record the number of downstream apps, flows, or copilots that use it.
- **Calculate development time saved**: Survey developers to estimate time saved by using existing integrations vs. creating new ones.
- **Surface the metrics**: Include value dashboard in your Center of Excellence reporting. Show not just integration usage but integration **impact**.

This data creates a compelling narrative for leadership – not just "we built this, look how many things we built" but "here's how often it's reused, here's how much time it saves, and here's the risk it helped us avoid."

Integration maturity **unlocks velocity**. As you begin to scale your AI solutions across departments and systems, **data is moving further, faster, and through more hands than ever before**. The challenge now shifts: How do you ensure the right guardrails are in place to support growth and not restrict it?

The next section moves from how data flows into your AI solutions to how you protect that data once it does. This stage is about **control, clarity, and confidence** at scale.

Establishing Data Protection and Security Practices

Guardrails for scale: Building security into everyday operations

By the time your organization reaches this stage of maturity, data security is no longer a theoretical discussion – it's embedded in the flow of daily operations. Whether it's a Power Platform solution built by a business team or generative AI agent ingesting sensitive documents. Every touchpoint must be governed by clear protections.

Here is the shift in mindset: protection is now about enabling safe innovation by making security visible, consistent, and automated. Your policies should now be working toward

- Being enforceable at the point of action
- Aligned to your integration and AI adoption strategies
- Built into the fabric of how your data moves, gets classified, and is accessed

Over the next few sections, we'll break down how to do this – focusing on practical tools, platform capabilities, and real-world implementation patterns you can embed across your ecosystem.

We'll explore

- Using Data Loss Prevention (DLP) policies to shape safe behaviors across makers, connectors, and environments
- How sensitivity labels and classifications offer context-aware security
- The role of Microsoft Purview as a scalable, integrated governance backbone

- Guidance on protecting AI-specific data risks, including personal and sensitive information
- Best practices for managing API permissions and external access

Data Loss Prevention (DLP) Policies

Data Loss Prevention (DLP) policies are one of the most powerful governance tools available within the Power Platform. But many organizations either underuse them or treat them as one-size-fits-all restrictions. DLP is a granular, adaptable guardrail mechanism that can grow with your organization's data maturity and AI adoption.

At its core, a DLP policy defines which connectors can be used together - and where. It draws a hard boundary between trusted (business) connectors and untrusted (nonbusiness) connectors, preventing unintentional data exfiltration or risky integration patterns.

This section explains how DLP works in Power Platform, including tenant and environment scopes, the role of connectors, how runtime and design-time enforcement work, what happens when multiple policies overlap, and how new capabilities like Advanced Connector Policies (preview at time of writing) offer even greater control.

For the emerging stage, we look to move beyond blanket rules. We're now looking to craft nuanced policies that reflect the sensitivity of your data, the risk profile of your integrations, and the context of your environments.

Why DLP Matters?

Data Loss Prevention is a mechanism that directly supports several responsible AI principles:

- **Privacy and security**: Controlling where data can be sent and how it's used
- **Accountability**: Ensuring traceability when sensitive connectors are invoked
- **Reliability and safety**: Preventing unexpected integrations or shadow systems
- **Transparency**: Making connector restrictions visible and explainable to makers

When applied thoughtfully, DLP makes AI adoption safer and more scalable by minimizing misuse and data leakage.

How DLP Works

At its core, DLP in Power Platform controls how connectors are used within an environment. A connector, whether to SharePoint, SAP, Outlook, or a custom API, represents a potential bridge between systems - and therefore, a potential risk.

DLP allows administrators to classify these connectors and apply rules about how they interact.

Two layers of control: Tenant vs. environment policies

DLP policies in Power Platform operate at two distinct scopes in Power Platform:

- **Tenant-level DLP**: Organization-wide policies that apply to all environments
- **Environment-level DLP**: Policies specific individual or grouped/selected environments

These can coexist - and when they do, **the most restrictive policy applies**. For example, if a connector is blocked in any applicable policy, it's blocked. If it's "business-only" in one policy and nonbusiness in another, its usability is limited to where it's grouped with other "compatible" connectors - more on that shortly.

Connector classification: Blocked, business, and nonbusiness

Connectors can be placed into three groups:

- **Business data only**: Approved for sensitive data and trusted interactions.
- **Nonbusiness data**: Intended for external, consumer-grade, or lower-trust use cases.
- **Blocked**: Cannot be used at all. This overrides all other classifications.

Important An app or flow can only use connectors that are classified in the **same group (business or nonbusiness)** *within the same DLP policy*. If a connector is classified differently across overlapping policies (e.g., tenant-level vs. environment-level), Power Platform evaluates them together and applied the most restrictive outcome. This enforcement helps prevent data from, say, a secure internal HR system in SAP, from flowing into a lower-trust destination like personal Dropbox – even if connectors are classified inconsistently across scopes.

Design-time vs. runtime

DLP enforcement happens in two phases:

- **Design-time**: Makers building apps or flows see connector restrictions immediately. If a connector is blocked or misclassified, the maker will receive an error and won't be able to save or publish.

- **Runtime**: Existing solutions are monitored and evaluated regularly. If a policy changes and renders a connector noncompliant, the flow or app will be suspended and the connection itself disabled until resolved.

This dual-layer model ensures that both new and existing resources remain compliant as policies evolve.

Policy overlap: What happens when multiple DLPs apply?

If several DLP policies apply to the same environment (e.g., one tenant-level and one environment-level), Power Platform evaluates them together and uses the most restrictive outcomes.

Where things get complex is when multiple policies classify the same connector differently - not as blocked but as business or nonbusiness. In these cases, Power Platform evaluates all policies and fragments connectors into separate groups. An app can only use connectors from a single group, meaning policy sprawl can unintentionally limit what can be built.

Governance tip: Keep the number of DLP policies per environment to a minimum. Overlapping policies increase complexity and can create unintended blockages due to fragmented connector groupings.

Advanced connector policies (ACP) [Preview]

A major evolution of DLP: Advanced Connector Policies provide greater granularity, environment group management, and action-level control - and are now available in preview for Managed Platform.

Key capabilities include

- **Allow listing**: Policies define what's allowed - everything else is blocked by default.
- **Trigger/action control**: ACP exposes connector internals, letting you block deprecated triggers or specific high-risk actions.

- **Environment group support**: Apply policies across related environments in bulk.
- **Mixed mode compatibility**: ACPs can work alongside classic DLP policies. The most restrictive rule (across both) applies at runtime.

ACP also supports automation via Power Platform APIs, enabling large-scale policy deployment and life cycle automation.

Governance Best Practices for DLP

- Start with core policies at the environment level, aligned to usage intent (e.g., production, sandbox, RAI sensitive, etc.).
- Minimize policy overlap to avoid complex group fragmentation.
- Adopt ACP for environments using advanced AI capabilities, where trigger level control is needed.
- Automate policy evaluation as part of solution readiness: include checks in your CoE Starter Kit flows.
- Audit connector usage regularly to identify policy violations and connector drift.

Understanding Connector Types

Connectors are the core building blocks that enable Power Platform apps, flows, and agents to connect with external systems. Whether your app is querying customer records from Dataverse, posting a message in Teams, or retrieving shipping data from SAP, it's the connector that brokers that interaction.

There are three primary connector types that appear in Power Platform governance tooling, and understanding how they work – both logically and architecturally – is essential to applying the right security policies.

1. **Certified connectors**

   Certified connectors are published and maintained by Microsoft or trusted partners. They undergo a review process to ensure they meet Microsoft's standards for performance, reliability, and security.

   A connector is only considered "certified" once it passes a structured approval process, which includes

   a. **Security review:** The connector must adhere to Microsoft's secure coding guidelines and demonstrate safe handling of data.

   b. **Functional testing:** Actions and triggers are tested against defined behaviors and edge cases.

   c. **Reliability standards:** Microsoft verifies uptime, error handling, and availability.

   d. **Documentation:** The partner must provide accurate descriptions of each action, parameter, and authentication method.

Tip If your organization develops a connector and wants it certified, you can follow the official Microsoft certification submission guide,[5] which outlines technical, policy, and branding requirements.

---

[5] https://learn.microsoft.com/en-us/connectors/custom-connectors/submit-certification

These connectors are built on a **common connector architecture**, with the following layers:

- **Connector manifest**: Defines actions, triggers, and parameters exposed to users (like Send Email, List Rows, or Detect Sentiment).
- **Runtime handler**: Handles authentication, request formatting, and response processing.
- **Connector gateway**: Routes the call securely to the target API (e.g., SharePoint Online, etc.), based on the privileges of the account that the connector is authorized with.
- **Cloud-hosted service**: The final API or endpoint the connector integrates with.

Security note: For most certified connectors using OAuth 2.0 authentication, the connector infrastructure never stores passwords. Instead, it securely stores and manages tokens, typically refresh tokens, which are used to request short-lived access tokens. This ensures secure, delegated access without exposing user credentials.

Governance insight: Even certified connectors can introduce risk depending on **how** and **where** they're used. For example, pushing HR data from Dataverse to Excel online may be compliant in one scenario but inappropriate in another. DLP policies allow fine-grained control over actions.

It's also worth noting connectors don't "move data" by themselves – they authorize, invoke, and broker communications with a service. The actual data exchange is handled by the Power Platform engine at runtime.[6]

2. **Custom connectors**

   Custom connectors allow you to wrap any external or internal API and expose it in Power Platform as if it were a standard connector.
   Under the hood, they still follow the same manifest-based structure as certified connectors, but

   a. You define the OpenAPI schema, authentication model, and UI definition.

   b. You're responsible for securing the endpoint, validating data formats, and managing performance.

   c. They can bypass traditional review processes and introduce sensitive data movement if not governed properly.

   Governance insight: Because custom-connectors are user-authored, they're a common vector for data exfiltration. If unmanaged, a single poorly secured connector can leak sensitive information to an external API with no audit trail. All custom connectors should be reviewed, classified, and – where appropriate – sandboxed or blocked.

[6] https://learn.microsoft.com/en-us/connectors/connector-architecture

3. **Virtual connectors**

   These are administrative or feature-based toggles surfaced as "connectors" in governance tools so they can be managed/blocked with DLP, but don't correspond to external APIs, for example, connectors that enable or disable Copilot Studio behaviors.

Connectors are the vessels that allow data to flow across systems, services, and environments. Governance of these vessels is essential - but what's being transported matters just as much.

If we can govern *how* data moves, we must also classify *what* is moving.

This is where data classification and sensitivity labeling come in. These are the foundational signals that tell your governance framework how to treat each data asset - whether to protect it more stringently, restrict its reuse, or monitor its downstream use in AI systems.

Sensitivity Labels and Data Classification

In any enterprise, not all data is equal.

Some datasets are harmless and public. Others carry regulatory obligations, personally identifiable information (PII), intellectual property, or latent risk when used in AI workflows. The more you can define and understand the sensitivity of your data, the more confidently you can unlock its value.

As data assets multiply and become increasingly woven into copilots, prompts, and automated workflows, understanding what you're working with - and how sensitive it is - becomes nonnegotiable.

This is where sensitivity labels and data classification frameworks come in.

What Sensitivity Labels Enable

It's important to clarify what sensitivity labels do – and what they don't.

Sensitivity labels are metadata tags defined in Microsoft Purview Information Protection and applied to files, data assets, emails, and more. They don't **move data, block access**, or **enforce policies** by themselves.

Who's responsible for sensitivity labels and Purview?

While sensitivity labels are used broadly across the organization, responsibility for defining, maintaining, and governing their usage typically falls into the following roles:

- **Information protection or compliance teams**

  Often housed within IT security or a central Risk and Compliance function, this team is typically the system owner for Microsoft Purview and leads the development of the data classification schema. They define what each sensitivity level means (e.g., Public, Internal, etc.) and own the organization's information protection strategy.

- **Security operations (SecOps)**

  These teams help operationalize classification decisions – ensuring labels trigger the right actions and integrate with Microsoft Defender, Microsoft Entra, or Intune if needed.

- **Platform governance or Power Platform Center of Excellence**

  For Power Platform and AI-specific scenarios, the CoE team takes responsibility for embedding label checks into environments – especially when scaling data reuse or enabling AI models to access shared sources.

Takeaway: Classification is a cross-functional responsibility - but ownership of Purview itself often sits within Information Protection or Security and Compliance teams. Power Platform CoEs should partner with them to ensure low-code assets respect and reflect those classifications.

As we've seen, sensitivity labels and classification frameworks are essential tools in shaping how data is understood and handled across the Power Platform. Tools, as you know by now, are not enough. Labels don't govern themselves, and risk doesn't pause between projects.

To scale data governance alongside the growing complexity of your AI solutions, you'll need more than metadata and policies - you'll need people, forums, and feedback loops.

This is where your governance operating model begins to evolve - shifting from a handful of good practices into an ecosystem of accountability, collaboration, and continual improvement.

Microsoft Purview As a Data Governance Backbone

Power Platform offers rich internal governance, but enterprise data doesn't live in a vacuum. It lives in SQL databases, shared drives, third-party SaaS, and cloud storage. This is where Microsoft Purview comes in - enabling classification, cataloging, and visibility across your entire data estate.

Purview allows organizations to apply sensitivity labels, discover data lineage, and establish data domains - even for sources Power Platform connects to indirectly. This becomes especially powerful when those sources are accessed by AI agents or used in model training.

Rather than duplicating classification logic across tools, CoE teams should integrate with the Purview framework. For example:

- SharePoint or Azure Blob documents used in agents can inherit sensitivity labels from Purview.
- SQL databases with classified columns (e.g., containing PII) can be discovered and labeled centrally.
- Purview's data map helps Power Platform teams understand upstream and downstream data flows – improving impact assessments for solution changes.

This isn't a Purview book – and this chapter won't attempt to fully explore its capabilities, but it's important to acknowledge that as your AI adoption grows, Purview becomes increasingly essential as the cross-platform governance system for your data estate.

Rather than duplicating classification, discovery, and cataloging logic in the CoE's own inventory, the operational approach is integration and coordination. This is where maturity becomes visible.

Core responsibilities for Power Platform teams working with Purview:

Task	Why it matters
Coordinate discovery scans on Key Power Platform sources (e.g., Dataverse)	Ensure that newly added Dataverse environments are classified
Use Purview lineage to assess downstream impact	Vital when agents train on or infer from structured datasets with upstream dependencies
Establish a contact point between the CoE and Purview administration	Enables ongoing governance tasks like classification alignment, data retention compliance, or audit responses
Reference Purview's sensitivity schema during RAI sensitivity scoring	Aligns security, regulatory, and AI-risk perspectives under a single labeling model

If you don't yet have a formal engagement between your Power Platform governance leads and your enterprise data governance team, here are some low-friction starting points:

- Attend a data governance council meeting as an observer.
- Request read-only access to Purview scans and metadata exports.
- Propose a joint process for labeling AI model inputs or high-risk data sources.

These small operational steps build shared language, visibility, and knowledge - making future governance coordination around AI dramatically more scalable.

Protecting AI-Specific Data Risks

AI introduces new categories of sensitive data - and new risks. A dataset that once powered reports may now influence decisions, generate predictions, or shape conversations. You need to deliberately protect against AI-specific harms, including

- **Latent bias**: Demographic proxies (as we discussed in the "Fairlearn" section), historical bias patterns, or one-sided training data
- **Privacy breaches**: Leaking of customer or employee data through agent responses
- **Regulatory violations**: Using regulated data (e.g., health records, financial identifiers) in ways not permitted under law or policy

Much of this can be mitigated by applying an RAI sensitivity evaluation through the agent inventory system you've already built. For example, if a dataset is reused more than "n" times or flagged as "High sensitivity," it can trigger a policy review or validation.

The governance tip here is that you are not just assessing data quality - you are **assessing data readiness**. What's safe for one use case may be dangerous when embedded into others.

Governing Connectors and API Permissions

Connectors are gateways - and every gateway needs a lock. Earlier, we explored DLP policies, connector classifications, and custom connector risks. Here, we look beyond classification into governing access and behavior at the API level.

Every connector - whether certified or custom - ultimately maps to an API or service. Governing that API means

- Enforcing authentication and authorization (e.g., via OAuth2, Azure Entra roles)
- Auditing access to sensitive or regulated endpoints

For internal APIs, consider hosting them behind Azure API Management (APIM), which offers

- Centralized gateway control, independent of the connectors' frontend behavior
- Built-in telemetry via Azure Monitor or Log Analytics
- Security policies, including IP filtering, JWT validation, and CORS control

The tip here is to treat connectors as privileged interfaces - not just configuration objects. Review them regularly, align them with RAI-sensitive use cases, and involve security architecture teams when exposing new APIs to low-code makers or AI agents.

- The role of Microsoft Purview as a scalable, integrated governance backbone
- Guidance on protecting AI-specific data risks, including personal and sensitive information
- Best practices for managing API permissions and external access

Evolving Your Data Governance Operating Model

In Part 2, we introduced the core structures and roles that make responsible governance possible - from CoEs to risk registers to role definitions. This section doesn't reopen those topics. Instead, it focuses on what happens next: how your governance model begins to scale, adapt, and reinforce itself as data use accelerates across the platform.

You're not reinventing governance - you're strengthening its feedback loop.

You'll find that as more assets are catalogued, reused, and embedded into AI-driven solutions, the weight of responsibility subtly shifts. More people touch sensitive data. More decisions happen faster. And the quality of your governance increasingly depends not on a single policy but on how well your organization is listening, surfacing issues, and adapting its practices in real time.

Scaling Stewardship

At the foundational stage, we focused on data stewards - the individuals responsible for maintaining metadata, enforcing lightweight policy, and acting as the first point of contact for governance questions.

Now, this responsibility needs to scale horizontally and vertically.

- **Data custodians emerge**: Individuals responsible not just for the metadata but for coordinating ongoing oversight and ensuring that datasets remain compliant, complete, and aligned with downstream AI use.
- **Data architects play a more prominent role**, especially when reuse expands across business units or environments. Their involvement ensures that data is modelled and shared in ways that optimize both value and control.

Implementation tip Start by reviewing the most used dataset. Assign a primary steward and supporting custodian role, ideally across technical and business lines. Custodians can begin by reviewing change logs and recent access patterns monthly/bimonthly.

Aligning Governance with the CoE

The CoE is not just a cheerleader for platform adoption. It must become a critical governance ally, responsible for bridging the data responsibilities across makers, IT, and compliance functions.

To evolve the operating model, bring the following into the CoE's remit:

- **Data risk escalation**: Create a simple process for data owners or stewards to raise concerns about datasets being reused in sensitive contexts (e.g., used in a public-facing agent).

- **Inventory contribution**: Stewards and developers should have clear paths to submit or update metadata records via the CoEs tools and guidance.
- **Sensitivity compliance reviews**: If sensitivity labels or classifications are being used, CoEs should embed checks into deployment workflows and maintain reporting on classification coverage.

Establishing Feedback Loops and Quality Signals

A strong operating model isn't just based on structure - it thrives on feedback.

Establish lightweight loops that allow governance signals to surface without relying on manual reviews or isolated incident reports.

Focus on the following:

- **Reuse telemetry**: Identify which data assets are being reused the most (reuse count in your inventory, Power Platform Admin Center). This gives you a strong proxy for impact and where governance attention should be focused.
- **Classification completeness**: Track how many datasets in your inventory are missing sensitivity evaluations or intended use metadata.
- **Data quality incidents**: Keep a basic log of data-related issues flagged during AI development, especially those impacting fairness, completeness, or reliability.
- **RAI sensitivity mismatches**: Review where high-sensitivity datasets are being reused without a corresponding risk evaluation.

Conclusion: Data Readiness and Security: Emerging Stage

Reaching the emerging stage is no small achievement.

It signals a lot more than just intent – it reflects a structured commitment to treating data not only as a technical enabler but as a governed, reusable, and trusted foundation for scalable AI. Unlike the foundational stage, where clarity and habit-building are the focus, the emerging stage introduces something much more demanding: **coordination at scale**.

In this chapter, we've introduced a series of practices and responsibilities that begin to connect platform decisions with broader organizational accountability:

- We've explored the scaling of internal governance – evolving the data inventory into a living system of ownership, validation, risk tagging, and quality oversight.
- We covered how to establish governance patterns for integrations – ensuring that data flowing between services, systems, and copilots is trackable, justifiable, and treated as a value stream.
- We unpacked data loss prevention in detail, clarifying the way connector access, runtime behaviors, and overlapping policies interact – including the introduction of Advanced Connector Policies.
- We emphasized sensitivity labels and classification as a bridge between traditional information protection and AI-aware data governance, calling out how these labels must not inform design-time and runtime decisions.

- We've focused heavily on aligning governance with value, and nowhere is that more evident than in the humble data inventory.

Seeing the Difference

If *Establishing Governance and Trust* was about creating the structures and principles that make safe, responsible platform use possible, then *Emerging: Data Readiness and Security* is where those principles start to operationalize across every asset, dataset, and integration point.

A note on the role of the CoE

Many organizations will find that direct data governance doesn't traditionally sit within the Power Platform Center of Excellence. That's a fair observation - but also one that's beginning to shift.

As the volume and velocity of AI-powered, data-driven apps continues to grow, CoEs are finding themselves responsible for more than just enablement and policy enforcement. They are becoming key stakeholders in governing how data is used, reused, and scaled across solutions - and that means at least some level of data governance awareness and engagement becomes essential.

No, your CoE doesn't need to take full ownership of classification, security controls, or Purview configuration - but they do need to understand

- What data is in use
- Who owns it
- How it's being used by AI
- And whether that usage aligns with both risk posture and reuse policies

Why the effort is worth it

Taking the time to build a proper onboarding process - even if it starts simply - is a strategic investment. While Microsoft Purview and other tools can offer powerful capabilities, they often require specialized knowledge and don't always align neatly with how Power Platform assets are created, reused, and governed. You know your platform best - and you know where governance friction lives.

A good inventory, with the right metadata, and a structured onboarding process

- Builds trust with compliance and data protection teams
- Enables makers to move faster, because expectations are clear
- Protects your organization from unexpected risks
- And increasingly, becomes the source of truth that determines which assets are eligible for agents, orchestration, or model training

You see, this isn't admin overhead - it's strategic governance, and it's exactly what this stage is about.

Emerging Stage Data Readiness Progression Checklist

Use this checklist to validate your progress and spot areas that may still need attention before moving forward:

Data inventory & ownership

- [ ] A centralized inventory exists and is regularly maintained across all environments.
- [ ] Metadata is consistently captured for all new assets (e.g., sensitivity level, owner, usage intent, integration method).
- [ ] A confirmation process exists to validate that asset creators are the correct owners (e.g., via adaptive cards or automated flows).
- [ ] Reuse metrics are captured or estimated to track data value creation over time.
- [ ] RAI sensitivity evaluations are applied where appropriate (e.g., custom prompts, AI models).

Quality & readiness processes

- [ ] Data quality policies are defined for core datasets (including completeness, timeliness and structure).
- [ ] Critical datasets have validation routines or acceptance criteria agreed with data owners.
- [ ] Checklists or gates exist in app/flow development pipelines to verify data readiness before release.
- [ ] Environment-specific data policies guide how data can be used in dev/test/prod.

Integration & connectors

- [ ] Common integration scenarios (e.g., ServiceNow, SAP, on-premises) are identified and governed.
- [ ] Connector usage is reviewed for risk, using DLP policies or Advanced Connector Policies.
- [ ] The impact of overlapping DLP policies is understood, and least-permissive behavior is expected.
- [ ] External API access and custom connectors are tracked in the inventory with defined ownership and risk assessments.

Security & classification

- [ ] Sensitivity labels are used in at least some Power Platform scenarios (e.g., Copilot knowledge sources).
- [ ] Purview (or another classification platform) is integrated with the broader data strategy.
- [ ] The CoE has a documented partnership with Information Protection, Security, or Compliance teams.
- [ ] Unlabeled or misclassified data assets are either blocked from use or flagged for review.

Governance operating model

- ☐ The Power Platform CoE has visibility into data use and reuse across AI-powered assets.
- ☐ Basic reporting on data reuse, value, and risk is shared with leadership or data governance forums.
- ☐ Emerging roles (e.g., data stewards, custodians) are acknowledged or assigned for key assets.
- ☐ Governance is no longer just a compliance function – it's embedded in the way solutions are delivered.

CHAPTER 11

Operationalized Stage: Data Readiness and Security

Strategic | Scalable | Integrated

By now, you've moved beyond spreadsheets, ad hoc approvals, and scattered ownership. Your data assets have structure. Your governance processes are recognized, repeatable, and respected. However, as data volumes and AI use cases scale, so too does the risk – and the value opportunity.

This chapter marks the transition from **manual oversight to automated orchestration.**

At the **operationalized stage**, you're no longer building the data governance framework – you're scaling it. This means automating previously manual controls, embedding governance into pipelines, surfacing risk signals in real time, and ensuring that every dataset, model, or connector is subject to intelligent, contextual policy enforcement.

It's also where **value creation becomes measurable.** Your data inventory no longer just helps manage risk – it becomes a tool for identifying and multiplying impact. The number of times a dataset is

S. Jeffery, *The Low-Code AI Maturity Model*, https://doi.org/10.1007/979-8-8688-1730-4_11

reused, where it's embedded, and how it performs over time are now part of your governance telemetry. This now goes beyond control – it's strategy.

You'll notice a shift in tone and structure in this chapter. We're not just formalizing practices from earlier stages. We're introducing **entirely new capabilities** – like automated model documentation and AI-specific policy branches.

It's also worth calling out a structural truth: **many of these practices weren't historically part of the Power Platform CoE's role**. The rise of AI, the shift to citizen-led development, and the sheer velocity of low-code adoption means that **data governance can no longer sit separately**. The CoE must engage – not just to protect the platform but to truly unlock its full potential responsibly.

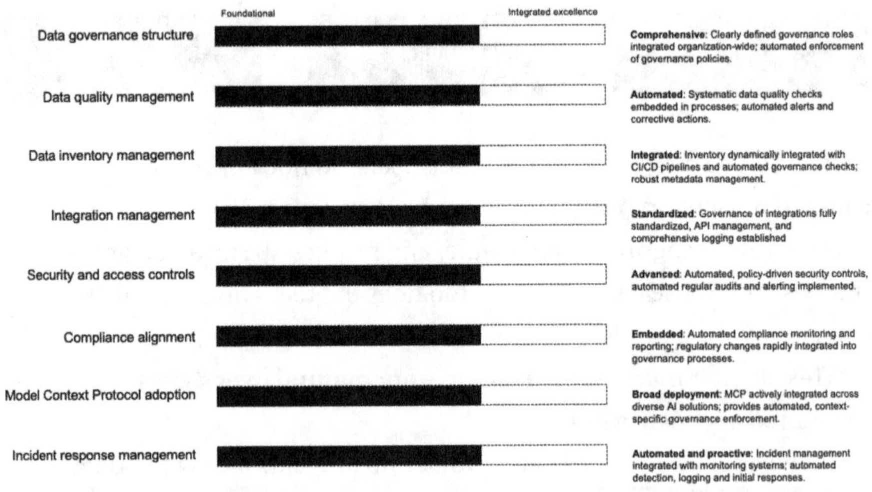

***Figure 11-1.** Typical characteristics of an organization at the operationalized stage*

If you've made it this far – and are actively working to mature your data governance to this level – you've already achieved what few organizations have. The rest of this chapter is designed to help you operationalize those efforts at scale, with confidence.

Orchestrating Data Life Cycle Governance at Scale

By the time you reach this stage of maturity, governance is no longer an aspiration or a set of scattered habits - it's a system. Data is flowing through apps, agents, prompts, flows, and APIs with increasing speed and scale, and your governance model must now do more than just observe and record - it must orchestrate.

At the operationalized stage, the challenge shifts from standing up basic practices to ensuring those practices scale across the full life cycle of a data asset: from its creation or integration, through validation and reuse, to eventual retirement or revision. The system you build here must be repeatable, visible, and - wherever possible - automated.

You've laid the groundwork. This section is about scale, orchestration, and making data governance sustainable - not just as a compliance tool but as a driver of value and trust.

Orchestrating Governance with Autonomous Agents

As your data landscape grows more complex and more AI-powered solutions move from experimentation to production, the limits of manual governance become apparent.

What once worked as a set of spreadsheets, owner reviews, and monthly data council meetings is no longer enough.

Now, we begin to introduce autonomous governance agents - small, domain-specific automation services that continuously monitor, evaluate, and act upon the data inventory. These agents do not replace people; they augment governance at scale, ensuring foundational controls are continuously applied and that risk signals are surfaced in real time.

From Inventory to Active Intelligence

In earlier stages, your data inventory acted as a system of record: a structured, trusted view of your data estate. Now, that same inventory evolves into a governance command center - where metadata not only describes your assets but also enables automated decisions and workflows.

Each autonomous agent specializes in a different aspect of governance. They act on the inventory as a source of trust, following defined policies and surfacing exceptions for human validation. This "agent pattern" mirrors the maturity of your platform: you no longer govern by spreadsheet or policy document but through distributed, intelligent execution.

Examples of autonomous governance agents:

Agent type	Purpose	Trigger/event source	Action taken
RAI sensitivity evaluation agent	Ensures every data asset has undergone responsible AI sensitivity evaluation.	New assets added to inventory or material changes to existing asset.	Sends adaptive card to data owner to complete classification form. Escalates unclassified assets after SLA.
Data quality monitor agent	Tracks whether asset validation dates are expired or incomplete. Manages data quality checks.	Schedules check across all inventory assets.	Flags when data quality is stale, notifies stewards, and recommends revalidation workflow.
Connector risk agent	Ensures that sensitive datasets aren't exposed to risky or deprecated connectors.	Policy updates or usage detected in new app/flow.	Flags policy violations and recommends mitigation actions.

Agent type	Purpose	Trigger/event source	Action taken
Metadata completeness agent	Audits inventory entries for missing or low-code quality metadata.	New assets created or regular scans.	Alerts CoE and requests data owners complete missing fields.
Life cycle agent	Monitors asset age, usage, and relevance for review or archival.	Aging thresholds reached.	Suggests review to owner and proposes status change to "deprecated," quarantined (apps), and disabled (flows).

These agents can be implemented using **Power Automate and Copilot Studio (autonomous agents)**, depending on where you're operating. The key is to ensure agents act on shared data - **your central inventory or metadata layer** - so governance is consistent and traceable across environments.

Architecture: Shared State, Decentralized Governance

Each agent follows a simple architectural principle:

- **Inventory as a source of truth**

  Agents query the inventory, not each other. They do not need to communicate peer-to-peer.

- **Single responsibility design**

Each agent handles one purpose: quality, classification, life cycle, risk.

- **Policy-triggered automation**

  Agents act on metadata signals and policy rules, minimizing the need for hardcoded logic.

- **Human-in-the-loop**

  Exceptions, escalations, and decisions (like overriding a classification or retiring a dataset) are directed to owners, stewards, or governance leads via Teams, email, or approval flows.

This creates a low-friction, high-visibility governance model, where people stay in control - but routine decisions and monitoring no longer require their constant attention.

Autonomous agents offer two major advantages:

1. **They allow you to scale governance without scaling teams.**

   Rather than growing oversight functions linearly with your adoption, you scale through automation, freeing up your CoE and governance leads to focus on exceptions, not routine enforcement.

2. **They embed governance into the life cycle - not as a gate but as a guide.**

   Agents prompt people when action is needed. They don't block - they support. This shifts governance from being perceived as a control function to a **value-adding operational practice**.

Autonomous agents don't just monitor what exists - they help guide where assets go next. Whether surfacing a stale dataset, flagging an unclassified model, or prompting a quality recheck, these agents signal when a data asset has reached a key inflection point in its life cycle.

To act on those signals consistently, you need a clear and shared understanding of how your data assets mature, evolve, and eventually retire. Governance doesn't stop at creation - it's a continuous loop. That's why the next step in operationalizing your data strategy is to **establish a structured life cycle model for your assets.**

Let's look at that next.

Establishing a Life Cycle Model for Data Assets

At the emerging stage, we focused on ownership, classification, and readiness. Now, with growing inventories, AI integrations, and autonomous agents in play, it's time to formalize how data moves through its life cycle - from introduction to sunset.

A life cycle model introduces clarity and consistency to how data assets are handled. It ensures each dataset, model, or integration follows a repeatable progression - and that interventions (such as quality checks, sensitivity reviews, or reuse decisions) happen at the right time, not just when someone remembers.

Let's walk through a practical model.

Five Phases of the Data Asset Life Cycle

1. **Introduction**
    a. Asset is first created or identified.
    b. Ownership assigned and basic metadata captured.
    c. Initial classification (e.g., sensitivity, risk level) applied.

d. Prompt or model type (if AI-related) recorded.

e. **Governance activity**: Inventory registration and RAI sensitivity evaluation.

   Agent responsibility: A registration agent checks metadata completeness and routes the data asset for RAI sensitivity evaluation.

2. **Readiness**

   a. Asset is validated for use in solutions.

   b. Data quality checks completed.

   c. Confirmed alignment with intended use case.

   d. Governance activity: Readiness checklist completed; eligibility for reuse unlocked.

   Agent responsibility: A data readiness agent ensures checklists are completed, prompts human review if necessary, and verifies all fields before approval.

3. **Active Use**

   a. Asset is integrated into one or more solutions (e.g., agents, apps, workflows).

   b. Reuse metrics tracked (how many consuming solutions, frequency of access).

   c. **Governance activity**: Regular reverification triggers based on sensitivity or usage volume (reuse count).

   Agent responsibility: A monitoring agent can trigger automated review tasks or alerts to human stewards.

4. **Review**

    a. Triggered by policy changes, low reuse, quality issues, or time (since last review).

    b. Autonomous agents can initiate review cycles.

    c. Human stewards assess value, validity, and alignment with current standards.

    d. **Governance activity**: Manual or automated interventions - reclassify, remediate, retire, or retrain.

    Agent responsibility: A review agent initiates the workflow, aggregates risk signals, and presents a summarized history for decision-making.

5. **Sunset**

    a. Asset is no longer in use or has failed governance checks without remediation.

    b. Removed from approved inventory.

    c. Access permissions revoked or deprecated.

    d. **Governance activity**: Logged as decommissioned, with audit trail maintained.

    Agent responsibility: Sunset agent initiates the decommission workflow, archives metadata, and notifies stakeholders.

A life cycle model is as much a "trust engine" as it is operational hygiene. Using agents drives consistency, and humans focused on exception handling, governance becomes faster, smarter, and more scalable.

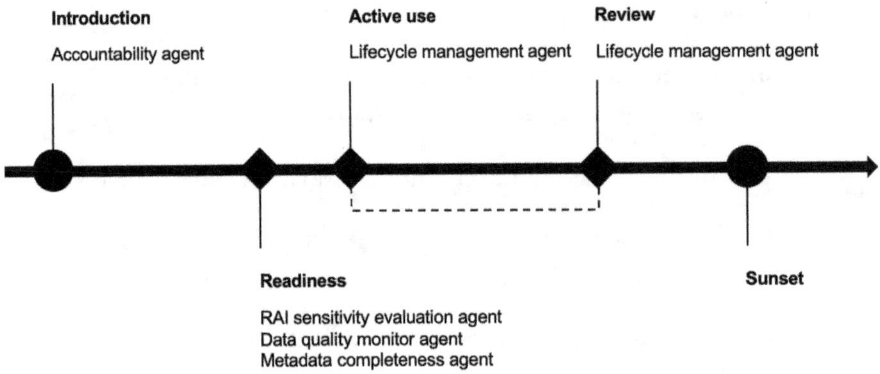

Figure 11-2. *Agentic approach to data life cycle management*

As the inventory grows, life cycle governance becomes the strongest defense against shadow AI, unverified reuse, and regulatory risks.

With a structured life cycle model in place - and autonomous agents ensuring data assets are managed through introduction, readiness, use, and eventually decommissioning - the next logical step is embedding governance earlier into the solution development process itself.

Rather than wait until an app, flow, or agent is live to spot issues, begin to bake governance into your pipelines. This means adding validation gates that automatically check for data readiness, risk classification, and RAI alignment - ensuring that nothing enters production without clearing the appropriate checks.

Integrating Validation Gates into Deployment Workflows

As you scale data governance alongside solution development, you'll eventually face a critical need:

How do you ensure that every deployed app, agent, model, or automation respects your data inventory and governance standards - without slowing down delivery?

In traditional software engineering, validation gates are built into CI/CD pipelines. In the Power Platform, especially at the emerging-to-operationalized stages, the same thinking applies - but adapted for the unique nature of low-code assets.

In this section, we'll explore how to embed a validation checkpoint into your deployment pipeline.

Specifically: ensuring that any new solutions moving between environments are registered, tracked, and evaluated in your data inventory - before they can be reused or consumed by agents, apps, or flows.

High-level process overview

Here's how the governance validation flow will work:

1. A new or updated solution is merged into GitHub.
    a. Someone submits a Pull Request (PR) and merges a solution export (a .zip file) from a development environment.
2. A GitHub Action triggers the deployment pipeline.
    a. This action initiates deployment to the target environment (e.g., Test, or UAT).
3. Before deployment happens, the action checks your inventory.
    a. It checks: Is this solution ID already registered in the inventory?
4. If the solution is NOT found:
    a. The system automatically registers the solution as "pending metadata."
    b. The submitter (the GitHub actor) is captured as the "provisional owner."

c. A governance adaptive card is sent to the submitted, requesting confirmation of metadata.

5. Once registration is completed:

a. The deployment proceeds, but metadata enrichment continues asynchronously (governance never blocks agility unless there's a severe violation).

To make this tangible and actionable, let's dive into a comprehensive, step-by-step guide for setting up a GitHub repository and embedding governance validation within your Power Platform solution deployment pipeline.

Comprehensive Guide to Setting Up a GitHub Repository and Governance Validation for Power Platform Solutions

This guide provides a comprehensive walkthrough to help you

1. Set up GitHub to manage Power Platform solutions using PAC CLI and pull request workflows.

2. Embed governance validation into the pipeline, ensuring that no solution moves forward unless it's registered in your inventory and enriched with required metadata.

You'll learn how to

- Export and commit Power Platform solutions to GitHub using PAC CLI.

- Trigger GitHub Actions that validate metadata presence in a custom inventory.

- Automatically register untracked solutions and notify the responsible owner via adaptive cards.
- Proceed with deployment only when governance checks pass - or escalate where needed.

Prerequisites

- A GitHub account.
- Git installed on your local machine.
- Power Platform CLI (PAC CLI) installed (install via "`npm install -g @microsoft/powerplatform-cli`" or download from `https://docs.microsoft.com/en-us/power-platform/developer/cli/introduction`).
- Access to Power Platform environments (Development, Test, UAT, Production).
- A local development machine with permissions to export solutions.
- The data asset inventory system to store solution metadata.
- A notification system (e.g., Microsoft Teams or email) for sending adaptive cards/actionable messages.
- Permissions to configure GitHub Actions and manage Power Platform environments.
- Basic familiarity with Git commands.

Part 1: Set Up the GitHub Repository and Sync Solutions

- Log in to github.com.
- Click the + icon in the top-right corner, and select **New repository**.

- Enter a repository name (e.g., "power-platform-solutions").
- Choose **Public** or **Private** based on your needs.
- Check **Add a README file** to initialize the repository.
- Optionally, add a ".gitignore" file (select the "Node" template to exclude files like "node_modules").
- Click **Create repository**.

Step 1.2: Clone the Repository Locally

- Copy the repository URL (e.g., "`https://github.com/your-username/power-platform-solutions.git`").
- Open a terminal on your local machine.
    1. On Windows, open Command Prompt or Git Bash (installed with Git for Windows)
    2. On Mac or Linux, open Terminal.
- Run:

```
bash
git clone https://github.com/your-username/power-
platform-solutions.git
```

Tip *If you're using Windows and don't have Git Bash or WSL, installing Git for Windows will give you Git Bash support out of the box.*

- Navigate to the repository folder:

```
cd power-platform-solutions
```

Step 1.3: Create a Folder for Solutions

- Create a directory to store solution export files:

```
mkdir solutions
```

Step 1.4: Export a Solution Using PAC CLI

- Authenticate with PAC CLI:

```
pac auth create --environment <ENVIRONMENT_ID_OR_URL>
```

  Replace "<ENVIRONMENT_ID_OR_URL>" with your environment's ID or URL (find it in the Power Platform admin center).

- List available solutions:

```
pac solution list
```

  – Note the "UniqueName" of the solution to export.

- Export the solution:

```
pac solution export --name <SOLUTION_UNIQUE_NAME>
--path ./solutions/<SOLUTION_NAME>.zip --overwrite
```

  – Replace "<SOLUTION_UNIQUE_NAME>" (e.g., "MyCustomSolution") and "<SOLUTION_NAME>" (e.g., "my-custom-solution").

- Verify the export in the **solutions** folder:

```
ls solutions
```

```
Or dir in Windows
```

Step 1.5: Sync the Solution File to GitHub

- Stage the solution file:

```
git add -v solutions/<SOLUTION_NAME>.zip
```

- Commit the changes:

```
git commit -m "Add solution export for <SOLUTION_NAME>"
```

- Push to GitHub:

```
git push origin main
```

- Verify on GitHub that the "solutions" folder contains the ".zip" file.

Step 1.6: Set Up Pull Request Workflow

- Protect the main branch:
  - In GitHub, go to repository **Settings ➤ Branches ➤ Add ruleset.**
  - Set **Branch name pattern** to "main".
  - Check **Require a pull request before merging** and **Require approvals** (e.g., one approval).
  - Save the rule.
- Create a development branch:

```
git checkout -b dev
```

- Export and commit to the "dev" branch:

```
pac solution export --name <SOLUTION_UNIQUE_NAME>
--path ./solutions/<SOLUTION_NAME>.zip --overwrite
git add -v solutions/<SOLUTION_NAME>.zip
git commit -m "Update solution export"
git push origin dev
```

- Create a pull request on GitHub:
    - Go to **Pull requests ➤ New pull request**.
    - Set **base: main** and **compare: dev**.
    - Add a description and create the PR.

At this point, you've created a secure process for managing changes using GitHub. The main branch is protected, and new changes are now routed through a pull request workflow. This ensures that updates to your Power Platform solutions are reviewed, traceable, and aligned with the team practices - a foundation that supports collaboration and maintains quality as you scale your development.

Part 2: Integrate Governance Validation in Deployment Pipelines

Next, you'll learn how to automate governance validation directly within your deployment pipeline. The goal is simple:

> *"No solution should move to a production-like environment without being registered, reviewed, and approved in your governance inventory."*

Let's walk through a GitHub-based implementation that allows you to

- Intercept solution deployments as they're pushed via PRs.
- Check for the solution's presence in the inventory.
- Register new solutions automatically and assign provisional ownership.
- Send adaptive cards to collect missing metadata and prompt RAI sensitivity evaluations asynchronously.
- Proceed with deployment once governance thresholds are met.

This is a major shift - embedding governance ensures compliance without blocking innovation. It's a model where policy enforcement is continuous, contextual, and powerful - executed not by people chasing approvals but by your **system** that is *guiding solutions toward readiness*!

Step 2.1: Creating Secrets for Use in GitHub Actions

Before you can make the HTTP call to Power Automate securely, you'll need to set up the necessary secrets in your GitHub repository.

- Navigate to your GitHub repository.
- Click **Settings ➤ Secrets and variables ➤ Actions.**
- Click New repository secret and add the following:

Secret name	Description
INVENTORY_FLOW_URL	The full HTTP POST URL of the Power Automate flow (copied from the flow's HTTP trigger)

- You can now reference these in your GitHub Actions YAML using ${{ secrets.INVENTORY_FLOW_URL }}.

Step 2.2: Create Cloud Flow to Query Inventory

- Go to Power Automate.
- Click **Create ➤ Instant cloud flow.**
- Choose "**When an HTTP request is received**" **as the trigger.**

  Security note: After selecting the "**When an HTTP request is received**" trigger, make sure to set "Who can trigger this flow" to whatever best suits your scenario. For improved security, consider using **Azure App Registration** and setting up authenticated access.

- Click **Create**.
- Define the input criteria:

  Click the HTTP trigger and paste in the following JSON schema:

  ```
  {
      "type":"object",
      "properties" : {
         "AgentId": {
            "type":"string"
         }
      },
      "required": ["AgentId"]
  }
  ```

- Initialize a variable for return value.
- Query the inventory:
  - Add a new step.
  - Choose Dataverse ➤ List rows.
  - Set:
    - **Table name**: DataInventory (or your inventory table name)
    - **Filter rows**: `yourprefix_agentid eq '@{triggerBody()?['AgentId']}'`
    - Set row count to 1 (for performance)

Note Replace "yourprefix_" with the actual logical name of the column in your Dataverse table. Logical names are typically lowercase and include a solution prefix (e.g., abc_agentid). You can find the logical name by opening the table and checking the "Name" column in the list of columns.

- Add a condition based on results:

```
If length(body('List_rows')?['value'])
is greater then 0
> Return "exists" : true
else
> Return "exists" : false
```

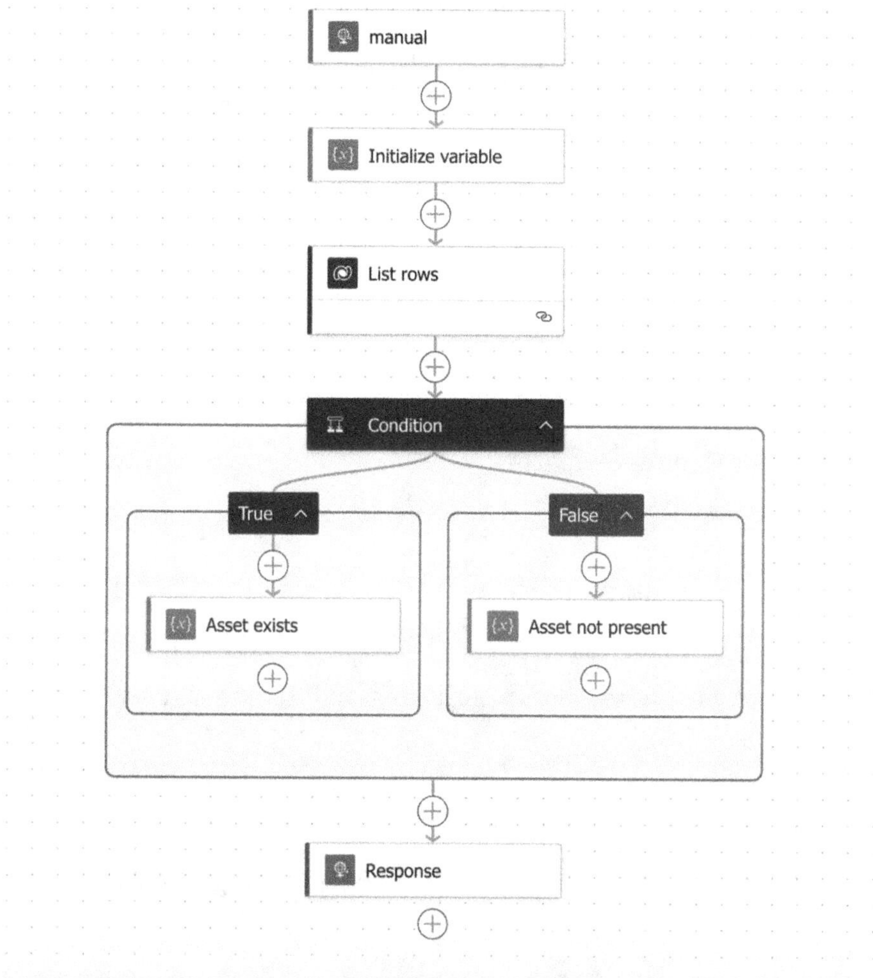

Figure 11-3. *Cloud flow checking for data asset in inventory*

Step 2.2: Create the GitHub Action

1. Create a workflow file (.github/workflows/deploy.yml):

```
name: Validate and Deploy Power Platform Solution

on:
  push:
    branches:
      - main

jobs:
  validate-inventory:
    runs-on: ubuntu-latest
    steps:
      - name: Checkout repo
        uses: actions/checkout@v3

      - name: Unzip solution and locate bot.xml
        run: |
          unzip -o solutions/*.zip -d extracted/
          find extracted/ -name "bot.xml" -print

      - name: Extract agent ID (applicationId)
        from bot.xml
        id: extract_agent
        run: |
          BOT_XML_PATH=$(find extracted/ -name "bot.
          xml" | head -n 1)
          echo "Found bot.xml at: $BOT_XML_PATH"

          AGENT_ID=$(grep -oP '"applicationId"\
          s*:\s*"\K[^"]+' "$BOT_XML_PATH" || echo
          "NOT_FOUND")
```

```
        echo "Extracted agent ID: $AGENT_ID"

        echo "AGENT_ID=$AGENT_ID" >> $GITHUB_ENV

    - name: Call Inventory Validation Flow
      id: call_inventory
      env:
        INVENTORY_FLOW_URL: ${{ secrets.INVENTORY_
        FLOW_URL }}
      run: |
        RESPONSE=$(curl -s -X POST "$INVENTORY_
        FLOW_URL" \
          -H "Content-Type: application/json" \
          -d "{\"agentid\": \"$AGENT_ID\"}")
        echo "Sent payload:
        {\"solutionId\":\"$AGENT_ID\"}"
        echo "Inventory check response: $RESPONSE"
        echo "flow URL used: $INVENTORY_FLOW_URL"
        EXISTS=$(echo $RESPONSE | jq -r '.exists')
        echo "EXISTS=$EXISTS" >> $GITHUB_ENV

    - name: Stop deployment if agent is unregistered
      if: env.EXISTS == 'false'
      run: |
        echo "Agent not found in inventory.
        Deployment halted until metadata is
        completed."
        exit 1

deploy-solution:
  needs: validate-inventory
  runs-on: ubuntu-latest
```

```
steps:
  - name: Deploy (Placeholder)
    run: echo "Proceed to deployment steps
    go here."
```

How to test the inventory validation

Before running this GitHub Action, ensure that

- You've **exported your solution** and **committed it to the repository**.
- Your solution includes a **Copilot Studio agent**, which appears in the bot.xml file. If the agent isn't included, validation will fail.

To confirm that the validation is functioning:

- Check that the job runs the "**Call Inventory Validation Flow**" step.
- **Even if the agent is found**, the log will stop deployment agent with a trailing (✓) to indicate the check passed and deployment was stopped intentionally.

Note This is expected behavior. The placeholder echo step halts deployment when the agent is registered, serving as a safety check before proceeding with the actual deployment steps.

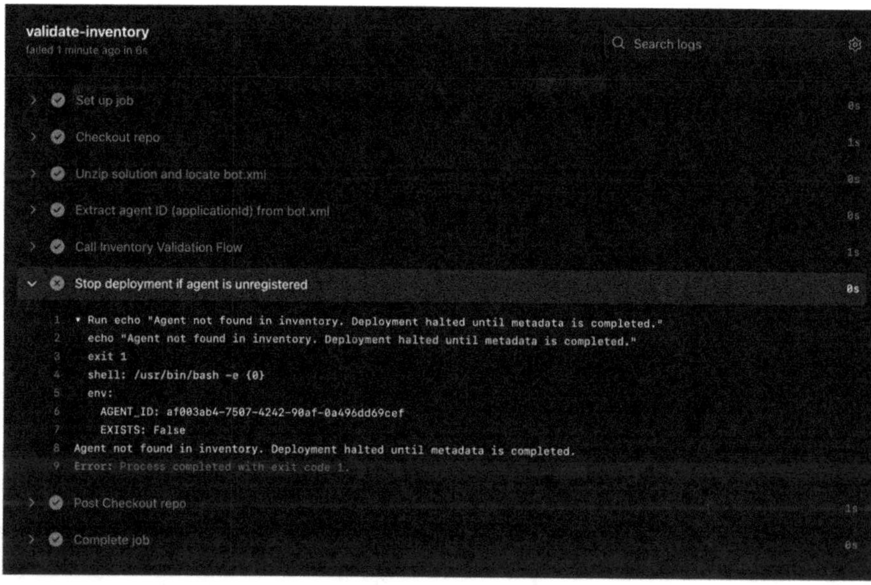

Figure 11-4. GitHub Action in process - failing because asset is not in inventory

By embedding governance validation directly into your deployment pipelines, you shift governance from a reactive checkpoint to a proactive enabler. Every agent, model, or solution must pass through a verification step - confirming ownership, metadata completeness, and responsible AI alignment - before being promoted. This doesn't slow down delivery; it safeguards it. And by relying on lightweight automation (like GitHub Actions and Power Automate), even complex governance scenarios become scalable and traceable.

Enhancing Reusability, Telemetry, and Inventory Intelligence

As organizations reach this stage, the role of data governance evolves dramatically. It's no longer enough to maintain a static inventory. The true power lies in unlocking the value hidden within it.

Every time a dataset, model, or integration asset is reused - across agents, apps, flows - it generates new returns on the original investment of time, effort, and governance. Reusable assets create compounding value: saving development time, accelerating AI adoption, reducing duplication, and building trust through proven quality.

This realization - compounding value - isn't going to happen passively. It requires intelligent inventory practices that surface insights - not just records.

It requires governance that prioritizes what matters, rather than applying the same scrutiny to everything.

Transforming your inventory from a compliance ledger into an active and intelligent asset means you have answers to critical questions:

- Which assets are driving the most innovation?
- Which assets pose the greatest risk if left unchecked?
- Where should our stewards, CoE teams, and governance resources focus their attention next?

Intelligent Surfacing of High-Value, High-Risk Assets

Earlier in this book, we introduced the concept of augmenting your inventory with an agent - a lightweight assistant that could interact with your data assets, surface insights, and guide better governance decisions.

At the time, the goal was simple: to make governance more **proactive and accessible**.

Now, the opportunity becomes far greater: your inventory agent can become a strategic command center for data value, risk, and life cycle management.

But what would that look like in practice?

Core Capabilities for an Inventory Agent

To move from a static registry to an intelligent, self-sustaining system, the following practical features need to be designed:

1. **Asset value surfacing**
    a. Metric-driven ranking of assets based on
        i. Number of reuse instances (e.g., how many solutions, agents, apps reference it)
        ii. Business impact (e.g., solutions linked to revenue, efficiency gains, or critical operations)
    b. Highlight "hidden gems" - assets with low reuse, but strong metadata.

    Why it matters: Focuses stewardship and platform growth efforts on the highest-value opportunities.

2. **Risk visibility and alerts**
    a. Identify assets with
        i. Outdated sensitivity labels
        ii. Missing or stale RAI sensitivity evaluations
        iii. Excessive reuse without recent quality review (e.g., reused >5× but last validated 6+ months ago)
    b. Trigger risk alerts and propose actions (e.g., review required, flag for RAI reassessment).

    Why it matters: Moves governance from reactive to preventive.

3. **Freshness and quality signals**

    a. Track when each asset was last verified or reviewed.

    b. Automatically flag assets that exceed a freshness threshold (e.g., no verification within 180 days).

    c. Allow stewards or agents to mark assets as verified through lightweight reviews.

    Why it matters: Maintains trust in fast-changing environments without needing manual effort.

4. **Governance life cycle management**

    a. Monitor assets transitioning through life cycle stages:

        i. Ready for reuse

        ii. Under review

        iii. At risk

        iv. Scheduled for sunset

    Why it matters: Avoids governance drift and ensures assets don't become abandoned liabilities.

5. **Human-in-the-loop stewardship**

    a. Allows stewards to ask the agent like

        i. "Which high-risk assets haven't been reviewed this quarter?"

        ii. "Show me top ten reused datasets in finance."

        iii. "Are there any expired data assets still being referenced?"

b. Enables approvals, reverifications, and risk escalations directly through agent interactions

Why it matters: Embeds governance into everyday platform usage - making it effortless, not an afterthought.

A Note on Implementation

This doesn't require massive technical investment upfront. Agents in Power Platform can deliver incredible value:

- Using Dataverse as your inventory back end.
- Power Automate flows trigger signals and alerts.
- Copilot Studio agents to interact conversationally or autonomously.
- Optionally layering in Azure OpenAI or Cognitive Services for more advanced natural language understanding.

From Intelligent Surfacing to Autonomous Inventory Governance

Simply cataloging data assets is a major step forward. Capturing ownership, sensitivity, and readiness helps shift data governance from reactive to proactive. However, we need to transform further.

Visibility is not enough. The inventory needs to become a governance system - autonomously surfacing insights, detecting risk, and driving stewardship actions at scale.

Intelligent Surfacing: Turning Visibility into Value

First, the inventory must serve as an intelligent advisor - highlighting where value is growing and where risk is emerging. This means layering metric-driven insights directly into your inventory system.

Key capabilities to implement

- **Value surfacing**
    - Track reuse metrics - how many apps, agents, and flows reference each assets.
    - Identify assets that generate the greatest downstream value (efficiency gains, cost savings, revenue impact).
    - Highlight "hidden gems" - assets with strong quality metadata but low reuse, ready to be promoted.
- **Risk visibility**
    - Detect assets with outdated sensitivity evaluations or missing readiness verifications.
    - Monitor assets whose classification does not match their usage patterns (e.g., unclassified assets fueling production solutions).
    - Identify heavily reused assets overdue for quality or RAI sensitivity review.
- **Freshness and quality signals**
    - Track the "last verified" date for every asset.
    - Flag assets that exceed freshness thresholds based on sensitivity (e.g., regulatory data requires revalidation every 90 days).

This transforms the inventory from a passive list into an active source of governance intelligence.

Autonomous governance: moving from insights to action

Insight alone doesn't protect or maximize value. Action does. You should automate stewardship behaviors based on the intelligence the inventory surfaces.

Key stewardship automations

- **Expiry tracking and escalation**
    - Trigger proactive notifications when an asset is nearing freshness expiration.
    - If no owner response, auto-escalate to governance leads for intervention.
    - If escalation fails, move the asset into a "restricted" or "under review" state automatically.
- **Quality degradation detection**
    - Detect data assets whose reuse volume surges (e.g., referenced in >10 solutions within a period, e.g., three months) without a corresponding revalidation.
    - Flag these assets for accelerated stewardship attention - automatically tagging them for review.
- **Risk auto-escalation**
    - If sensitivity labels are missing, outdated, or contradict asset usage patterns, auto-flag the asset for RAI reassessment.
    - Agents or flows can initiate lightweight reassessment workflows.

- **Governance life cycle tracking**
  - Maintain clear status tracking for each asset (ready, under review, at risk, scheduled for sunset).
  - Trigger transitions based on metrics, not just manual decisions.

Human-in-the-loop remains essential - but only where judgement is needed. The system now carries the operational burden, surfacing *where* stewardship effort is best directed.

Conclusion: Data Readiness and Security – Operationalized Stage

At the operationalized stage, governance moves from principle to practice - embedded not only in policies but in the workflows, pipelines, and tools your organization uses every day. Manual reviews are replaced by automated validation gates. Governance agents watch over data quality, life cycle progression, and responsible AI compliance - surfacing risks and prompting human review when needed. Your inventory evolves from a static registry into a living intelligence layer, driving both protection and performance.

Crucially, this is where governance becomes measurable. You begin to capture reuse metrics, track policy adherence, and quantify the value and risk of your assets. What was once invisible - like stale datasets silently reused by agents or unclassified prompts deployed into production - now becomes actionable. And what was once friction - governance processes that slowed teams down - becomes a flywheel for smarter, safer, more scalable AI adoption.

Operationalized Stage Data Readiness Progression Checklist

Use this checklist to help you ensure your data practices support enterprise-grade AI and Power Platform adoption, with governance, automation, and life cycle management embedded throughout.

Lifecycle governance

- [ ] Every dataset in use is registered in a centralized inventory with ownership and classification metadata.
- [ ] Autonomous governance agents monitor asset status, prompting reviews based on lifecycle stage, quality, or reuse volume.
- [ ] A formal data lifecycle model exists and is consistently applied across business units.

Value and usage telemetry

- [ ] Data reuse is tracked across environments and solutions, with visibility into where and how data is applied.
- [ ] Dashboards exist to report on reuse frequency, time to first use, and governed activation needs.
- [ ] High-value datasets are prioritized for maintenance and protection based on usage telemetry and risk profiles.

Policy enforcement and validation

- [ ] DLP policies are implemented at both tenant and environment levels, with ACPs used where fine-grained control is needed.
- [ ] Sensitivity labels and classification metadata are consistently applied to data assets and enforced in workflows.
- [ ] Deployment pipelines include validation gates to check for inventory registration, classification completeness and RAI sensitivity alignment.

AI-specific data readiness

- [ ] AI-sensitive datasets (used in prompts, models, agents) are reviewed for fairness, representativeness and compliance before deployment.
- [ ] Model Context Protocol or equivalent structures are used to control what structured data is passed into LLMs.
- [ ] AI Builder model usage is monitored using platform telemetry, with consumption trends available to admins and governance leads.

Cross-functional governance

- [ ] A data governance council is active, with participation from CoE, IT, Legal, Compliance and business units.
- [ ] Policy evolution is reviewed quarterly, incorporating insights from telemetry and incident response reviews.
- [ ] Data governance efforts are aligned with strategic business outcomes, not just risk avoidance.

CHAPTER 12

Integrated Excellence Stage: Data Readiness and Security

Pioneering	Sustainable	Influential

By the time you arrive at this stage – integrated excellence – your organization has done the hard work. You've established data governance not as a one-off project but as a living practice. You've scaled oversight across agents, apps, workflows… You've moved from manual reviews to life cycle models, from risk identification to automation, from tactical value to measurable ROI.

Governance moves beyond existing as just an internal function. It's part of how your organization competes, leads, and grows. You are no longer just managing risk – you're shaping how AI is used responsibly across your industry. You're not only tracking value, but you're also building systems that multiply it, track it, and communicate it with precision.

S. Jeffery, *The Low-Code AI Maturity Model*, https://doi.org/10.1007/979-8-8688-1730-4_12

So, in this final chapter, we shift the lens outward and upward. It's about

- Stepping into **industry leadership** by publishing, contributing, and influencing the standards of responsible AI and low-code governance
- Ensuring your platform governance amplifies cross-functional value and speaks to the broader enterprise strategy - not just technical metrics
- Embedding a culture where governance is just how things work, not conversations

This chapter is about **elevating your data operating model into a leadership position** - where your readiness and governance infrastructure prove, scale, and signal responsible AI to customers, regulators, and partners.

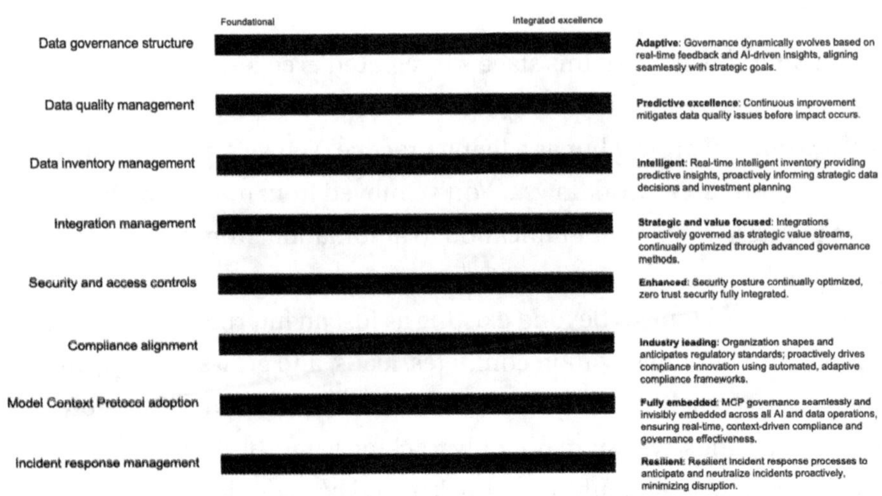

***Figure 12-1.** Characteristics of an organization at the integrated excellence stage*

Responsible AI at the Strategic Level

Your organization has reached a stage where responsible AI is no longer an isolated practice managed by a small team of experts. It becomes a visible, integrated, and verifiable component of how your organization develops, deploys, and measures AI - at every layer of the stack.

The organization shifts from internal assurance to external confidence. This shift marks the maturation or responsible AI into a strategic asset. RAI is no longer confined to ethics officers or compliance teams. It is now embedded into business strategy, product development, and enterprise architecture. Responsible practices shape not only the way AI is built but also where and why it is deployed. In short, RAI should be viewed as a lever for innovation, trust, and long-term value creation.

From Principles to Embedded Practice

Many organizations have published AI principles, and they tend to converge around familiar themes: fairness, safety, transparency, accountability, privacy, and inclusiveness. But declarations of intent are not the same as embedded practice. What distinguishes this level of maturity is the systematic way these principles are translated into policy, tooling, and decision-making frameworks across the organization.

For example, Accenture, one of the early leaders in this space, has created an internal responsible AI framework[1] that doesn't just guide development - it shapes client engagement, hiring practices, model procurement, and risk assessments for third-party tools. Their public documentation outlines clear roles and accountabilities, including RAI leads embedded within delivery teams and an ethics advisory council that governs high-risk solutions. This integration of ethics and operations is critical - and difficult to fake.

[1] https://www.accenture.com/us-en/case-studies/data-ai/blueprint-responsible-ai

Similarly, Microsoft's internal RAI governance includes mandatory reviews for sensitive use cases, tooling to assess model impact, and detailed design guidelines that balance human-AI interaction, error handling, and explainability. These are not optional. Teams must document their rationale, mitigation plans, and measurement strategies for fairness, privacy, and safety as part of the development life cycle. As such, responsibility becomes an integrated feature of the system - not a wrapper applied after the fact.

Beyond Mitigation: Responsibility As Differentiation

Organizations realize that responsible AI is not only about avoiding harm - it is about earning trust. Trust is what enables adoption at scale. It's what allows a financial institution to roll out AI-enabled loan adjudication without public backlash. It's what lets a healthcare provider deploy generative AI into sensitive clinical settings, because patients and regulators believe their data will be handled with care.

Trust doesn't come from promises. It comes from performance - and transparency about how that performance is governed.

Strategic organizations begin to publish impact assessments, fairness audit summaries, and model cards not because they are required to but because they understand that disclosure strengthens relationships with users, regulators, and society at large. Responsible AI moves from an internal safeguard to an external statement of intent - and it starts to differentiate companies in the marketplace.

BCG[2] (Boston Consulting Group) has emphasized that investors and partners are increasingly applying pressure on companies to demonstrate responsible AI governance. As metrics evolve to include digital ethics

[2] https://www.bcg.com/capabilities/artificial-intelligence/responsible-ai

and AI safety, RAI becomes part of an organizations strategic positioning, reputation, and even valuation.

Some organizations move beyond internal controls altogether - they bring in external auditors, publish voluntary disclosures, or participate in cross-industry standards efforts. Google, for instance, shares updates on its AI principles and associated progress, while PwC and Deloitte now offer independent responsible AI audits as part of their enterprise assurance services.

This not only supports stakeholder trust but also prepares organizations for the regulatory environments of tomorrow. Companies that demonstrate control, fairness, and accountability today will face lower barriers when formal audits or legal mandates arrive.

How to apply this in your organization

- Begin documenting ethical considerations and value tradeoffs alongside technical decisions - make the invisible visible.
- Build fairness and explainability into solution reviews, not just post-deployment testing.
- Partner with legal, branding, and corporate responsibility teams to align RAI with your public commitments and market strategy.
- Encourage transparency internally - build dashboards or tools that make model behavior legible to business leaders and subject matter experts.

Responsibility at this level is a differentiator. It doesn't slow you down - it strengthens your license to operate, grow, and lead.

External Reporting and Transparency

Many organizations view governance and responsible AI as internal functions. But the most mature and most trusted recognize that reporting how AI systems behave, how data is used, and how decisions are governed is part of their responsibility to customers, regulators, partners, and society at large. You could say transparency is strategic communication.

Why it matters

- **Trust is the currency of scalable AI**: Users need to understand - and believe in - the systems that impact their experiences, data, or outcomes. Internal assurance is no longer enough.
- **Regulators are watching**: The EU AI Act and various sector-specific guidelines all point to the same future: explainability, auditable processes, and risk classification must be provable, not aspirational.
- **Talent and partnerships align to values**: Organizations that demonstrate ethical use of data attract stronger partners and more motivated internal teams.

What transparency involves

An external-facing data or AI governance website. For example:

- Salesforce provides public documentation on its Ethics by Design framework[3] and shares the "trusted AI and agents impact report."[4]

[3] https://www.salesforce.com/uk/news/stories/salesforce-technology-ethics/

[4] https://www.salesforce.com/en-us/wp-content/uploads/sites/4/assets/pdf/reports/salesforce-trusted-ai-and-agents-impact-report.pdf

- Microsoft regularly publishes updates through its AI transparency center[5] and responsible AI standard. These include design guidance, toolkits, and real-world application of its principles.

What does this tell Salesforce and Microsoft's customers and regulators?

When organizations like Microsoft and Salesforce share detailed, structured disclosures about how their AI products are designed and governed, the signal is loud and clear: **We take responsibility for what our technology does, and we can prove it.**

To customers, it shows

- **Accountability is operational**: The company is not just making promises about ethics or fairness; it has embedded responsible AI into its development life cycle, review process, and product design.
- **AI is being managed intentionally**: Customers, especially in regulated industries, gain confidence that the products they are using have been evaluated for bias, risk, and misuse.
- **Partnership goes beyond licensing**: Vendors are not simply providing tools; they are co-owning the governance responsibility. This makes adoption safe and more scalable.

[5] https://www.microsoft.com/en-us/ai/responsible-ai?activetab=pivot:impact-tab

To regulators, it signals:

- **Preparedness for scrutiny**: The organization is not waiting to be forced into transparency. It has built systems and documentation that align with future regulatory frameworks - such as the EU AI Act or US Algorithmic Accountability Act.
- **Maturity of controls**: The presence of internal audits, ethics councils, and toolkits like model cards suggests that compliance isn't an afterthought. Risk is being proactively managed.
- **Leadership, not just compliance**: These disclosures indicate a company that is helping shape the future of AI governance - not only trying to meet the minimum bar.

In both cases, transparency serves as a proxy for trustworthiness. It reduces friction in procurement, eases concern from data protection officers, and accelerates enterprise adoption. Customers and regulators are far more likely to invest in or approve systems that are visibly governed, not just technically functional.

Getting started in your organization

If you're considering moving to external transparency, start small:

- **Pilot internal transparency first**: Create a dashboard that exposes key RAI and data governance metrics to senior stakeholders.
- **Choose strategic KPIs**: Not everything needs to be public - but some metrics offer low-risk, high-signaling starting points.
- **Build stories, not just numbers**: Pair your metrics with real-world narratives that illustrate impact, challenge, or leadership decisions.

- **Review with legal and communications**: Transparency should be proactive, not risky. Work with internal legal and communications teams to align your messaging and clarify your goals.

Transparency is not a compliance requirement. It's a competitive edge. Mature organizations use it to differentiate themselves, shape public policy, and build reputational resilience - all while improving the quality of their internal practices.

Maximizing Value and ROI from Your Data Estate

The objective now shifts from minimizing risk to maximizing value. This means unlocking the strategic potential of your data estate through reuse, telemetry, and alignment with enterprise outcomes. Data governance shifts to a business capability that drives measurable ROI from a compliance requirement.

From data storage to strategic activation

While most enterprises have invested in data lakes, warehouses, and platforms, few systematically measure how often data is reused, how quickly it creates value, or ho responsibly it is governed. This is why the focus has been on creating such an environment, where the following principles are applied:

- **Data reuse**: Encouraging and tracking reuse of verified datasets, prompts, and models across solutions
- **Usage telemetry**: Capturing how datasets are used (and where), to quantify value and detect issues
- **Life cycle intelligence**: Identifying when data is no longer trustworthy or relevant, ensuring governance guides sunset decisions

These aren't abstract concepts. Organizations like AstraZeneca have implemented data marketplace models to scale reuse across their global business. AstraZeneca's[6] FAIR data model[7] has been described as a foundational step in scaling data-driven research across departments. In this book, we've focused deeply on the importance of developing a data inventory and considering how governance and value interweave with the data life cycle process.

Data governance as a value multiplier

Organizations that integrate telemetry into their Power Platform CoE gain the ability to track not just usage but the performance of data assets over time. The Power Platform Admin Center provides insights into where apps and flows are deployed and reused, there is also the CoE Starter Kit and the Microsoft Copilot Studio Kit (both developed by the Power CAT team) that both enable you to build dashboards to see how your users' contributions are used and how they drive impact across the enterprise.

Incentivizing responsible reuse

ROI doesn't come from governance alone - it also comes from behavior. Some companies now reward data sharing and stewardship in the same way they recognize innovation or sales performance. For example: Airbnb developed an internal data quality platform (MIDAS)[8] to verify data sources and track usage.

They treat governance not as gatekeeping but as an enablement infrastructure - ensuring quality, promoting reuse, and increasing confidence.

---

[6] https://www.astrazeneca.com/what-science-can-do/topics/data-science-ai/unlocking-potential-data-ai-driven-drug-discovery-development.html#!

[7] https://www.go-fair.org/fair-principles/

[8] https://medium.com/airbnb-engineering/data-quality-at-airbnb-870d03080469

Metrics that matter

One reuse is tracked and metadata is enforced, organizations can begin to measure ROI more directly using:

- **Reuse frequency**: How often a dataset is reused across different apps or copilots
- **Time to first use**: How long it takes for a dataset to be used after it's registered
- **Governed activation rate**: Proportion of reused datasets that have completed governance checks
- **Value-to-risk ratio**: Comparing reuse volume to risk classification – highlighting high-value, low-risk assets and flagging dangerous overuse.

Advanced metrics are possible when governance and usage data are integrated. This allows leadership to prioritize which data assets deserve attention, investment, or deprecation – turning governance into an ROI compass.

How to apply this in your organization

- **Capture reuse**: Covered deeply in the previous chapters; extend your inventory with fields for "reused," "last verified," and "value indicator."
- **Expose usage metrics:** Build a dashboard to show leaders which datasets drive the most value and where gaps exist.
- **Recognize contributors**: Implement a lightweight reward or recognition system for teams that contribute widely used and well-governed data assets.
- **Make ROI part of your story**: Governance teams should report not only on compliance but also on the **value their policies protect and enable.**

Aligning data strategy with enterprise objectives

Data strategy is not an isolated IT function but a core component of the organization's overall strategy. Data initiatives are aligned with business goals, ensuring that data efforts directly contribute to enterprise objectives. For example, Unilever has integrated its data strategy with its sustainability goals[9] - enhancing its supply chain transparency and sustainability efforts.

How to implement in your organization

- **Establish cross-functional governance**: We already covered this, but forming or joining a data governance council with representatives from various departments ensures there is alignment between data initiatives and business objectives.
- **Define clear KPIs**: Set measurable goals for data projects that reflect their impact on enterprise objectives, such as customer satisfaction, operational efficiency, or revenue growth.
- **Regular strategy reviews**: Periodically review and adjust the data strategy to remain aligned with evolving business goals and market conditions.

Cultivating a data-driven culture

A data-driven culture empowers employees at all levels to make decisions based on data insights. This cultural shift involves training, accessibility, and encouragement to use data in daily operations. We covered this in part one of this book, but as a reminder, here are some useful tips:

[9] https://diginomica.com/unilever-is-using-google-cloud-to-help-it-create-1-billion-one-to-one-relationships-with-its-consumers

Implementation Guidance

- **Invest in training**: Provide ongoing education and resources to improve data literacy across the organization.
- **Enhance data accessibility**: Implement user-friendly tools and platforms that allow employees to access and analyze data without technical barriers.
- **Recognize data usage**: Acknowledge and reward teams and individuals who effectively utilize data to drive business outcomes.
- **Knowledge sharing**: Promote the dissemination of best practices and lessons learned throughout the organization.

The tenets of success in this area are to ensure your user community have the resources and incentives they need to buy-in to the data culture that you are trying to build. Some areas of the business will naturally adopt and in many cases thrive at this opportunity, but as with any form of tech adoption, look for the areas of your organization that lag. Talk with their leaders; get their support and buy-in. This is why value is so closely related to governance in this regard. Demonstrating the value of reusable data is going to help.

Closing Reflections: Governing for Impact, Building for the Future

Your organization by now has moved far beyond foundational controls. What began as an effort to bring visibility to your data estate has evolved into something much greater: a system of intelligence that supports growth, trust, and long-term value.

You've progressed from building inventories and setting policies to orchestrating autonomous agents, integrating governance into pipelines, and aligning data strategy with business outcomes. Security is no longer siloed. Data is no longer just a risk to be contained - it is a strategic asset to be nurtured, reused, and trusted.

In many ways, the journey of data readiness and security mirrors the broader evolution of your organization's maturity.

As AI evolves and regulatory expectations increase, the organizations that thrive will not be those who waited to be told what to fix - but those who already built the systems that make trust measurable, governance automatic, and value scalable. You govern now with conviction, because it is the right thing to do. You share your practices because they strengthen the ecosystem. And you invest in people, systems, and standards - not just for today's requirements but for the leadership demands of tomorrow.

This is integrated excellence.

Integrated Excellence Checklist: Data Readiness and Security

Data readiness is your competitive advantage. Your systems, culture, and metrics are harmonized to drive value, uphold trust, and lead your industry in responsible AI and Power Platform adoption. This checklist below is a useful way to document progress:

Strategic lifecycle management

- ☐ Autonomous agents continuously manage and monitor data lifecycle across all environments – from introduction to decommissioning – with minimal intervention.
- ☐ Lifecycle reviews are triggered by usage telemetry, sensitivity changes, or contextual shifts (e.g., business rules, regulatory updates).
- ☐ Asset value and risk are measured over time, and decisions to retire, retrain, or prioritize assets are based on strategic outcomes – not operational fatigue.

ROI-driven data strategy

- ☐ Data reuse, impact, and contribution to AI or business outcomes are routinely measured and reported.
- ☐ Metrics such as value-to-risk ratio, governed activation rate, and contribution to key business KPIs are part of leadership scorecards.
- ☐ Contributors and stewards of high-value, well-governed data assets are recognized, rewarded, and featured in internal knowledge-sharing efforts.

External transparency & assurance

- ☐ Public-facing RAI or data transparency pages exist – showcasing governance principles, model assurance processes and accountability frameworks.
- ☐ The organization participates in cross-industry initiatives, standards development, or third-party responsible AI audits.
- ☐ Regulatory readiness is built-in, not bolted on – audit trails, fairness metrics, explainability tooling, and incident response processes are standardized and externally verifiable.

Advanced AI integration practices

- ☐ Structured, governed data is delivered to generative AI systems using controlled protocols with policy-bound context delivery.
- ☐ AI-specific risk factors (e.g., hallucination, misuse of prompts) are actively managed through monitoring and mitigation workflow.
- ☐ High-risk AI use cases undergo formal impact assessments, with disclosures available to partners or customers as needed.

Embedded governance culture

- [ ] Data governance is fully embedded in organizational culture – seen not as a control function, but as an enabler of trusted digital innovation.
- [ ] Governance teams are strategically placed in key planning, architecture, and executive decision-making forums.
- [ ] Training and awareness on Responsible AI and data ethics are part of onboarding, leadership development and product launch workflows.

CHAPTER 13

The Journey Forward: Leadership in the Age of Low-Code AI

When I first began helping organizations adopt Power Platform, AI wasn't part of the conversation. The questions were about forms, workflows, and how to empower business users without overwhelming IT.

It was exciting but also constrained – governance was reactive, and data was often treated as someone else's responsibility.

Then everything changed.

As generative AI moved into mainstream platforms like Power Platform and Microsoft Copilot Studio – I saw a shift – not just in capability but in risk, value, and complexity. Suddenly, the same tools that enabled citizen developers to build apps could also generate decisions, suggest actions, or consume sensitive customer data.

It wasn't enough to teach people how to use these tools. We had to teach **how to use them responsibly**, how to **govern them at scale**, and how to **realize their full value without losing trust**.

That's why I wrote this book.

S. Jeffery, *The Low-Code AI Maturity Model*, https://doi.org/10.1007/979-8-8688-1730-4_13

Not because I have all the answers - but because I saw how many organizations were asking the same questions.

- *"How do we get started without losing control?"*
- *"What does 'responsible AI' actually mean when you're working with low-code?"*
- *"How do we build governance that helps, not hinders?"*

This book is my attempt to answer those questions - not as a prescription but as a maturity journey: a way to assess, grow, and adapt practices over time.

It's based on patterns I've seen succeed, challenges I've helped resolve, and lessons learned the hard way in real organizations, with real complexity and real consequences.

If there's one thing I hope you take away, it's this:

You can lead. You don't need perfect tools or executive mandates to begin.

Governance is not about slowing down - it's about doing the right things faster.

Responsible AI is not a luxury - it's the foundation for trust, credibility, and long-term impact.

Thank you for reading. Thank you for doing the hard work. The future of AI-powered low-code isn't being decided by labs and launch events - it's being shaped inside organizations like yours, every day.

Let's make it count.

Index

S. Jeffery, *The Low-Code AI Maturity Model*, https://doi.org/10.1007/979-8-8688-1730-4

D

F

G, H

K

L

N

O

T, U, V, W, X, Y, Z

GPSR Compliance
The European Union's (EU) General Product Safety Regulation (GPSR) is a set of rules that requires consumer products to be safe and our obligations to ensure this.

If you have any concerns about our products, you can contact us on

ProductSafety@springernature.com

In case Publisher is established outside the EU, the EU authorized representative is:

Springer Nature Customer Service Center GmbH
Europaplatz 3
69115 Heidelberg, Germany

www.ingramcontent.com/pod-product-compliance
Lightning Source LLC
LaVergne TN
LVHW052208200926
848481LV00001B/54
9798868817298